Royal British Columbia Museum

Memoir No. 1

Buttercups, Waterlilies, and Their Relatives:

(The Order Ranales)

in

British Columbia

BY

T. CHRISTOPHER BRAYSHAW

ILLUSTRATIONS BY THE AUTHOR

ROYAL
BRITISH
COLUMBIA
MUSEUM

Province of British Columbia
Ministry of Municipal Affairs, Recreation and Culture
Hon. Rita M. Johnston, Minister

Published by

The Royal British Columbia Museum, Victoria

First Printing..1989

Canadian Cataloguing in Publication Data

Brayshaw, T. Christoper, 1919-
    Buttercups, waterlilies and their relatives

    (Royal British Columbua Museum memoir, ISSN
0843-5383 ; no. 1)

    Bibliography: p.
    ISBN 0-7718-8739-6

    1. Ranunculales. 2. Water-lilies - British
Columbia. I. Royal British Columbia Museum. II.
Title. III. Series.

QK495.A12B72 1989   583'.111   C89-092056-7

ABSTRACT

This contribution to the flora of British Columbia describes 98 plant species in the families Ranunculaceae, Nymphaeaceae, Ceratophyllaceae, and Berberidaceae, comprising the representatives of the Englerian order Ranales that have been found growing wild, native or naturalized, in British Columbia.

The species described are illustrated, and their distributions are indicated by maps based on herbarium records. Diagnostic keys are provided to aid in the identification of species and subspecific categories. Some cases of questionable identity are discussed in detail. One hybrid is described, and six other nomenclatural innovations are made.

Key Index Words:- BERBERIDACEAE, BRITISH COLUMBIA, BUTTERCUPS, CERATOPHYLLACEAE, NYMPHAEACEAE, RANALES, RANUNCULACEAE, WATERLILIES.

DEDICATED TO THE MEMORY OF DR. MARGARET L. HEIMBURGER

Plant geneticist and authority on *Anemone*, formerly at the University of Toronto; whose studies on *Anemone* and *Ranunculus*, continued in her retirement as a Research Associate at the Royal British Columbia Museum, made a significant contribution to the research on which this volume is based.

PREFACE

The family Ranunculaceae and its relatives have for many years stood at the centre of interest for the author.

Work toward the production of a handbook on this family in British Columbia, begun in 1968, soon had to be given up in favour of other, more pressing duties; and it was not taken up again seriously for another ten years. This delay is now not regretted, in view of the amount of knowledge that has been gained through field work, and through herbarium research, in particular on the genera *Ranunculus* and *Anemone* by the author in partnership with the late Dr. Margaret L. Heimburger, during the intervening years.

The manuscript was completed initially during the summer of 1984; but the prospect of publication at that time appeared so remote that the work was left on a shelf for over a year. In 1986 and early 1987 an extensive revision was made, partly in response to the reviews of the original manuscript, and partly in an effort to keep the work up to date. This revision was completed in February, 1987.

The author is grateful to the curators of the various herbaria that he visited for their assistance in providing the records that he sought, to Dr. Peter W. Ball, of the University of Toronto, and to the late Dr. Heimburger, for reviewing the original manuscript, and also to the staff of the Royal British Columbia Museum for assistance in many aspects of the research and the preparation of this volume.

T. Christopher Brayshaw
March, 1987

CONTENTS

INTRODUCTION

Definition of the group

The order Ranales, as treated here, is taken in the broad sense defined by Engler and Prantl (1887–1915). Their system of classification, known as the Englerian system, was almost universally accepted when published, and is familiar to most botanists. It has been commonly followed by authors of floristic manuals, such as Fernald (1950) and Hitchcock and Cronquist (1964) in major part. The work of Dalla Torre and Harms (1900–1907, re-edited 1963), a comprehensive register of the known genera of seed plants, using the Englerian system, assigned a number to every genus. Its alphabetical index ("register"), published separately in 1958, is in widespread use as an index to large herbaria.

Though the Englerian system is now recognized as not completely natural in that it does not precisely reflect the currently recognized relationships among the major plant categories, its wide familiarity has favoured its continued use. Most of the families and many orders, including the Ranales, are clearly natural groups, and have been retained in rearranged positions in the many diverse classifications that have been published in more recent times.

In the recent classification by Takhtajan (1961) and Cronquist (1968, 1981) the old order Ranales, as defined above, with the addition of a few other, smaller orders, has become the subclass Magnoliidae, which is now subdivided into several, individually more closely circumscribed, orders. Differences in anatomy, pollen type, and embryo development provide ample justification for this reorganization, which reflects what we now understand about the divergent evolutionary trends among the members of this ancient group.

Of the sixteen families in the Englerian order Ranales, only four, divided between two orders by Cronquist (1981), are represented in the flora of British Columbia. These orders, with their families are: the order Nymphaeales, including the families Nymphaeaceae (Waterlily Family) and Ceratophyllaceae (Hornwort Family), and the order Ranunculales, including the families Ranunculaceae (Buttercup Family) and Berberidaceae (Barberry Family).

The order Ranales is difficult to define and to distinguish from other orders, because of the diversity of anatomy and floral structure among its members: a diversity that yet is associated with a near absence of most of the specializations that distinguish other and more advanced orders in the class Dicotyledonae. In a sense, members of this order are distinguished as much by the features that they do not share as by those they do share. Basically: our Ranales are mainly herbaceous plants that have radially symmetrical flowers that bear completely separate organs; often in rather indefinite numbers, and at least partly spirally, rather than entirely cyclically, arranged. The perianth often is not differentiated sharply into sepals and petals, and the petals and stamens may intergrade. In fact, the petals are thought to have originated from sterile stamens. The stamens are usually numerous and spirally arranged; and the carpels also are commonly indefinitely numerous, free from each other, and spirally arranged on prominent receptacles. The fruits of many of the species are follicles — the most primitive kind of fruit — or types, such as achenes, that are formed by direct modification of follicles. Apart from the herbaceous habit, all the characteristics mentioned above are regarded as primitive in flowering plants.

To each of the above-mentioned characteristics, a few genera show exceptions, while still displaying their affinity to the order by general agreement with the overall complement of its characteristics.

Brief overview of the families of Ranales in British Columbia

The families treated here may be briefly characterized as follows:—

Nymphaeaceae: The Waterlily Family: Aquatic perennials with submersed stems and roots, alternate, long-petioled, floating leaves, and usually conspicuous floating or emergent flowers, commonly with indefinite numbers of spirally arranged and weakly differentiated floral organs except for the often coherent carpels. Petals not bearing distinct nectaries. Pollen grains monosulcate (with a single furrow-like aperture) or with an encircling groove.

Ceratophyllaceae: The Hornwort Family: Rootless, totally submersed perennials, with whorls of forked leaves and unisexual flowers; each flower type consisting of a single carpel or an anther (or cluster of anthers?) subtended by a whorl of small appendages that may represent a perianth. Pollen grains thin-walled, and without special apertures: pollination taking place beneath the water surface.

Ranunculaceae: The Buttercup Family: Terrestrial, amphibious or aquatic, alternate-leaved herbs or opposite-leaved vines. Petals, when present, bearing nectar glands. The stamens, and usually the carpels, indefinite in number and separate; the carpels forming follicles or achenes, rarely berries or capsules. Pollen grains with three furrows or many small apertures.

Berberidaceae: The Barberry Family: Terrestrial shrubs or herbs with alternate leaves. Flowers with sepals, petals, and usually stamens, in multiple whorls of three each. Petals, when present, bearing distinct nectar glands. Stamens commonly in two whorls of three each; the anthers opening by uplifting valves. Carpel one per flower, forming a berry or an achene. Pollen grains with three to many furrows or a spiral groove.

Material and observations

The research on this group has included the examination of the specimens in the herbarium of the Royal British Columbia Museum (V), and of many more filed at the University of British Columbia, Vancouver (UBC), at the National Museum of Natural Sciences (CAN), and the Biosystematics Research Centre of Agriculture Canada (DAO), in Ottawa. Smaller numbers of specimens have been borrowed for examination from the University of Toronto (TRT), the University of Calgary (UAC), the University of Washington, Seattle (WTU), the University of California, Berkeley (UC), the University of Idaho, Moscow (ID), Washington State University, Pullman (WSU), Pomona College, Claremont, California (POM), the New York Botanical Garden (NY), and the Gray Herbarium of Harvard University, (GH), in Cambridge, Massachusetts.

It has not been practical to examine all the specimens that have been the bases of published or written reports of species in British Columbia. Such specimens are widely scattered in herbaria across all of North America and Europe.

Field work in various parts of British Columbia has been carried out during the past few summers with the object of obtaining fuller collections of species in the Ranales. Particular efforts were made to obtain representative material of the aquatic members of this order, since, at the outset, such plants were not well represented in the Royal British Columbia Museum herbarium.

Research in the genera *Ranunculus* and *Anemone* has been a joint project by the late Dr. Margaret Heimburger, formerly of the University of Toronto, and the author. Dr. Heimburger's many years of experience in investigating the genetics of these two genera were of immense value in understanding their interspecific and intraspecific relationships.

While in the greater part of the order, examination of the visible morphological features has been adequate for the resolution of specific distinctions, in some of the more difficult groups in the genus *Ranunculus*, such as the section *Batrachium* and the *R. flammula* and *R. occidentalis* complexes, more detailed examination has been necessary. Dr. Heimburger carried on breeding experiments with *R. occidentalis* and *R. californicus*, and Drs. Adolf and Oldriska Ceska made analyses of the flavonoid chemistry of material of the section *Batrachium*, in efforts to resolve the questions of identity that many of these plants present.

Descriptions

In order to make this work usable by a wide range of readers, the descriptions have been written in such a way as to avoid, where possible, much reliance on highly technical terminology. However, in the interest of verbal precision and economy, the use of some technical terms is practically unavoidable. Terms used are defined in a glossary starting on page 188; with the types of structures illustrated in Figure 105.

In addition to the gross, externally evident structural features that distinguish families and species, certain microscopic features are often important. Of these may be mentioned the pollen grain structure, and such internal characters as the structure and arrangement of the vascular (conductive) tissues. Further, on the subcellular scale, the complement of chromosomes has become recognized as of fundamental importance as a determinor, and thus as an indicator, of genetic relationships; since the chromosomes carry the genetic code that determines the expression of every character manifested by an organism.

Each body cell contains in its nucleus duplicate sets of a complement of chromosomes that is fundamental and characteristic for a species or variety. This duplicate, or diploid, number is signified by the abbreviation 2n, and is mentioned in the species descriptions where it is known.

In flowering plants, in the cell divisions that lead to the formation of the pollen or of the microscopic female plants in the ovules, the equivalent duplicate chromosomes join in pairs and then separate, one complete set, comprising what is known as the haploid number (n), going to each pollen grain and sperm, or to each female plant and egg. Subsequent fertilization restores the diploid number by addition of the parental haploid complements.

A genus, section or group of species is commonly characterized by the possession of a fundamental haploid number of chromosomes, designated by x, which is represented in all the member species. Species in the group may be diploid with respect to this basic number, in which case n = x. However, species or individuals are often found having in their cells four, six, or even eight copies of the fundamental complement for the genus, in which case n = 2x, 3x, or 4x.

Subspecific variation

All species show variation to some degree, from individual to individual, and from one local population to another. In order to accommodate variation within a species, the categories of subspecies (ssp.), variety (var.) and forma (f.) are used, in descending order of rank. Many such variations have been described in our species; and of these, some appear to be of major significance while others appear trivial and questionably deserving of formal recognition. The decision on whether to recognize a given variation is often a difficult and highly subjective one.

The author has here followed a policy of mentioning all those variations that he could identify; and of providing the information whereby the reader can make his own judgement. The author may then register his own impression on the matter when it differs from that indicated by the publishers of the named variations.

Distribution

Maps showing the distribution in British Columbia of the records of all the species described in this work are included in an Appendix. All the records shown in these maps represent specimens that have been examined by the author in, or on loan from, the herbaria mentioned above.

It is important to understand that the patterns of records on these maps cannot just be assumed to display proportional representation of any species by region. For most species, the incidence of records in any region is a product of both the incidence of the species and the incidence of interested botanists.

British Columbia is becoming better known with every passing year, as remote areas become increasingly accessible with the building of roads. Yet the coverage of the province by plottable plant records remains extremely uneven, and large areas are still unsampled. The map published by the author (Brayshaw, 1976, p. 9), outlining the main botanically sampled areas in British Columbia, remains, in general, valid today (1986), save for recent intensive sampling on western Vancouver Island. In areas such as southeastern Vancouver Island, the southern Okanagan Valley, and the Kamloops district, the records are replicated and crowded until their density exceeds the capacity of a small scale map to indicate all their locations clearly. Yet, even in such places, the most abundant and dominant species often tend to be overlooked, or ignored as too commonplace to excite interest. Other areas, empty of mapped records, all too often are recording the absence of botanists sampling them rather than the absence of particular species. The distribution patterns on the maps should be interpreted in the light of these considerations.

Forest regions identified by name are those described by Rowe (1972).

CHECK-LIST OF FAMILIES, GENERA AND SPECIES OF RANALES IN BRITISH COLUMBIA

NYMPHAEACEAE
Brasenia schreberi Gmelin
Nuphar polysepalum Engelmann
N. variegatum Engelmann *ex* Clint.
Nymphaea mexicana Zuccarini
N. odorata Aiton
N. tetragona Georgi

CERATOPHYLLACEAE
Ceratophyllum demersum L.
C. echinatum Gray

RANUNCULACEAE
Aconitum columbianum Nuttall *in* Torrey &
    Gray
A. compactum (Reichenbach) Gayer
A. delphiniifolium DC.
Actaea rubra (Aiton) Willdenow
Anemone canadensis L.
A. cylindrica Gray
A. drummondii S. Watson
A. lyallii Britton
A. multifida Poiret
A. narcissiflora L.
A. occidentalis S. Watson
A. parviflora Michaux
A. patens L.
A. piperi Britton
A. richardsonii Hooker
A. riparia Fernald
Aquilegia brevistyla Hooker
A. flavescens S. Watson
A. formosa Fischer
A. vulgaris L.
Caltha biflora DC.
C. leptosepala DC.
C. natans Pallas
C. palustris L.
Cimicifuga elata Nuttall *in* Torrey & Gray
Clematis ligusticifolia Nuttall *in* Torrey &
    Gray
C. occidentalis (Hornemann) DC.
C. tangutica (Maximowicz) Korshinsky

C. vitalba L.
Coptis aspleniifolia Salisbury
C. occidentalis (Nuttall) Torrey & Gray
C. trifolia (L.) Salisbury
Delphinium ambiguum L.
D. bicolor Nuttall
D. burkei Greene
D. glaucum S. Watson
D. menziesii DC.
D. nuttallianum Pritzel
Isopyrum savilei Calder & Taylor
Myosurus aristatus Bentham & Hooker
M. minimus L.
Nigella damascena L.
Ranunculus abortivus L.
R. acris L.
R. alismifolius Geyer *ex* Bentham
R. aquatilis L.
R. bulbosus L.
R. californicus Bentham
R. cardiophyllus Hooker
R. cooleyae Vasey & Rose
R. cymbalaria Pursh
R. eschscholtzii Schlechtendal
R. eximius Greene
R. ficaria L.
R. flabellaris Rafinesque
R. flammula L.
R. gelidus Karelin & Kirilov
R. glaberrimus Hooker
R. gmelinii DC.
R. hyperboreus Rottboell
R. inamoenus Greene
R. lapponicus L.
R. lobbii (Hiern) Gray
R. macounii Britton
R. nivalis L.
R. occidentalis Nuttall *in* Torrey & Gray
R. orthorhynchus Hooker
R. pedatifidus J.E. Smith *ex* Rees
R. pensylvanicus L. f.
R. pygmaeus Wahlenberg

R. repens L.
R. rhomboideus Goldie
R. sardous Crantz
R. sceleratus L.
R. subrigidus W.B. Drew
R. sulphureus Solander
R. testiculatus Crantz
R. trichophyllus Chaix *in* Villars
R. uncinatus D. Don *in* G. Don
R. verecundus Robinson
Thalictrum alpinum L.
T. dasycarpum Fischer, Meyer & Ave-Lallemant *in* Fischer & Meyer

T. occidentale Gray
T. sparsiflorum Turczaninow *ex* Fischer & Meyer
T. venulosum Trelease
Trautvetteria caroliniensis (Walter) Vail
Trollius laxus Salisbury

BERBERIDACEAE
Achlys triphylla J.E. Smith
Berberis aquifolium Pursh
B. nervosa Pursh
B. vulgaris L.

EXCLUDED SPECIES

Anemone deltoidea Hooker

Records of this species, reported by Macoun (1883) from the Salmon River, and by Raup (1934) from Dease Lake, have been found by Scoggan (1978) to be based on misidentified material. This species is not known at present to grow in British Columbia.

Clematis hirsutissima Pursh (*C. douglasii* Hooker)

This species was reported for southern British Columbia by Hitchcock and Cronquist (1964, 1973); however, no records substantiating this statement have been found by the author.

Paeonia brownii Douglas *ex* Hooker

This species was reported by Macoun (1883) and Henry (1915) in the Ranunculaceae for Vancouver Island. This occurrence is quite unlikely, since the natural range of this plant is east of the Cascade Range from Washington State to California. It is not known to occur anywhere in British Columbia. It is now segregated in a distinct family, the Paeoniaceae.

NOMENCLATURAL INNOVATIONS IN THIS VOLUME

p. 38   *Anemone multifida* Poiret var. *saxicola* B. Boivin forma *hirsuta* (C.L. Hitchcock) T.C. Brayshaw *stat. nov.* (based on *Anemone multifida* var. *hirsuta* C.L. Hitchcock *in* Hitchcock, C.L., and Arthur Cronquist: Vascular Plants of the Pacific Northwest: Part 2: 327. 1964).

p. 110   *Ranunculus* X *heimburgerae* T.C. Brayshaw *nom. nov.* (*Ranunculus occidentalis* Nuttall X *californicus* Bentham).

p. 124   *Ranunculus eschscholtzii* Schlechtendal ssp. *suksdorfii* (Gray) T.C. Brayshaw *stat. nov.* (*Ranunculus suksdorfii* Gray; Proc. Am. Acad. 21:371. 1886.)

p. 145   *Ranunculus gmelinii* DC. var. *hookeri* (G. Don) L. Benson forma *prolificus* (Fernald) T.C. Brayshaw *comb. nov.* (*R. purshii* Richardson var. *prolificus* Fernald; Rhodora 19:135; 1917)

p. 170   *Thalictrum occidentale* Gray var. *breitungii* (B. Boivin) T.C. Brayshaw *stat. nov.* (*Thalictrum breitungii* B. Boivin; Canadian Field-Naturalist 62(6):168; 1948).

p. 179   *Achlys triphylla* (J.E. Smith) DC. ssp. *californica* (Fukuda & Baker) T.C. Brayshaw *stat. nov.* (*Achlys californica* Fukuda & Baker *in* Fukuda, Ichiro; Taxon 16:315; 1967).

p. 182   *Berberis aquifolium* Pursh ssp. *repens* (Lindley) T.C. Brayshaw *stat. nov.* (*Berberis repens* Lindley; Bot. Reg. 14:pl.1176; 1828).

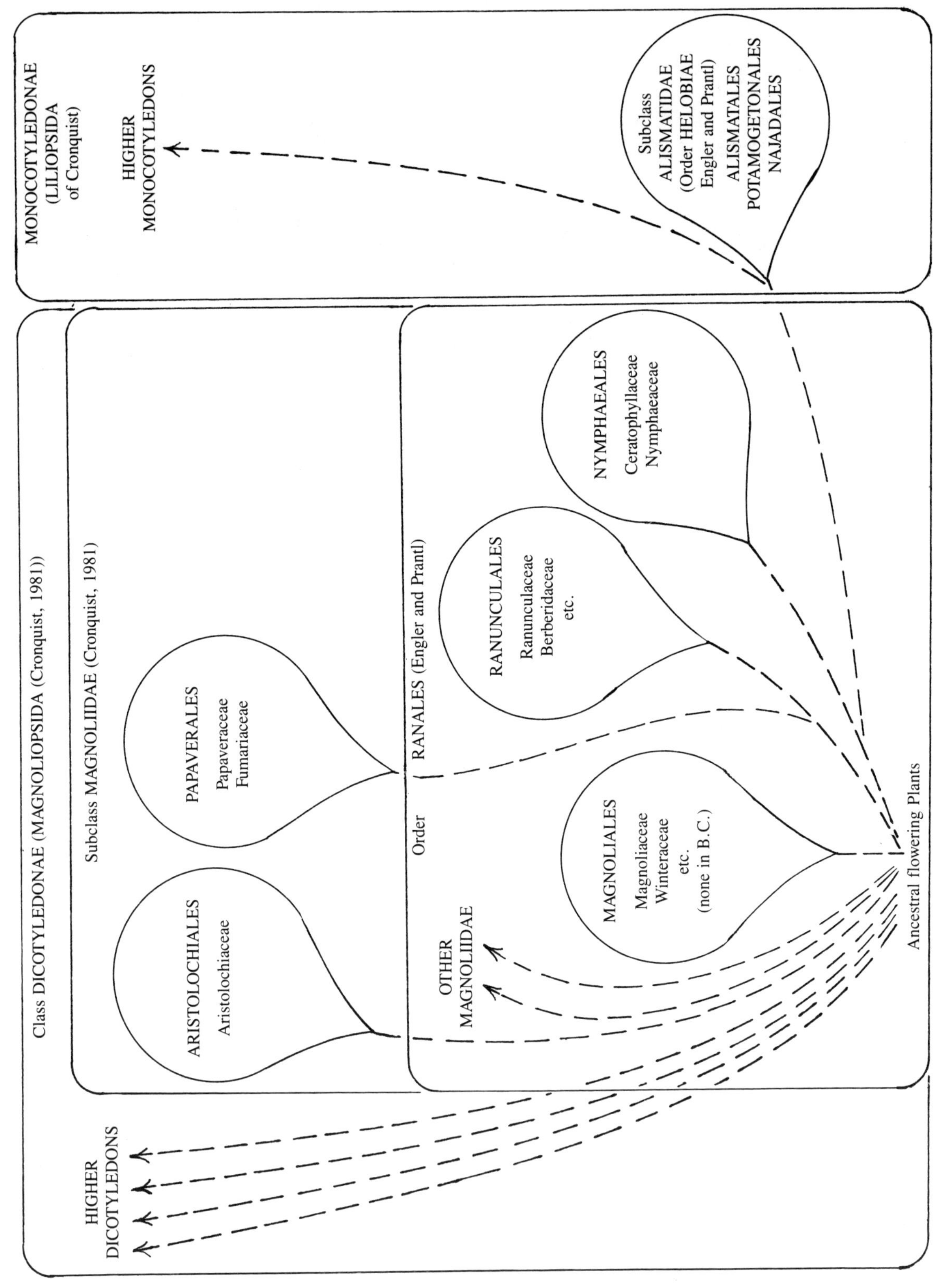

Figure 1. Diagram of the putative evolutionary relationships among British Columbian families of the Ranales and their relatives; showing their relationships to the two systems of classifications described in the text. Orders named in balloons are as used by Cronquist (1981).

EVOLUTIONARY POSITION OF THE RANALES

The Order Ranales is considered to be the most primitive order of the Flowering Plants (Angiosperms), because it has retained more features that are believed to have characterized the first, ancestral Flowering Plants than have any other surviving orders in this phylum.

No single species or genus of the Ranales displays all the features that are primitive in Flowering Plants. In every existing member, even those that appear the most primitive, we see, juxtaposed to a series of primitive characteristics, some that are evidently derived and indicative of specialization along some direction.

Although looked upon as the most primitive in the phylum, the forms that we see today cannot be considered as ancestral to other orders of Flowering Plants. It is better to regard them as the surviving ends of long lines of descent that have evolved in their own diverse ways from ancient ancestral groups, that were also fore-runners of other, more progressive orders of Dicotyledons, and of the Monocotyledons, as well as being more closely related to each other than are their modern descendants. Each surviving line-end in this order retains its own array of primitive characteristics that have been superseded in other lines that have progressed further in other directions. The complete 'living fossil', bearing all the primitive characteristics of its group, but that survives in spite of having been left behind in an evolutionary sense, is a rare phenomenon indeed, and is not known in this order.

The Ranales thus present us with reminiscences of the ancestral stock of the Flowering Plants, not that stock itself, which has long since disappeared.

In addition to the Dicotyledons, the ancestral Monocotyledons are thought to have evolved from the early Ranales (more specifically, the Magnoliales of Cronquist's (1981) system) via a stock that also was ancestral to the present day *Nymphaeaceae*; with which the most primitive Mono-cotyledons share a number of characteristics of the anatomy and pollen.

Figure 1 indicates, in simplified schematic form, the putative evolutionary relationships between our members of the order Ranales and related orders of flowering plants; and how the two systems of classification mentioned above are related, with respect to the families treated here.

KEY TO FAMILIES OF RANALES IN BRITISH COLUMBIA

1. Plant totally submersed, rootless, with whorled, sessile, forked leaves. Flowers axillary, minute, apetalous, sessile or subsessile .... **CERATOPHYLLACEAE**

   Plant with roots and alternate or opposite leaves, or, if leaves whorled, plant terrestrial **2**

2. Aquatic plants with submersed rhizomes. Leaves mostly floating, long-petioled, with elliptic to orbicular blades unlobed except for basal angles, or peltate. Petals without distinct nectar glands .... **NYMPHAEACEAE**

   Plants terrestrial, or, if aquatic, with dissected or linear to lanceolate leaves. Petals, when present, with distinct nectar glands **3**

3. Shrubs with alternate leaves. Yellow flowers with perianth parts in multiple whorls of 3. Anthers opening by uplifting flaps. Carpel one. Fruit a berry ....
   .... *Berberis* in **BERBERIDACEAE**

   Herbs of diverse form, or woody vines with opposite leaves. Perianth parts not in multiple whorls of three **4**

4. Scapose herbs with large (over 10 cm wide) trifoliolate leaves with sinuate margins. Flowers in spikes, white, apetalous. Carpel one. Fruit a crescentic achene ....
   .... *Achlys* in **BERBERIDACEAE**

   Diverse herbs or opposite-leaved vines. If scapose and trifoliolate, leaves small (less than 5 cm wide). Flowers various, but if white and apetalous, with two or more, usually separate, carpels, or borne on leafy-stemmed plants. Anthers opening by longitudinal slits
   .... **RANUNCULACEAE**

DESCRIPTIONS OF FAMILIES, GENERA AND SPECIES

The Family NYMPHAEACEAE
Water-Lily Family

Aquatic perennial herbs, commonly with submerged or buried rhizomes and alternate leaves with elongate petioles and broad floating blades (except in *Cabomba*, with mostly submersed, opposite, dissected leaves). Conductive system commonly, but not always, of scattered slender vascular strands that are incapable of increase in thickness. There is no cambium. Xylem tissues of the vascular system greatly reduced and largely replaced by xylem canals, entirely without vessels (tracheae), except in the roots of *Nelumbo*.

Flowers solitary on elongate pedicels. Floral appendages in whorls of 3 or 4; or sometimes indefinite in number, spirally arranged, and tending to intergrade in form. Anthers introrse, i.e. opening inward, toward the axis of the flower, except in *Nelumbo*. Pollen grains are monocolpate,[1] i.e. have a single furrow-like aperture, except in *Nelumbo*, which has tricolpate pollen, with three furrows, and *Nymphaea*, which has *zonosulcate* pollen, (with an encircling groove). Carpels three to many, separate or connate in a whorl. Ovules one to many in each carpel, attached to the lateral or dorsal sides of the locules, not to the ventral suture as in most flowering plants.

Fruit a follicle or capsule (or achenes in *Nelumbo*) decaying to release the seeds, which contain endosperm, except in *Nelumbo*.

Embryo with two distinct cotyledons, or sometimes apparently one by the lateral union of the cotyledons on one side.

The first leaf produced by the seedling, after the cotyledons, has a linear, flattened form, and represents a petiole, without the differentiation of a distinct expanded leaf blade. Subsequent leaves possess very small, lanceolate to ovate blades; and thence successive leaves show increased size and transitional forms until the adult form is attained. The primary root of the seedling is small and short-lived, and is replaced early by the adventitious roots that grow out from the developing rhizome and are the characteristic roots of the adult plant.

A rather small, worldwide, freshwater family of 8 genera.

Because of the exceptions to the family characteristics mentioned above, *Nelumbo* is commonly separated as a distinct family, the *Nelumbonaceae*; and less often, the genera *Brasenia* and *Cabomba* are separated by some authors as the family *Cabombaceae* (Cronquist, 1981).

KEY TO GENERA

1. Leaves peltate, elliptic .... ***Brasenia***
   Leaves with a basal sinus **2**

2. Leaves elliptic, with rounded basal lobes, sepals orbicular .... ***Nuphar***
   Leaves broadly elliptic to orbicular, with angled basal lobes, sepals lanceolate ..... ***Nymphaea***

BRASENIA Schreber
Water Shield

Perennial herbs with slender rhizomes and ascending leafy branches bearing alternate, long-petioled, centrally peltate floating leaves with elliptic blades 3–8 cm long, without stipules, and solitary flowers on obliquely axillary pedicels (arising to one side of the leaf axil). Vegetative branches are axillary or obliquely axillary. The stem contains one or a pair of double vascular strands, the phloem of each on either side of a central xylem canal. A number of air canals surrounds the central strand. Sclereids are absent. The younger submerged parts of the plant are covered in a layer of mucilage secreted by special unicellular glandular hairs.

---

[1] Monocolpate pollen is characteristic also of the Magnoliales, the most primitive order of Dicotyledons, and the Piperales, and of the majority of the Monocotyledons.

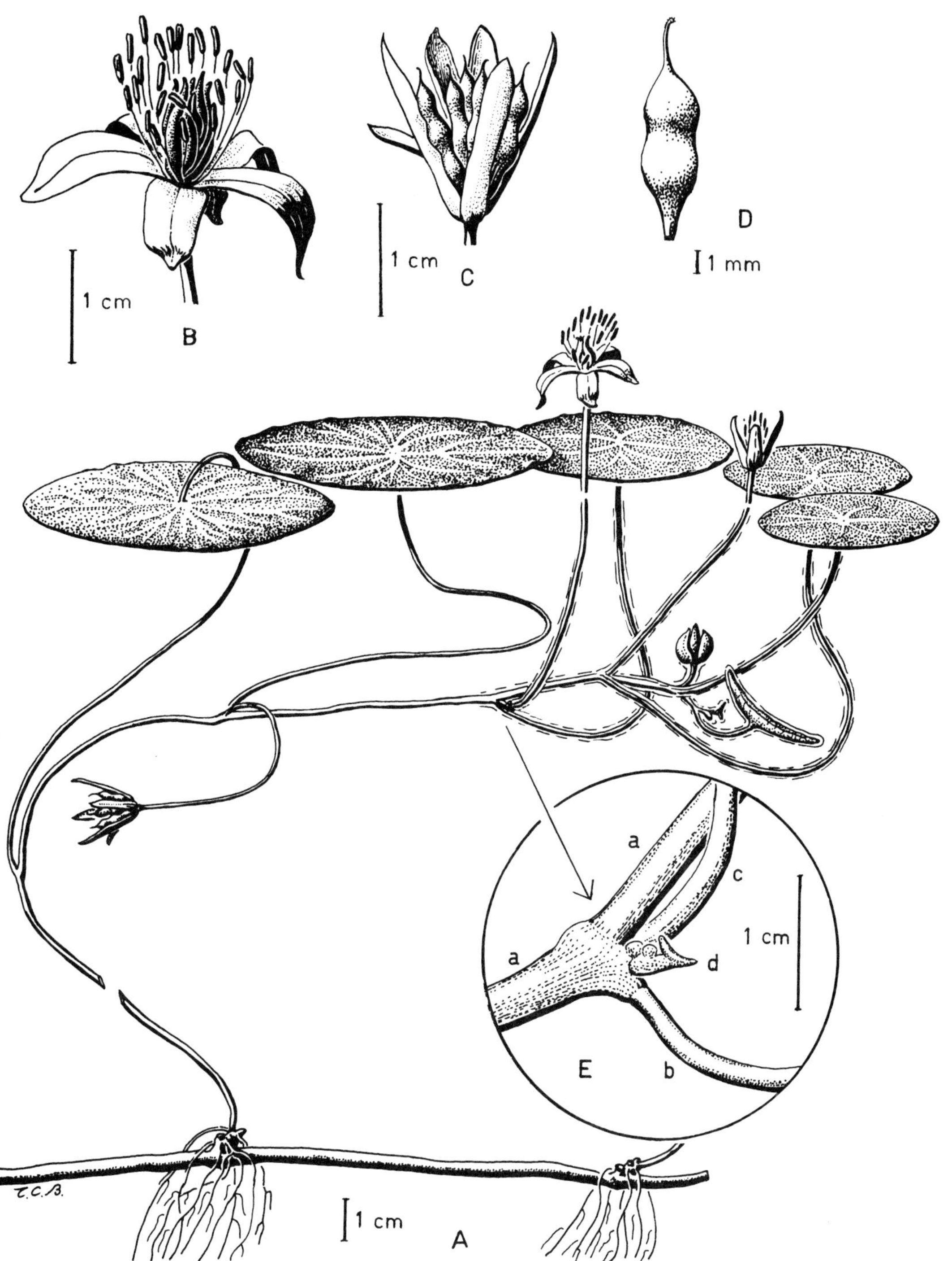

Figure 2. *Brasenia schreberi*

A. flowering and fruiting plant.
B. flower.
C. 'flower' at fruiting stage.
D. fruit (follicle).

E. branching pattern at node:
   (a) main stem.
   (b) leaf stalk (petiole).
   (c) flower stalk (pedicel).
   (d) young branch.

Flowers small, dull purple, with 3 sepals, 3 petals, 12 to 20 or more stamens with slender filaments and introrse anthers. Carpels 4–18, free from each other and from other floral parts, with linear stigmas, each containing 2–4 ovules attached to the inner carpel wall at or near the dorsal suture.

Fruit a follicle ripening under water and decaying to release the seeds. $2n = 80$.

Brasenia, with one cosmopolitan species, together with the mainly tropical genus *Cabomba*, is placed by Cronquist (1981) and others, in a separate family: the Cabombaceae.

Brasenia schreberi Gmelin forms extensive colonies in still water 1–2 (occasionally to 3) metres deep, covering the surface with its floating leaves.

NUPHAR J.E. Smith (*nom. cons.*) (*NYMPHOZANTHUS* L.C. Richard)

Rhizomatous aquatic perennial with sparingly branched rhizomes anchored to the bottom by adventitious roots arising in clusters below the leaf bases; and with conspicuous raised leaf scars and pedicel scars. Branching sclereids abundant in the tissues.

Leaves alternate, without stipules, of two kinds: submersed, membranous, short-petioled, spring leaves, and thicker, long-petioled, floating or emersed summer leaves, with rounded basal lobes in ours.

Flowers solitary on pedicels that arise from the rhizome in place of, rather than in the axils of, leaves.

Flowers with several to many broad, thick, concave sepals, many inconspicuous fleshy petals, and many stamens with introrse anthers. Ovary of many fused carpels, quite free from the perianth and stamens; the stigmas as radiating lines on the concave, sessile apical stigmatic disc. Ovules many, scattered on the lateral walls of the locules.

The fruit is an ovoid longitudinally ribbed capsule, ripening at the water surface in ours, and decaying and rupturing to liberate the numerous seeds in a mucilaginous matrix. Seeds without arils.

Internally, the petioles, pedicels, and the bulk of the rhizome consist largely of spongy lacunate tissue containing multitudes of very fine air passages partly obstructed by the many branching sclereids. Vascular strands are slender and scattered through these tissues (Ogden, 1974).

Our two species are weakly differentiated. In areas where their ranges meet, intermediate forms (hybrids?) are not uncommon; raising the question as to whether they are truly distinct species or, as some consider, mere subspecies of one circumboreal species (*N. lutea* (L.) Sibthorp & Smith) (Calder & Taylor, 1968).

KEY TO SPECIES

1. Sepals 8–17, commonly 9. Petiole terete .... *N. polysepalum*
   Sepals 6–8. Petiole flattened, with lateral ribs .... *N. variegatum*

Nuphar polysepalum Engelmann   **Yellow Water-Lily**
 Nymphozanthus polysepalus (Engelmann) Fernald   **Yellow Pond-Lily**
 Nuphar lutea (L.) Sibthorp & Smith ssp. *polysepalum* (Engelmann) Beal

Coarse perennial with a massive log-like rhizome 8 to 15 cm thick by up to 5 m long. The rhizome glabrous, bearing conspicuous raised scars in spiral series: the scars of petioles with lateral projecting angles, and accompanied on their basal sides by double rows of adventitious roots or root scars. The scars of pedicels scattered in the same spiral series, smaller, elliptic, rounded, and devoid of associated roots. Internally, the rhizome is a mass of spongy tissue traversed by scattered slender vascular strands.

Summer leaves with soft petioles, elliptic and rounded in cross section except at the flanged bases; the blades 15–50 cm long, elliptic, with rounded basal lobes, pinnately veined. The blades floating on deep water, but commonly emergent, even to ½ metre or rarely to 1 metre, in shallow water commmunities where competition is intense.

Flowers solitary, often standing a few centimetres above the water, on long pedicels arising, in place of petioles, directly from the rhizome; the flowers 5–8 cm wide, with 8–12 or more fleshy concave sepals, the outer green, and the inner larger and yellow, often 'fading' to orange. Petals several to many, small and inconspicuous, little if at all longer than the adjacent stamens but rather wider, and rectangular; often reddish tinged, fleshy. Stamens many, yellow, reddish apically, at first aggregated closely against the base of the ovary; but recurving as they mature in succession, and then lying on and concealing the petals. Anthers linear, on and partly embedded in the ventral surface of the wide flat connective. Appendages transitional in form between sepals, petals and stamens sometimes occur. Pollination is carried out by small flies.

Ovary superior, of many coherent carpels, with a terminal and central funnel-shaped depression bearing stigmas as radiating lines near the upper edge.

Fruit an ovoid, ribbed, green capsule, 5–9 cm long, ripening at the water surface. Seeds ovoid, 3–5 mm long, light brown when ripe. 2n = 34.

Range from Alaska to California, and east to Colorado and South Dakota in lakes and sluggish streams generally with muddy bottoms, in water to 3 metres deep or more, or on wet mud.

Nuphar variegatum **Engelmann** *ex* **Clint.**   **Yellow Water-Lily**
Nymphozanthus variegatus **(Engelmann) Fernald**
Nuphar lutea **(L.) Sibthorp & Smith ssp.** *variegatum* **(Engelmann) Beal**

Coarse rhizomatous perennial similar in aspect to *N. polysepalum*; but differing in the following respects:

Rhizome, petioles and pedicels on the average more slender. Petiole rather flattened, with lateral ribs, firmer in texture than that of *N. polysepalum*, but similar in internal structure. Summer leaves floating or sometimes emersed in shallow water.

Flowers averaging slightly smaller than those of *N. polysepalum*, 3–6 cm wide, with 6–8 sepals. Stamens yellow. Pollination is mainly by beetles and flies (Leppik, 1964).

Fruit similar to that of *N. polysepalum*, but commonly rather smaller and relatively more slender, 3–5 cm long. 2n = 34.

Nuphar variegatum ranges from the interior of Yukon and British Columbia eastward through the forested parts of Canada and the northern United States to Newfoundland. It is found in habitats similar to those of *N. polysepalum*. Forms (hybrids?) between this species and *N. polysepalum* occupy a narrow geographic belt that delineates the line of contact between these species in British Columbia (see map).

NYMPHAEA L.

Rhizomatous perennials, with rhizomes attached to the bed of the lake or pond by adventitious roots arising in groups below the leaf bases.

Leaves alternate on the rhizome, with axillary ligules apparently formed of stipules joined across the leaf axils, long petioles and floating blades, which are not peltate in our species. The petioles are traversed by 4 large and several slender air channels; the tissues containing branching sclereids.

Flowers on long pedicels that arise from the rhizome in place of, not axillary to, leaves.

Sepals 4, greenish in our species. Petals and outer stamens commonly not sharply differentiated, but intergrading in form; their bases adnate to the ovary. There are usually many of both petals and stamens. Anthers linear, on the ventral surface of the wide, flat connective.

Ovary partly inferior among the adnate bases of the petals and stamens, composed of many laterally coherent, many-ovuled carpels; the ovules scattered on the lateral walls of the locules. Stigmatic disc sessile, concave, the margin projecting into incurved appendages opposite each radiating stigmatic line.

Fruit a capsule containing many seeds, each seed in a mucilaginous aril. In some species the pedicel coils up after pollination, drawing the fruit beneath the water for ripening.

Internally, the petiole and pedicel are similar, containing 2–4 conspicuous air channels and sometimes a few smaller ones, scattered sclereids and with one central and several peripheral vascular strands (Odgen, 1974). The rhizome, which is often hairy, contains a layer of lacunate tissue with air passages, and a central vascular system with branches leading to the petiole and root bases.

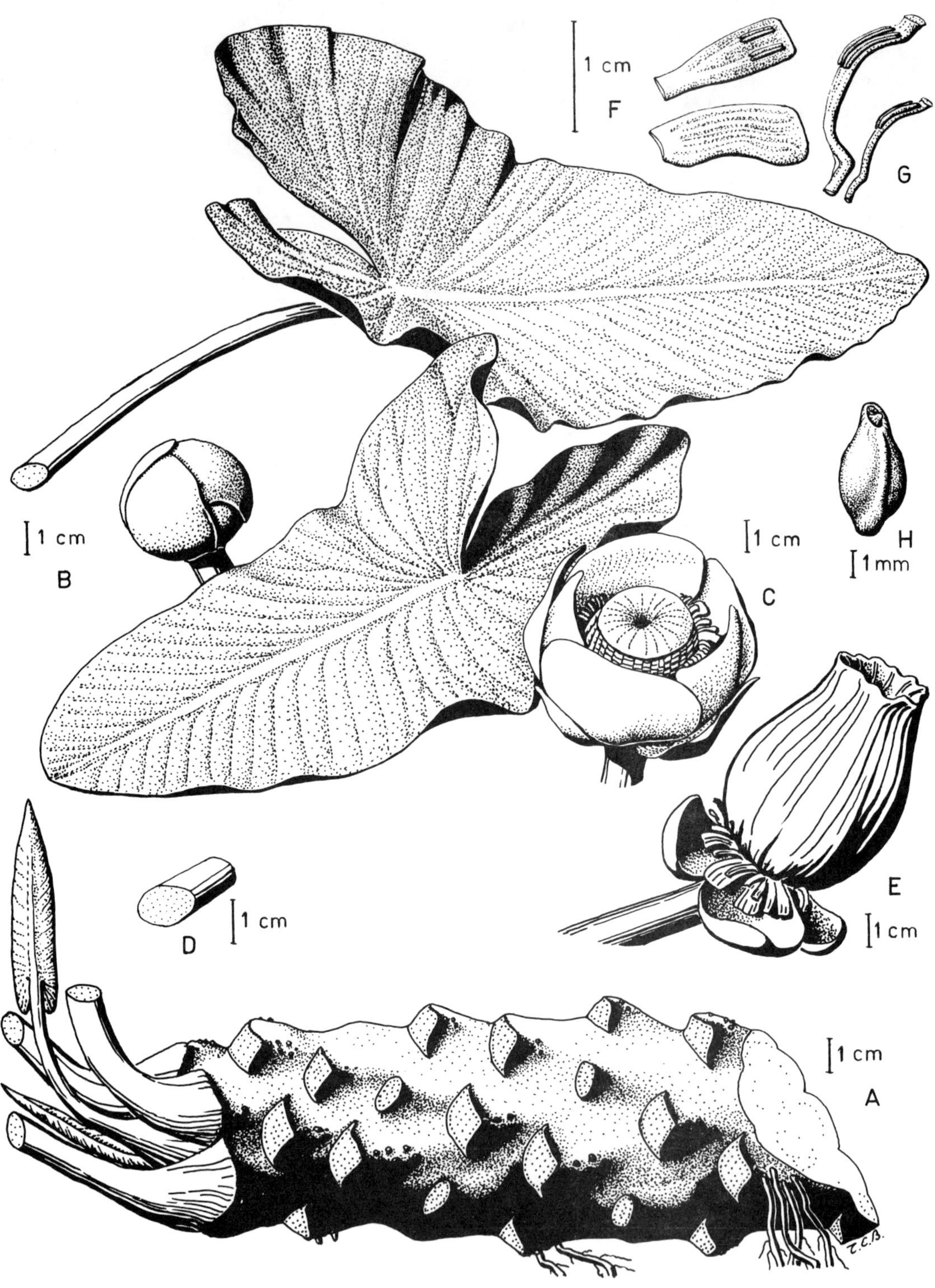

Figure 3. *Nuphar polysepalum*

A. rhizome and terminal bud.
B. leaves and flower bud.
C. flower.
D. cross-section of petiole.

E. fruit.
F. petal (lower) and petal-stamen transition (upper).
G. upper and lower stamens (same enlargement as F).
H. seed.

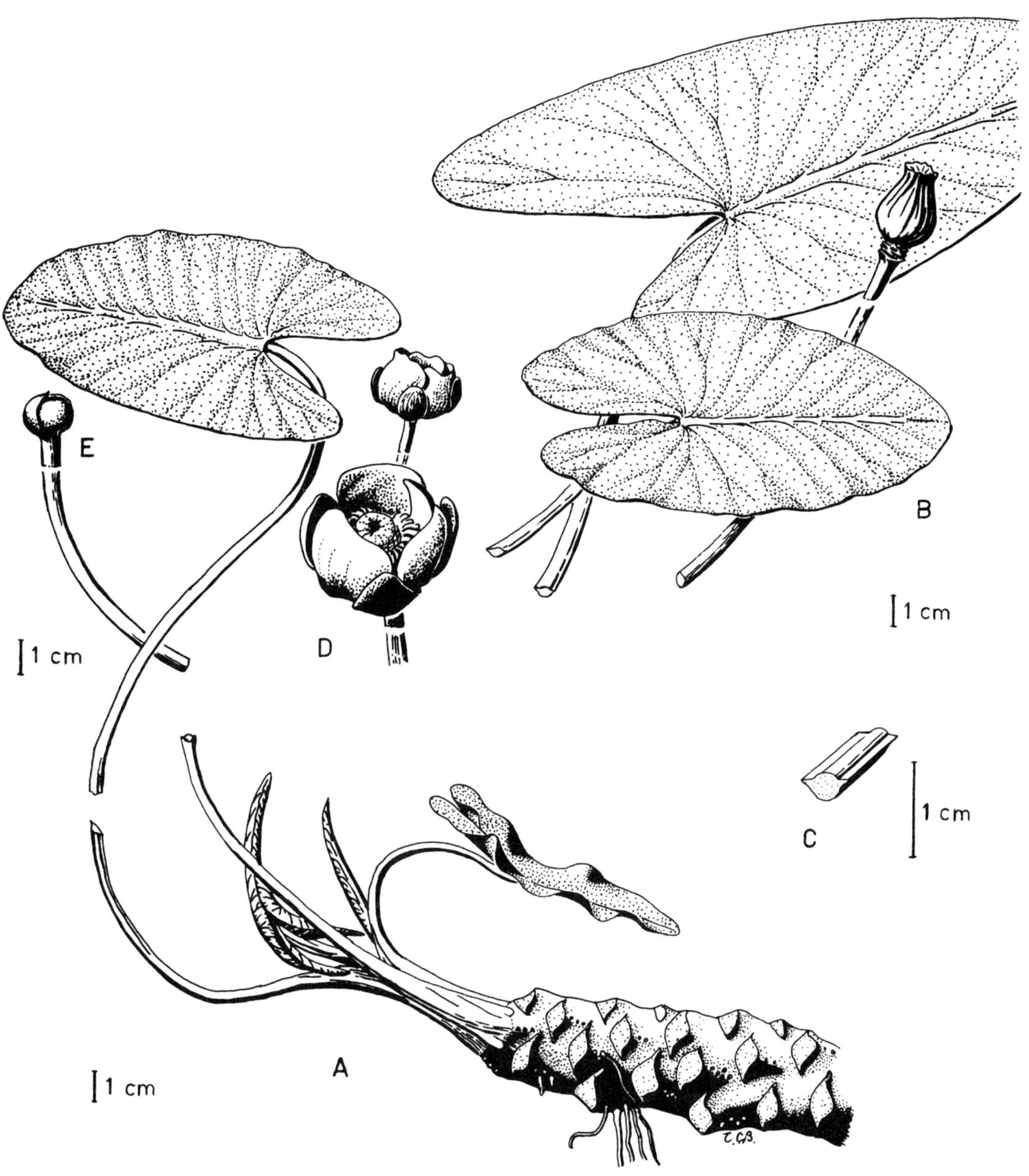

Figure 4. *Nuphar variegatum*

A. rhizome and submerged leaf.
B. floating leaves and fruit.
C. cross-section of petiole.

D. flowers.
E. flower bud.

KEY TO SPECIES

1. Leaves oval, 4–8 cm long. Flowers 3–6 cm wide, with 7–15 petals. Rhizome short, erect.. ***N. tetragona***

   Leaves and flowers larger, petals more numerous (12 or more). Rhizome horizontal........... **2**

2. Flowers yellow, leaves variegated, with red or purple streaks and blotches beneath, and sometimes above... ***N. mexicana***

   Flowers white, pink or red, leaves not variegated.. **3**

3. Petals 12–20.. ***N. alba***

   Petals 20 or more.. **4**

4. Petals oblanceolate to spatulate, obtuse to rounded, widest above the middle. Rhizome with short, detachable, tuber-like branchlets. Leaves green beneath.......................... ***N. tuberosa***

   Petals lanceolate, acute, widest at or below the middle. Rhizome branches not tuber-like. Leaves commonly purple beneath .... ***N. odorata***

Only *N. tetragona* is native to British Columbia; but several exotic species and their ornamental varieties and hybrids have been introduced into lakes, and may persist at the sites of introduction. Of these exotic species, only *N. odorata* appears to be able to spread beyond the sites of its introduction, and invade native aquatic plant communities.

Nymphaea tetragona Georgi   **Small White Water-Lily**

Perennial with erect, short, tuber-like rhizome densely covered by multicellular brownish hairs among the roots and petiole- and pedicel-bases.

Leaves with long, very slender petioles and elliptic floating blades 4–8 cm long. Petioles with 2 wide and 2 or 4 very fine air channels. Pedicels with 4 nearly equally wide air channels.

Flowers 3–6 cm wide, with 4 sepals often with projecting basal angles, giving the bud and closed flower a 4-angled base. Petals 7–15, commonly 9 or 10, white. Stamens 30–45, yellow or the outermost with wide, petaloid, white filaments.

Ovary partly inferior with 6–9 stigmas and as many incurved sterile appendages.

Fruit globose, about 1 cm in diameter, surrounded by the erect persistent sepals, the petals and stamens decaying away. Seed 1–2 mm long, in an aril. 2n = 112.

Circumboreal, widely scattered; native in the northern half of British Columbia, in small lakes and sheltered bays; in water ¼–2 m deep.

Nymphaea odorata Aiton   **Large White Water-Lily**

Horizontal creeping and branching rhizome 2–3 cm in diameter, the younger parts tomentose with brownish hairs. Petioles and pedicels are alternate in spiral series, the petioles accompanied at their bases by broad, glabrescent axillary ligules (connate stipules?) that leave crescent-shaped scars when shed, and by clusters of adventitious roots. Pedicels not associated with ligules or roots.

Leaves with long petioles and orbicular floating blades with angular lobes beside the basal sinuses, purple beneath.

Flowers on pedicels equal to the depth of water, floating, with 4 greenish sepals, 20–30 petals, lanceolate, broadest about the middle or below, acute to somewhat rounded at the tips, white, or sometimes pink or red in forma *rubra* Guillon, 30–100 stamens, the outer with petaloid, and sometimes white, filaments, the inner progressively smaller and more slender, the innermost with yellow filaments narrower than the yellow anthers. Ovary with several stigmas with projecting and incurving appendages.

Fruit 1–2 cm in diameter, covered by the persistent perianth or the outer sepals and petals falling away. Seeds 2–3 mm long with longer arils. 2n = 84.

Native in eastern North America, this species is widely cultivated elsewhere. In British Columbia it is spreading from introductions in small lakes and sheltered bays, in water ½ to 2½ metres deep; competing with more or less success with the native yellow water-lilies. Pollination is

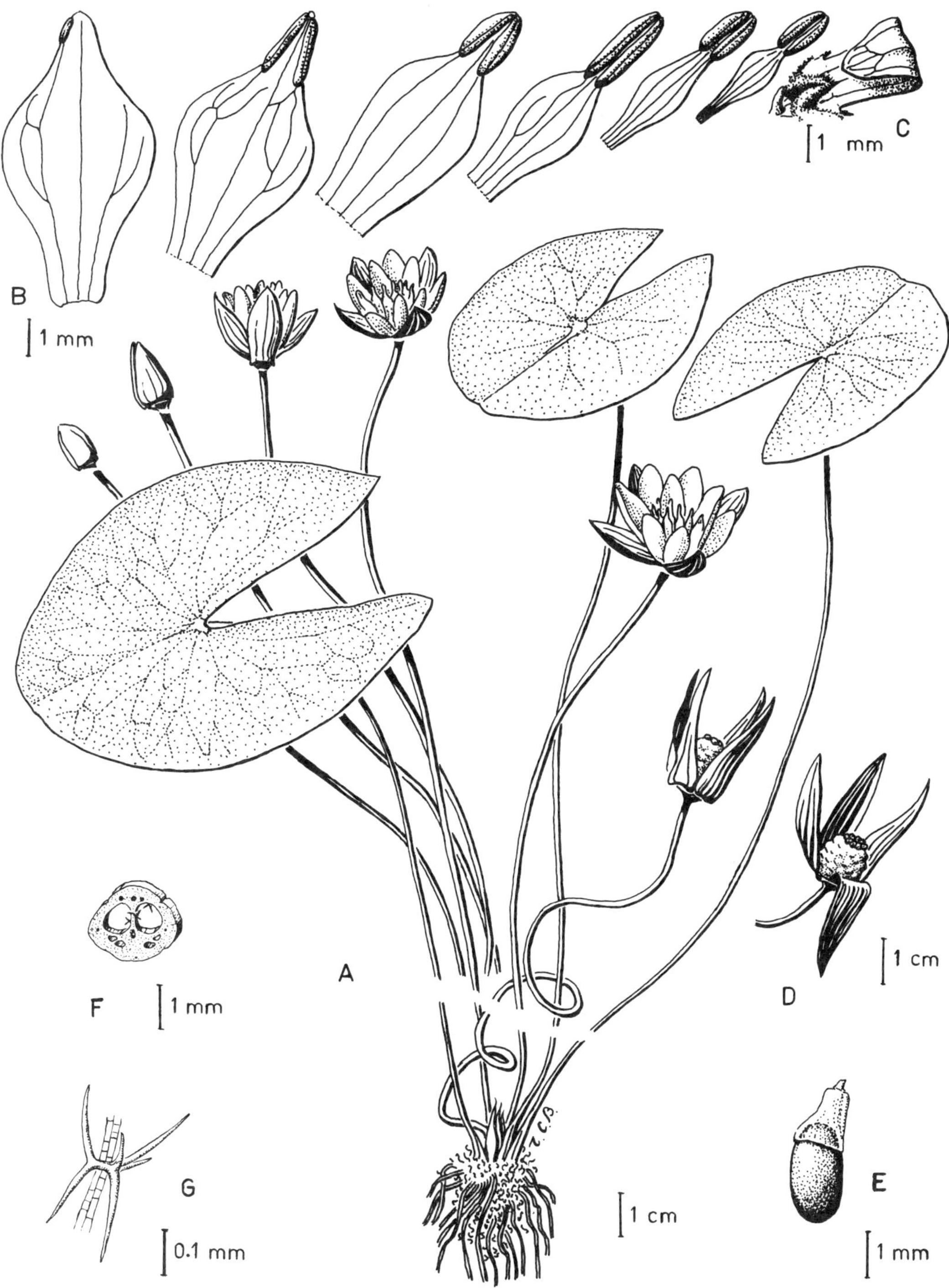

Figure 5. *Nymphaea tetragona*

A. flowering and fruiting plant.
B. 6 stamens: from outer (transitional to petals) to inner.
C. sector of top of ovary, with one stigma and its sterile appendage.
D. fruit, with one sepal turned down.
E. seed with aril.
F. cross-section of petiole.
G. sclereid.

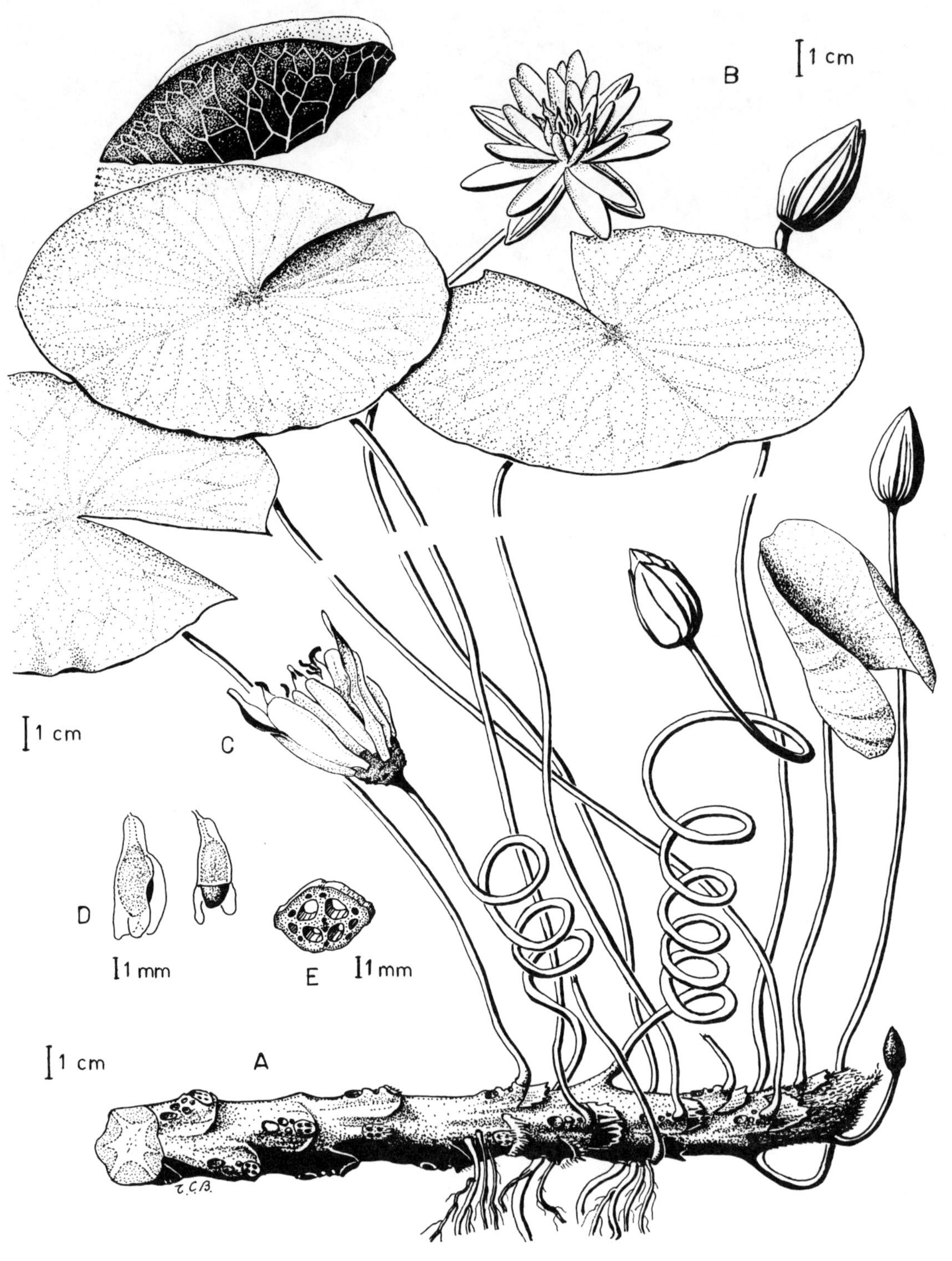

Figure 6. *Nymphaea odorata*

A. rhizome, with buds and fruit (C).
B. flower and foliage.
C. fruit.

D. seeds with arils.
E. cross-section of petiole, showing air-passages.

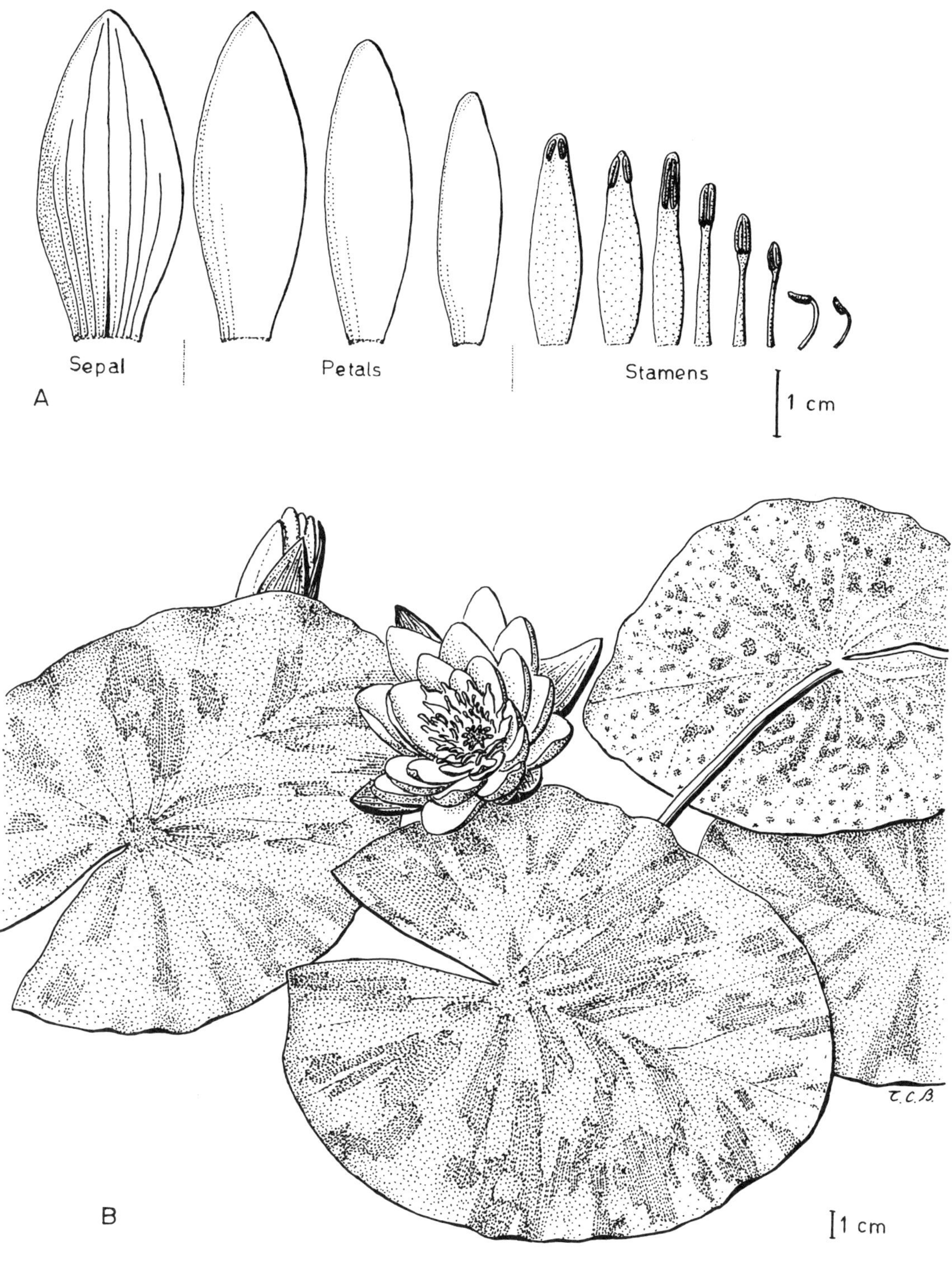

Figure 7.

A. *Nymphaea odorata*: selected series of floral appendages, from sepal through petals and stamens, showing transitions.
B. *Nymphaea mexicana*: flower and foliage.

performed mainly by beetles, according to Leppik (1964), but bees and bumblebees have also been seen by the present author to visit the flowers.

Nymphaea mexicana Zuccarini

Rhizomatous and stoloniferous plant with the general aspect of *N. odorata*. Leaf blades broadly elliptic to nearly orbicular, variegated with red spots and blotches beneath, and also often streaked with red above. Flowers floating, yellow, with many petals and stamens. $2n = 42$.

Native to Mexico and southernmost United States; sometimes cultivated in British Columbia, and persisting at points of introduction.

Nymphaea tuberosa **Paine,** of eastern North America, and *N. alba* **L.,** of Europe, are occasionally planted, and may persist at their points of introduction. They resemble *N. odorata* in general appearance, differing as indicated in the key. Pink and red flowered forms of *N. alba* are also known.

The Family CERATOPHYLLACEAE

Hornwort Family

Rootless, monoicous, submersed aquatic herbs with slender, weak stems. The stem with one central vascular strand, the xylem of which is without vessels and unlignified, being partly specialized for starch storage, and partly replaced by an air channel. Leaves forked, in regular whorls. Branches are axillary, and usually only one per leaf whorl.

Flowers apetalous; a flower (or inflorescence, according to interpretation) surrounded by a whorl of 8–15 basally connate appendages variously interpreted as bractlets or sepals, and terminating a short peduncle arising between adjacent leaf axils.

Staminate inflorescence comprising 8–18 stamens with no filaments, the anthers, with 2 pollen sacs, crowned by spongy, inflated connectives. Pollen grains thin-walled, without apertures. Pistillate inflorescence normally one-flowered; the single carpel terminated by a style and a linear (sometimes divided) stigma, and containing one pendulous ovule.

Fruit a biconvex achene crowned by the persistent style and furnished with lateral spines. The seed contains an embryo with almost no discernible root, but with a well-developed young shoot with three or four nodes and internodes.

The seedling possesses a rudiment of a root, which, however, makes no further development. Above the linear cotyledons, the first true leaves are an opposite pair, but all subsequent leaves are whorled.

A small family of one cosmopolitan genus. It is thought to be a highly specialized offshoot from the ancestral Nymphaeaceae.

CERATOPHYLLUM L.

Hornwort

Characters of the family; a widespread genus of nine species.

KEY TO SPECIES

1. Leaf divisions usually rather stout and stiff and distinctly toothed. Fruit with a pair of basal lateral spines, seedling with the first true leaves undivided.................................... *C. demersum*

   Leaf divisions fine and thread-like; teeth, if present, few and microscopically fine; fruit with 3–4 pairs of lateral spines. Seedling with all true leaves divided................. *C. echinatum*

Ceratophyllum demersum L.   **Hornwort, Coon-Tail**

Extensively growing perennial herb with leaves 6–12 in a whorl, each leaf once to thrice dichotomously forked, the divisions toothed on their outer and dorsal sides.

Pistillate and staminate inflorescences separate, minute and inconspicuous, terminating usually solitary short peduncles in different leaf whorls, each arising above a gap between adjacent leaves rather than in the axil of a leaf.

Each inflorescence has an encircling involucre of 8–15 minute, bristle-tipped, basally connate bractlets. Ovary normally solitary; if more than one, each additional ovary with its own circle of appendages within the overall involucre, suggesting that the complete structure is an inflorescence rather than simply a flower. The ovary contains one pendulous ovule and is beaked by a style and a usually undivided stigma. Anthers commonly 8, about 2 mm long, with minutely apically toothed connectives.

When the pollen is ripe, the anthers become detached, and float to the water surface, there to shed their pollen, which sinks slowly to make contact with the stigmas of the pistillate flowers.

Fruit a flattened nutlet about 5 mm long, tipped by a spine about 7 mm long, developed from the style, and bearing also a pair of baso-lateral spines. Fruit is sparingly produced, and not often collected. $2n = 24$.

The first few leaves of the seedling are thread-like and undivided, with an opposite pair at the first node, and a whorl at the second node. Divided leaves first appear at the third or fourth node of the stem (Les, 1985).

This plant may anchor itself to the bottom by colourless branches that grow downward into the mud; but it often may be suspended freely in the water, sometimes forming extensive mats in eutrophic lakes.

The foliage is not covered with a mucilaginous layer, and with its stiff, toothed leaves, it has a harsh, rough feel when handled.

Perennation is by means of shoot-tips that are scarcely specialized enough to be called winter buds, but which carry out the same function. These shoots, developed in the Autumn, show a distinctive form, with exceptionally stout, often dilated leaves with conspicuously toothed, incurved divisions. These shoots become detached, and sink to the bottom for the winter (Muenscher, 1944).

A cosmopolitan species, extending from arctic to tropical latitudes, it is found all over British Columbia in still ponds and lakes. This is our common species of this genus, becoming especially abundant in some polluted lakes. Floating or suspended mats may be independent of water depth, but plants may grow attached to the bottom in water to 4 metres deep.

Ceratophyllum echinatum Gray

A perennial herb similar to *C. demersum*, but finer and softer textured. Leaves forked 2 or 3 times into very slender, thread-like, divisions, but the sub-basal segments inflated. Teeth, if present, few and microscopically fine.

Fruit with a row of 3–5 spines along each edge. Seedling with all leaves forked; the first pair of seedling leaves once forked.

Ceratophyllum echinatum is widespread in the eastern half of North America, from about latitude 50° North southward. The western population, occurring in British Columbia and the coastal areas of the states of Washington and Oregon, appears to be separated from the eastern range by a wide region from which this species is absent. A map of its distribution has been published by Les (1985). This species is comparatively scarce in British Columbia.

The Family RANUNCULACEAE

Buttercup Family

Mostly perennial herbs of diverse aspect, usually with alternate leaves. Stipules may be present, or absent as such, and replaced by flaring and sheathing dilations of the petiole base.

Flowers complete and perfect or unisexual and dioicous, hypogynous, the parts nearly always separate; the symmetry usually radial. Sepals and petals whorled; stamens and carpels few and whorled or many and spirally arranged. Sepals (3–) 5–20, inconspicuous and often shed early; or large, coloured and petaloid; the latter kind especially distinctive when the petals are absent or reduced to nectaries. Sepals sometimes spurred. Petals (2–) 5–20, more persistent, large, coloured

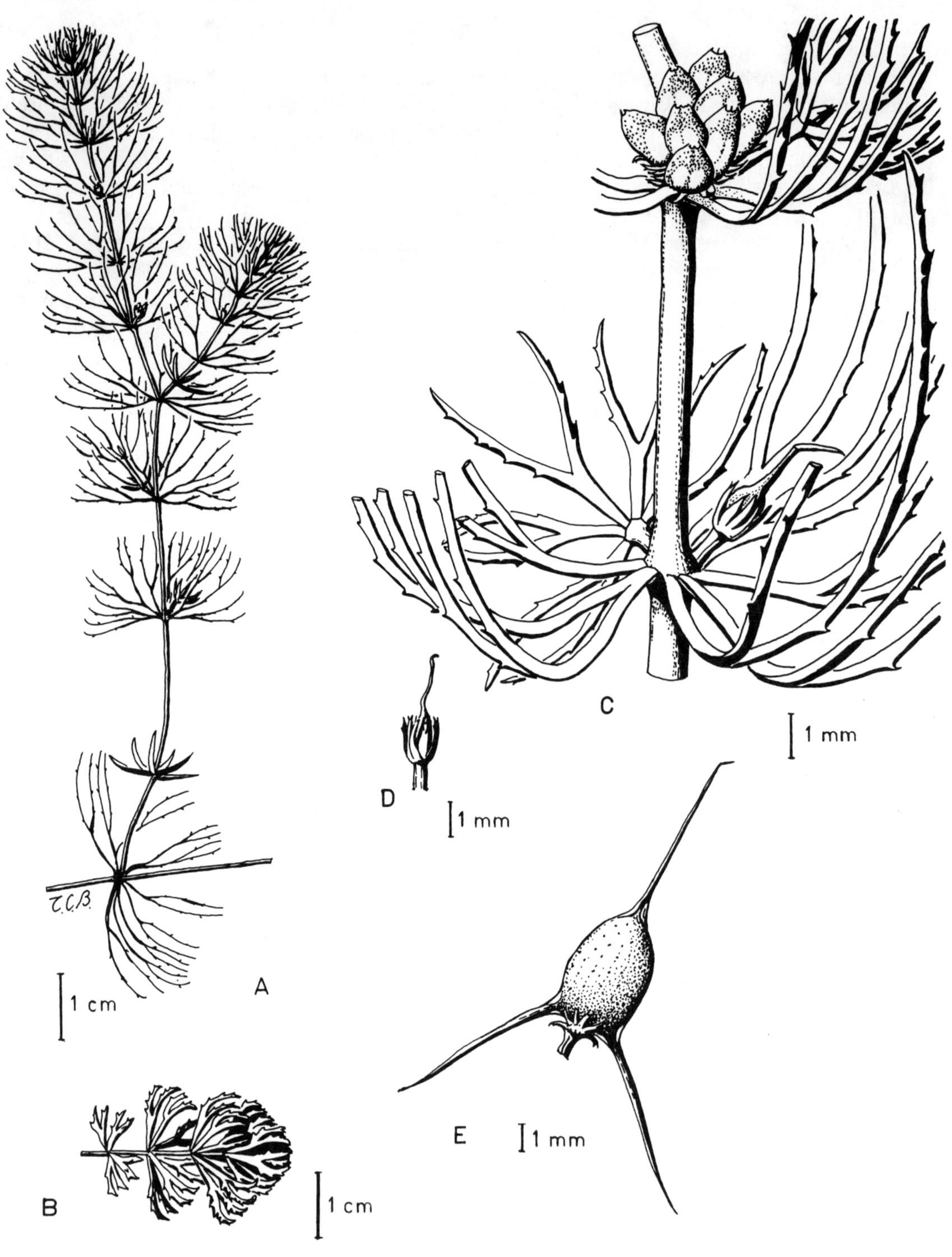

Figure 8. *Ceratophyllum demersum*

A. branch.
B. overwintering shoot-end.
C. two nodes, with staminate (upper) and pistillate (lower) inflorescences.
D. pistillate flower.
E. fruit (achene).

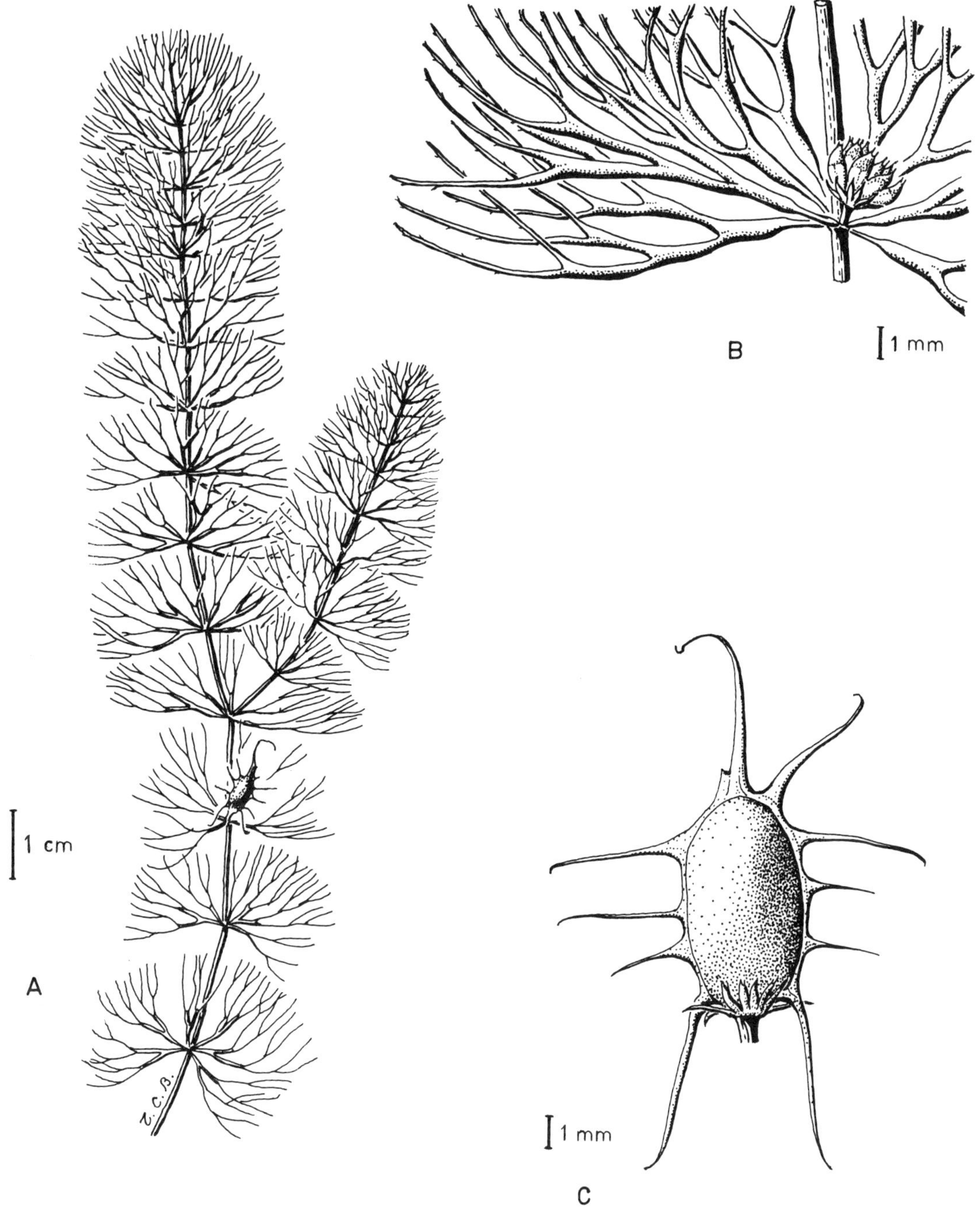

Figure 9. *Ceratophyllum echinatum*

A. fruiting branch.   B. leaf whorl and staminate inflorescence.   C. achene.

and showy, bearing nectaries near the base, or with reduced blades and essentially consisting of stalked nectaries, or they may be absent. Stamens few to many, usually spirally arranged; the anthers opening by longitudinal slits extrorsely, i.e. away from the floral axis. Pollen grains with 3 to many germination pores. Carpels few to many, typically separate, uniovulate or multiovulate, borne on receptacles that may become elongated in fruit.

The fruit produced from a flower is commonly a cluster, in the form of a head or spike, of achenes or follicles; but in *Actaea* it is a berry, and in *Nigella* it is a capsule.

The Ranunculaceae is a fairly large family; containing around fifty genera, depending on interpretation, and about 2000 species. The family is of practically worldwide distribution; but its member species are most common in temperate regions of the Northern Hemisphere, many of them being associated with moist or aquatic habitats.

Most members of the Ranunculaceae are poisonous to some degree. In species of *Anemone*, *Caltha*, and *Ranunculus* the harmful ingredient is a blistering irritant: protoanemonin, or its precursor, ranunculin. Toxic alkaloids are present in *Aconitum* and *Delphinium*; and the latter genus is a noted cause of livestock poisoning. An essential oil with toxic properties is present in *Actaea*, including its berries. The subject is well reviewed by Kingsbury (1964). The irritant properties have been exploited, commonly for counter-irritant purposes, in traditional medicine in both the New and Old Worlds (Turner, 1984.)

Classification of the Ranunculaceae.

This family, with its wide diversity of generic characteristics, has been a classifiers' puzzle since it was first recognized as a family, and a persistent source of disagreement among botanists concerned with its classification. In assigning genera among subfamilies and tribes the classical approach has emphasized fruit types, followed by flower and leaf structures as the fundamental distinguishing characters.

This approach has led to the division of the family, as represented in our flora, into two or three tribes or subfamilies: the Helleboroideae, with multiseeded follicles, berries or capsules, and the Ranunculoideae, (or Anemonoideae), with achenes. The further subdivision of these subfamilies into tribes varies among authors according to the degree of importance assigned to flower structure, leaf form and arrangement, etc. Leppik (1964) produced the scheme shown in Table 1.

TABLE 1:   CLASSIFICATION OF BRITISH COLUMBIAN GENERA OF RANUNCULACEAE AFTER LEPPIK (1964)

Subfamily	Tribe	Genera
HYDRASTIDOIDEAE	Hydrastideae	*Hydrastis, Glaucidium,*
PAEONIOIDEAE		*Paeonia*
ANEMONOIDEAE	Clematideae	*Thalictrum, Anemone, Pulsatilla, Clematis, Hepatica*
	Ranunculeae	*Myosurus, Ranunculus, Trautvetteria*
HELLEBOROIDEAE	Isopyreae	*Cimicifuga, Coptis, Isopyrum, Actaea, Aquilegia*
	Trollieae	*Caltha, Trollius, Nigella, Aconitum, Delphinium*

Botanists investigating the cellular mechanisms of the plants found fundamental differences in form and numbers of the chromosomes among the genera. This is well expressed by Gregory (1941), who, postulating that achenes had originated more than once by reduction from follicles, proposed a classification using chromosome forms and basic numbers for genera as the first decisive characters. Table 2 shows a classification of the genera in British Columbia into tribes, based on Gregory's (1941) paper:—

HELLEBOREAE: Large, 'Ranunculus-type' chromosomes; basic number $x = 8$ (6 in most *Nigella*). Fruit: follicles, berry, or capsule.

Trollius	*Cimicifuga*	*Delphinium*
Caltha	*Actaea*	*Aconitum*
Nigella		

RANUNCULEAE: (Anemoneae by Gregory: Ranunculeae is correct according to Article 19 of International Code of Botanical Nomenclature). Large, 'Ranunculus-type' chromosomes; basic number $x = 8$, sometimes 7. Fruit: achenes, usually one-seeded; seed attached basally or usually low on a ventral placenta.

Anemone	*Ranunculus*	*Trautvetteria?*
		(not placed by Gregory)
Clematis	*Myosurus*	

THALICTREAE: Small, kidney-shaped, 'Thalictrum-type' chromosomes; basic number $x = 7$. Fruit: follicles or achenes. Leaves bipinnately or biternately compound, the leaflets with rounded outline or lobes.

(a) Fruit: follicles.

| *Isopyrum* | *Aquilegia* |

(b) Fruit: achenes, one seeded; seed attached apically.

Thalictrum

COPTIDEAE: Small, slender, 'Coptis-type' chromosomes; basic number $x = 9$. Fruit: follicles.

Coptis

It can be seen that this scheme cuts across groupings based on fruit and flower morphology. The chromosome complement carries the code directing the chemical processes that ultimately determine all the morphological characteristics of the organism. The author therefore feels that its characteristics are of basic importance as indicators of the most natural, if not the most obvious, relationships among species; and prefers it over other classification schemes proposed.

KEY TO GENERA

1. Flowers in racemes **2**

    Flowers solitary or in cymes, panicles or umbels **5**

2. Flowers bilaterally symmetric, blue, rarely white. Leaves simple or rarely palmately divided **3**

    Flowers radially symmetric, white. Leaves bipinnately compound **4**

3. Upper sepal spurred. Petals 4 or 2, visible externally .... ***Delphinium***

    Upper sepal hooded. Petals usually 2, concealed in hood .... ***Aconitum***

4. Petals several, small. Carpel 1. Fruit a berry. Plant 1 metre or less tall. Widespread .... ***Actaea***

    Petals none in ours. Carpels 1–4. Fruit follicles. Plant 1–2 metres tall. Local in Cascade Range .... ***Cimicifuga***

5. Woody vines with opposite compound leaves .... ***Clematis***

    Herbaceous plants **6**

6.   Tufted, linear-leafed scapose annuals with spurred sepals & elongate receptacles bearing quadrilateral achenes... *Myosurus*

      Mostly perennials with broad leaves; or if linear-leafed, with arching, stolonate stems.... **7**

7.   Stem leaves opposite or in whorls. Petals none. Sepals showy. Fruit achenes......... *Anemone*

      Stem leaves alternate or none... **8**

8.   Flower with an involucre of finely dissected bracts. Carpels joined. Fruit a capsule... *Nigella*

      Flower without an involucre. Carpels separate.. **9**

9.   Petals present, bearing nectaries.. **10**

      Petals absent or very small and inconspicuous.. **13**

10.   Petals with conspicuous hollow spurs. Leaves biternately compound................. *Aquilegia*

      Petals not spurred.. **11**

11.   Petals conspicuous, yellow or sometimes white. Fruit achenes.................... *Ranunculus*

      Petals often inconspicuous or concealed. Fruit follicles.. **12**

12.   Petals concealed under stamens. Leaves simple but palmately lobed. Sepals large, white.. *Trollius*

      Petals visible, though reduced, shorter than sepals. Leaves compound.................. *Coptis*

13.   Leaves compound.. **14**

      Leaves simple.. **16**

14.   Plant scapose. Fruit follicles.. *Coptis*

      Plant with leafy stems.. **15**

15.   Flower solitary, with large white sepals. Fruit follicles................................... *Isopyrum*

      Flowers in panicles. Sepals small. Fruit achenes.. *Thalictrum*

16.   Leaves crenate-margined, not lobed. Fruit follicles.. *Caltha*

      Leaves both lobed and sharply toothed.. **17**

17.   Sepals large, white, showy. Flower usually single. Fruit follicles....................... *Trollius*

      Sepals small, shed as flower opens. Stamens white: the conspicuous floral parts. Flowers in a terminal cyme. Fruit achenes... *Trauvetteria*

ACONITUM L.

Monkshood

Perennial, non-rhizomatous herbs with compact fibrous or fleshy roots, alternate, palmately lobed leaves, and terminal, and sometimes axillary, racemes of bilaterally symmetric flowers followed by follicular fruits.

Flowers with five showy coloured sepals: the upper sepal large and forming a hood, often with a small apical beak, the upper laterals (partly covered by the hood) oval to kidney-shaped, the lower laterals narrower. Petals concealed under the hood, the upper two always present, with long stem-like basal claw and a small apical blade and large nectar-gland; the lower three petals absent or vestigial. Stamens many, with dilated filament bases. Carpels 3 to 5, distinct, multi-ovulate, tipped by styles and terminal stigmas.

Fruits a follicle from each carpel, opening along the ventral suture. Seeds angled and sometimes winged or otherwise appendaged.

A northern temperate zone genus of 50 or more species, most in Asia; two species native in British Columbia. Pollination is performed by bumblebees.

KEY TO SPECIES

1. Leaves somewhat crowded beneath the inflorescence. Blades divided quite to base into 3–5 leaflets... ***A. compactum***

   Leaves basal and evenly spaced along the stem. Blades shallowly to deeply lobed; seldom divided to base... **2**

2. Hood higher than long.. ***A. columbianum***

   Hood not higher than long, broadly crescent-shaped as seen from the side.................... ***A. delphiniifolium***

Aconitum columbianum **Nuttall** *in* **Torrey & Gray**

Tall (up to 2 m) perennial with one to several stems from a tuberous root; the stem thick and hollow, glabrous below. Leaves mainly on the stem, up to 2 dm wide, not dissected to base, the lobes broad or narrow, and variously toothed or lobed. Petioles long on lower leaves, but very short on the upper leaves.

Flowers in an elongate terminal raceme, sometimes with axillary racemes in vigorous plants. Peduncles and pedicels pubescent and sometimes glandular. Sepals blue or rarely white, in our material puberulent to glabrous; the hood high-crowned, and often beaked in front. The two nectariferous petals erect, with a stout claw and hooked nectary, other petals not developed. Carpels 3, glabrous to sparsely puberulent.

Follicles 3, one to two cm long. Seeds 3–4 mm long, with a longitudinal keel and numerous transversely arranged fringes. 2n = 16, 18.

Moist woods, commonly in alluvial situations, in the Subalpine and Columbia forest regions from southern British Columbia and Alberta southward along the mountain ranges to California and Colorado.

Aconitum delphiniifolium **DC.**

Tap-rooted perennial of very variable size from 1 to 10 dm tall, with a single slender, hollow stem; finely pubescent to nearly glabrous.

Leaves 3–12 cm wide, palmately cleft almost to base into divisions that on the average are narrower than in *A. columbianum*, the divisions pinnately divided with linear lobes, leaves tending to be more concentrated at the base than in *A. columbianum*. Petioles long on lower leaves, but very short on the upper leaves.

Flowers deep blue, rarely white, narrow, and nearly circular as seen from the side. Hood broadly crescent-shaped as seen from the side, and longer than deep. Upper lateral sepals nearly circular, and the lower ones narrowly elliptic to lanceolate. Upper petals with long, arching claws, hollow spur-like or sac-like nectar glands, and narrow, recurved blades. Other petals vestigial or absent. Stamens many, recurving. Carpels 3, or sometimes 4 or 5, pubescent to glabrescent.

Follicles up to 27 mm long, seeds about 3 mm long by half as wide with several oblique membranous fringes. 2n = 16 in North America, 32 in Asia.

This species ranges from the Mackenzie Valley, N.W.T. westward across Alaska and eastern Siberia, and southward to around Mount Robson in British Columbia. It is found in subalpine forest and upward onto alpine tundra. Its size reflects its environment; and woodland plants may be a metre tall with many-flowered racemes, while plants of exposed rocky tundra may be only 10 cm tall with one or two flowers.

The great range of individual variation in this species has led to the publication of subspecies based on very weak criteria, such as stature and leaf width; both of which respond to environmental conditions. Only the following distinction is recognized here:

KEY TO SUBSPECIES

1. Nectary hooked.. **ssp.** *delphiniifolium*

   Nectary very short, not hooked... **ssp.** *paradoxum*

25

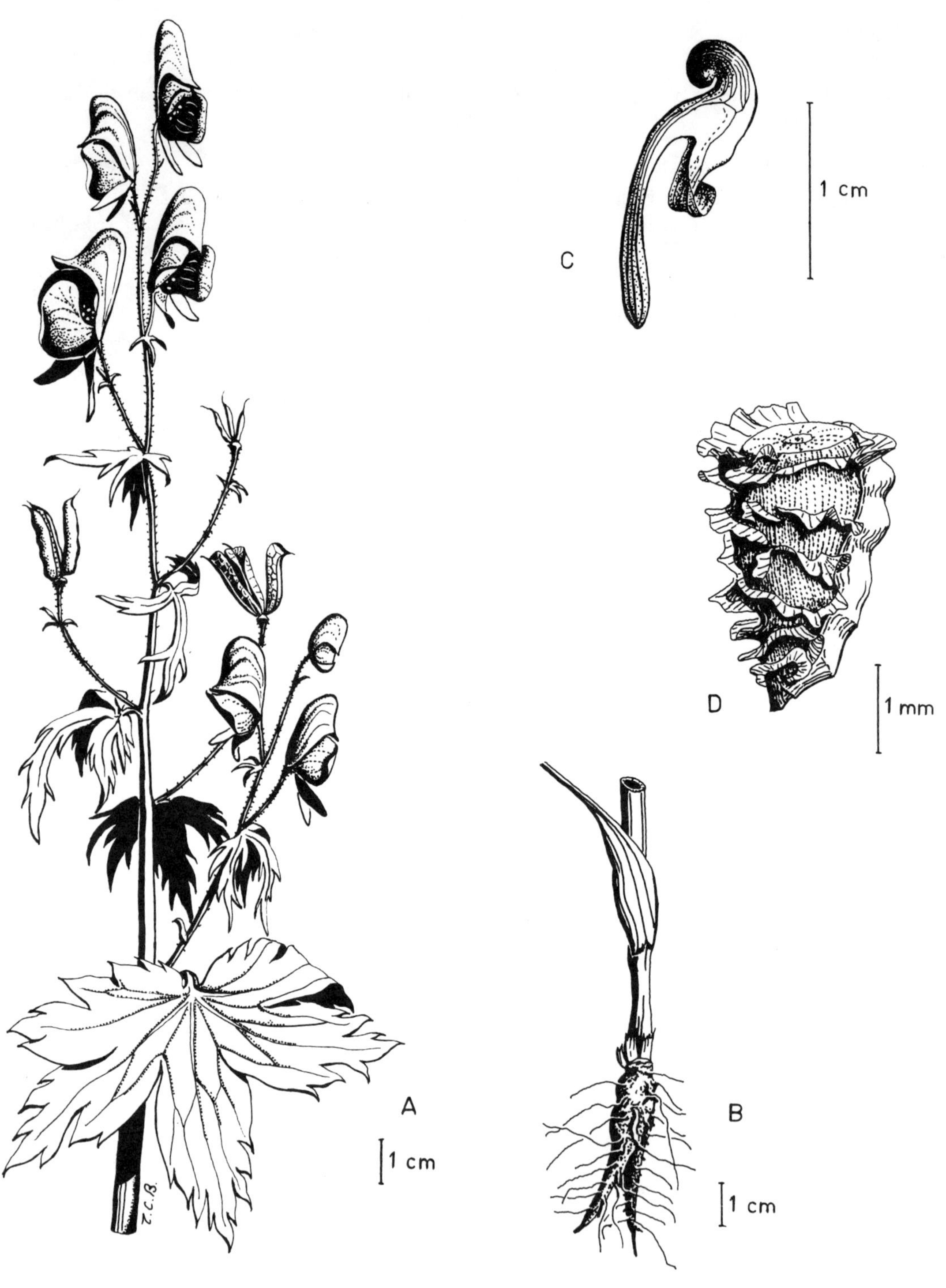

Figure 10. *Aconitum columbianum*

A. top of plant.
B. base of plant and root.

C. nectariferous petal.
D. seed.

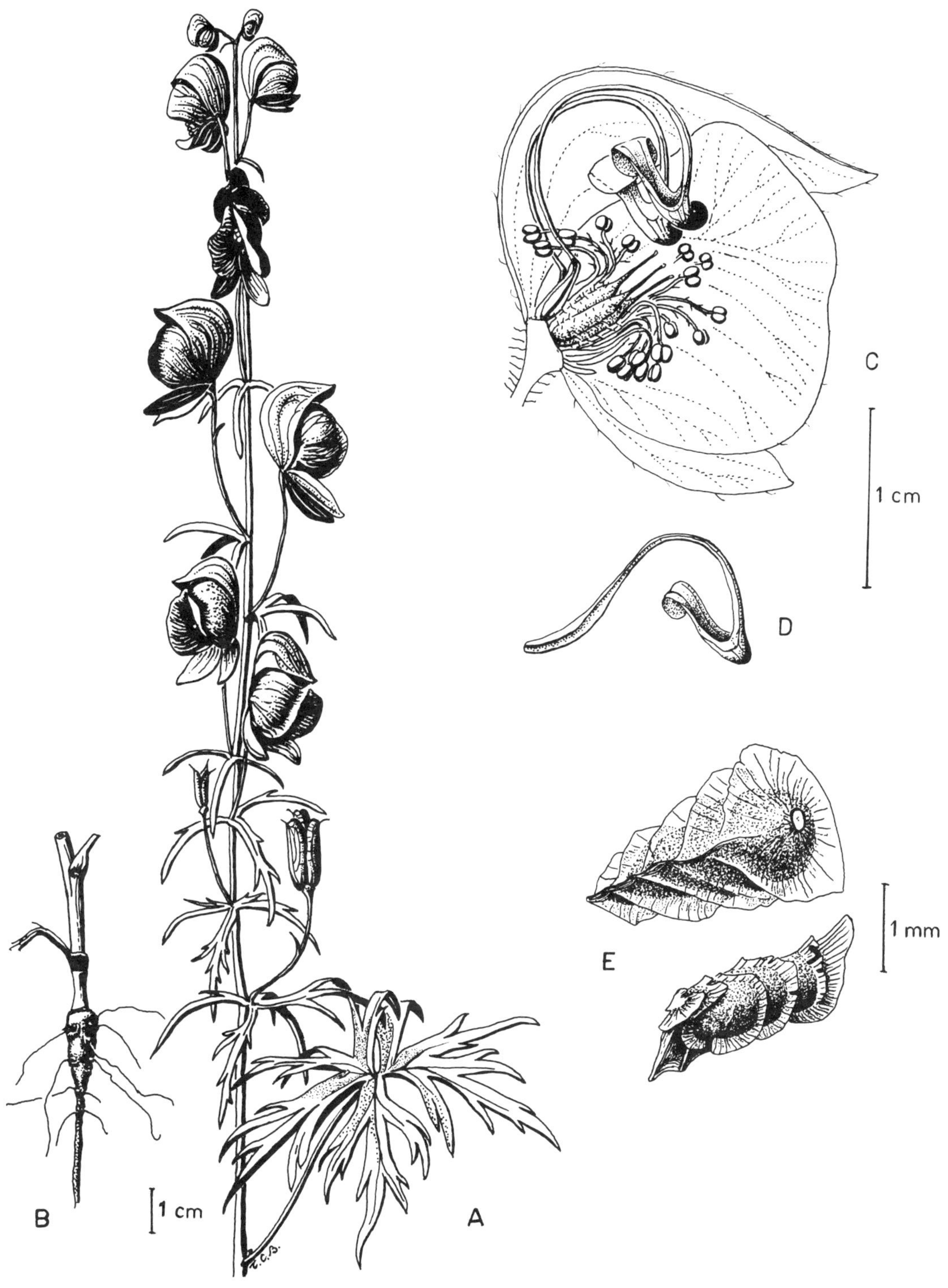

Figure 11. *Aconitum delphiniifolium*

A. top of plant.
B. stem base and root.
C. longitudinal section of flower of ssp. *delphiniifolium*, showing petals.
D. petal of ssp. *paradoxum*.
E. Seeds.

Subspecies *paradoxum* (Beal) Hulten, described as a very short plant with one to only a few flowers, and growing on the arctic tundra and very scattered further south, is known with us from Rocher Deboule, near Hazelton.

Aconitum compactum (Reichenbach) Gayer
? *A. napellus* L. var.?

Resembling *A. delphiniifolium*, but the leaves are larger (12–15 cm across), and rather crowded beneath the raceme; the blades divided completely to base into 3–5 leaflets that in turn are deeply divided into linear, acuminate lobes. Pedicels short (ca. 1 cm), stout, and ascending close to the axis of the raceme.

This member of the *A. napellus* L. complex of closely related species is a native of western Europe that is sometimes cultivated here. It has been found once, growing as a garden escape on land-fill at Cadboro Bay, near Victoria.

ACTAEA L.

Baneberry

Perennial herbs with compound leaves and terminal racemes of small flowers. Flowers perfect, usually complete, with many stamens and one multiovulate carpel. Fruit a several-seeded berry.

A genus of about 6 species of north temperate latitudes. One species in British Columbia.

Actaea rubra (Aiton) Willdenow
Red Baneberry
A. spicata L. var. *rubra* Aiton
A. arguta Nuttall
A. rubra ssp. *arguta* (Nuttall) Hulten
A. neglecta Gillman (=*A. rubra* forma *neglecta* (Gillman) B.L. Robinson)
A. eburnea Rydberg (=*A. rubra* forma *neglecta*)

Sparingly branched or simple-stemmed plant up to a metre tall, sparingly pubescent; commonly with one basal (or near-basal) leaf, and one or two cauline leaves.

Basal leaf (actually a very low cauline leaf) long-petioled, ample, biternate and pinnate or triternate; the leaflets lobed and toothed with usually sharp teeth. Mid-cauline leaf similar but smaller, the upper cauline leaf reduced and with fewer leaflets.

Flowers small, white, in a short raceme 2–6 cm long; the axis of the raceme elongating to 8 to 15 cm long as the fruit matures; the pedicels also elongating. Sepals 3–5, 2–4 mm long, deciduous as the flower opens, or shortly afterward. Petals 5–10, much shorter than the sepals, spatulate, white. Stamens many, white, longer than the perianth parts. Carpel one, with a small terminal, sessile stigma.

Fruit a shiny, several-seeded berry; usually red but occasionally white. 2n = 16.

The western form of this species may be distinguished as ssp. *arguta* (Nuttall) Hulten on the basis of its berries, which are slightly smaller (5–8 mm in diameter) than those of the eastern North American ssp. *rubra* (10–11 mm in diameter). The distinction appears to be weak:- some specimens from Vancouver Island have fruit that scarcely differs in size from some from the Province of Quebec.

The berries are normally red. The occasional white-fruited plants may be distinguished as forma *neglecta* (Gillman) Robinson.

Actaea rubra is widespread across North America; from Alaska to Newfoundland, and southward in the west to California. It is found in woods, especially in rich alluvial sites. The whole plant, including the berries, is poisonous.

ANEMONE L.

Anemone, Wind-Flower

Perennial herbs with stems and basal leaves clustered on a branched subterranean caudex, or scattered on a creeping rhizome. The stem with a false whorl ('involucre') of 2, 3, or more leaves

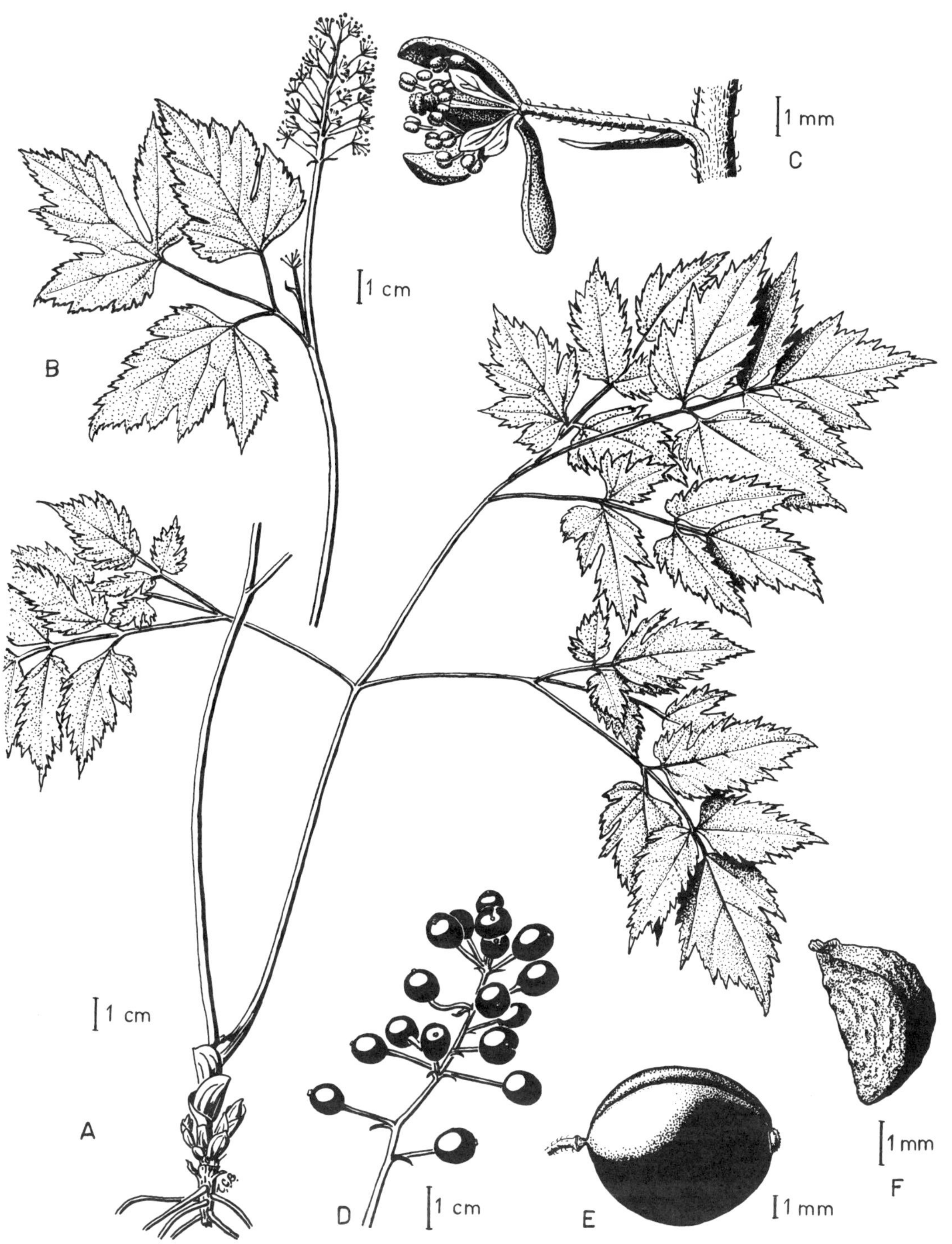

Figure 12. *Actaea rubra*

A. base of plant, with basal leaf.
B. top of flowering plant.
C. flower, newly opened.
D. fruiting raceme.
E. berry.
F. seed.

(bracts) at the base of the inflorescence, and secondary whorls ('involucels') in several-flowered inflorescences. These stem leaves are basically alternate, but appear whorled or opposite owing to the failure of some internodes to elongate. Leaf blades variously dissected.

Inflorescence basically cymose, but commonly umbel-like owing to the non-elongated internodes at the base of the inflorescence, or often, single-flowered. Flowers lacking petals, but the sepals large and coloured, and overlapping in bud. Stamens many. Carpels many, single-ovuled. Fruit a head or spike of achenes. $x = 7, 8$.

A genus of about 100 species, mostly of northern temperate latitudes. Pollination is effected by a variety of insects, including bees and beetles, and bumblebees in section *Pulsatilla*. The achenes of many species are provided with long feathery styles, or are covered with dense, long, woolly hairs, and appear to be adapted for dispersal by the wind; but some species produce glabrous achenes.

Like most other members of this family, *Anemone* species are poisonous.

Our twelve species of *Anemone* are classified among two subgenera and six sections, as in Table 3. The descriptive treatment follows the species order in that table.

TABLE 3: SUBGENERA, SECTIONS, AND BRITISH COLUMBIAN SPECIES OF *ANEMONE*

Subgenus *Anemone*
    Section *Anemone* $x = 8$
        A. cylindrica
        A. riparia
        A. drummondii
        A. multifida
        A. parviflora
    Section *Sylvia* $x = 8$
        A. lyallii
        A. piperi
    Section *Anemonidium* $x = 7$
        A. canadensis

    Section *Omalocarpus* $x = 7$
        A. narcissiflora

    Section *Rivularidium* $x = 7$
        A. richardsonii

Subgenus *Pulsatilla*
    Section *Pulsatilla* $x = 8$
        A. occidentalis
        A. patens

KEY TO SUBGENERA, SECTIONS AND SPECIES

1. Achenes with long, plumose styles 2 cm or more long... Subgenus ***Pulsatilla***; Section ***Pulsatilla***   **2**

   Achenes with short styles less than 7 mm long and not plumose ...... Subgenus ***Anemone***   **3**

2. Leaves ternate, the divisions again ternate. Sepals blue to violet. Fruiting receptacle globose.. ***A. patens***

   Leaves ternate and the divisions pinnately multifid. Sepals white. Fruiting receptacle elongate .... ***A. occidentalis***

3. Leaves of involucre usually sessile, simple and broadly though often deeply lobed............ **4**

   Leaves of involucre stalked, and compound or deeply multifid into linear divisions............ **6**

4. Achenes glabrous, with the stylar beak reflexed. Flowers yellow .... Section ***Rivularidium***; ***A. richardsonii***

   Achenes normally with at least some hairs. Flowers white, blue or purple............................ **5**

5. Flowering stems 20 cm or more tall, with more than one flower. Leaves more than 5 cm wide. Mature achenes sparsely hispid, keeled marginally... Section ***Anemonidium; A. canadensis***

   Flowering stems usually less than 15 cm tall; one-flowered. Leaves 4 cm wide or less. Mature achenes copiously woolly... Section ***Anemone; A. parviflora***

6. Achenes glabrous, distinctly keeled marginally .... Section *Omalocarpus; A. narcissiflora*
   Achenes hairy **7**

7. Plant with creeping rhizome. Basal leaves separate from flowering stem. Achenes finely
   pubescent or hispid .... Section *Sylvia* **8**
   Plant with erect caudex. Basal leaves clustered around base of flowering stem. Achenes densely
   woolly .... Section *Anemone* **9**

8. Rhizome pale, horizontal. Sepals 4–8 mm long. Stamens 12–30. Coast to Cascade
   Range.. *A. lyallii*
   Rhizome dark brown, oblique to ± vertical. Sepals 8–16 mm long. Stamens 35–60.
   Kootenay region.. *A. piperi*

9. Achenes in elongate spikes 2–5 times as long as thick. Leaves with broad cuneate divisions **10**
   Achenes in ovoid to globose heads. Leaves with ± linear divisions.................................. **11**

10. Involucre of usually 3 leaves. Involucels regularly present and conspicuous, placed well
    above the involucre. Achene beak about 1.5 mm long .... *A. riparia*
    Involucre of 4–9 leaves. Involucels, if present, set down among the involucral leaves; all
    pedicels appearing leafless. Achene beak 1 mm or less long................................ *A. cylindrica*

11. Achene beaks 2–4 mm long, slender. Flower one per stem. Sepals often blue dorsally
    .... *A. drummondii*

    Achene beaks 1–1.5 mm long, relatively stout. Flowers often more than one per stem.
    Sepals white or ivory-coloured to purple or pink, rarely blue .... *A. multifida*

Anemone cylindrica Gray   **Thimbleweed**

Non-rhizomatous perennial, with an upright caudex, often branched, giving rise to one or more
greyish hirsute to glabrescent stems and leaves. The stems 3–7 dm tall, normally bearing one
involucre and 3 or more flowers on apparently bractless pedicels.

Basal leaves clustered around the stem, long-petioled; the blade 4–10 cm wide, pentagonal in
outline, and palmately divided into 3, or occasionally 5, sessile, rather concavely cuneate-based,
rhombic leaflets that are lobed and toothed. Involucral leaves 4 –9, on petioles 1–5 cm long; with
blades similar to those of the basal leaves. Involucels, if present, usually small and placed
inconspicuously, low down among the involucral leaves.

Flowers 2–7 per stem; their pedicels appearing to lack involucels. Sepals normally 5, narrowly
elliptic, 10–15 mm long, white, tomentose dorsally, obliquely ascending, the flower commonly
rather cup-like. Stamens many. Carpels very many, in a column, on an elongate, hairy receptacle.

Achenes in a cylindrical head or spike 2–4 cm long by 7–10 mm thick at maturity. The achene
body elliptic, 2–3 mm long, tapering into a short (ca. 1 mm or less long), glabrous or very short-
haired beak with a minute recurved tip; the body almost concealed among the long woolly hairs that
arise mostly near its base. 2n = 16.

The plant is more readily recognized in fruit than in flower; as no other *Anemone* in this region
has such elongate and slender spikes of achenes.

Anemone cylindrica ranges over much of North America, from eastern British Columbia across
the Prairie Provinces, to southern Quebec and Maine; and south to New Mexico. In British
Columbia it is found in the Peace River basin, and, uncommonly, in a few localities west of the
Rocky Mountains. It is typically a plant of open prairie, but is sometimes found in dry open woods.

Anemone riparia Fernald   **Thimbleweed**

Robust perennial, generally similar to *A. cylindrica*. Stems arising from a caudex, up to over a
metre tall, sparsely villous to glabrous, bearing 3–10 flowers.

Basal leaves with blades 8–20 cm wide, divided into 3–5 leaflets that are obovate to
oblanceolate or rhombic, rather more broadly cuneate and straighter-margined at base than in

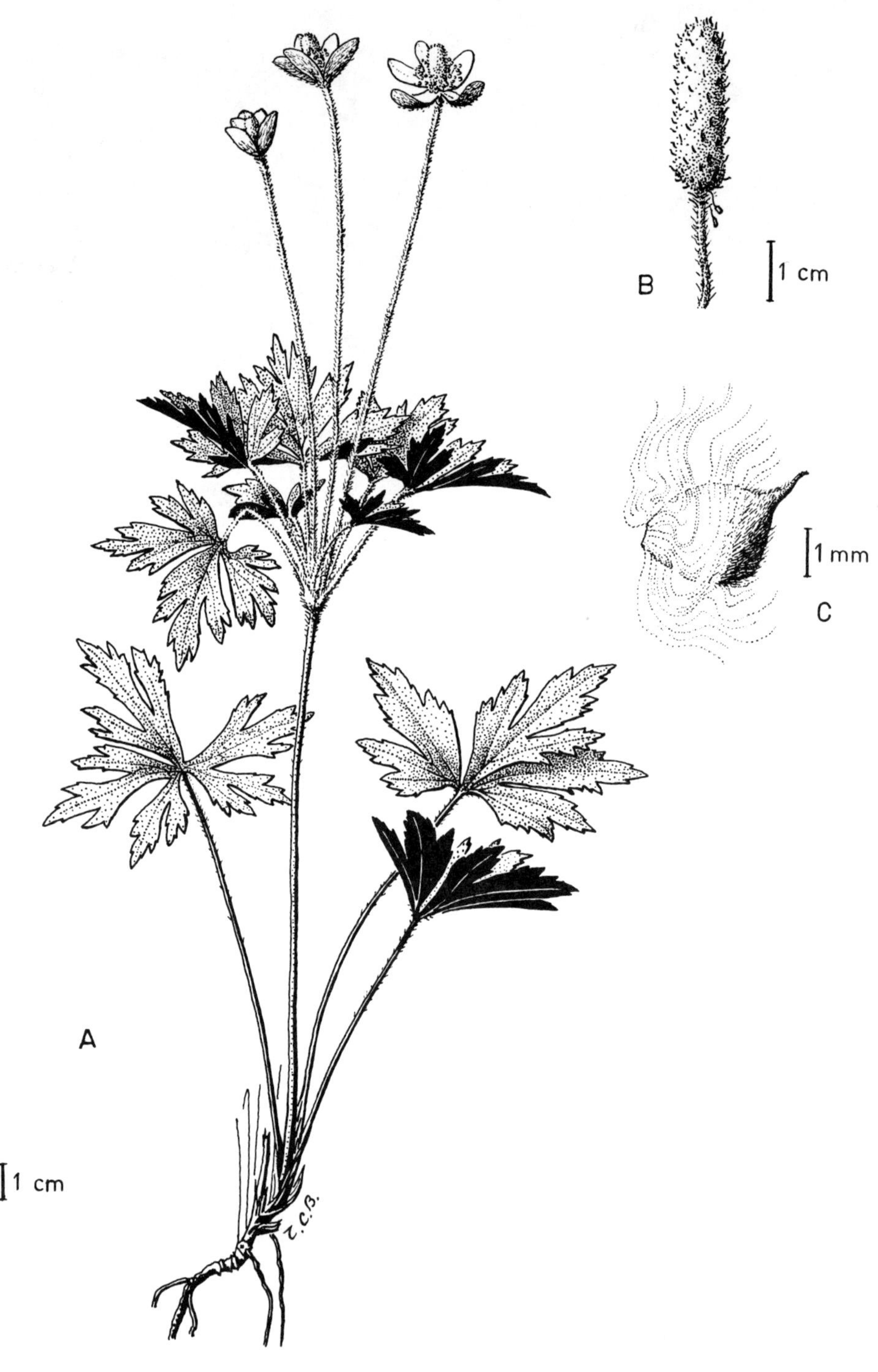

Figure 13. *Anemone cylindrica*

A. flowering plant.   B. spike of achenes.   C. achene.

Figure 14. *Anemone riparia*

A. flowering plant.   B. head of achenes.   C. achene.

33

A. cylindrica, sharply lobed and toothed, and sparsely pubescent. Involucral leaves normally three, short-petioled, the blades similar to those of the basal leaves. Involucels present, clearly above the involucres, of normally two, rather smaller leaves; the peduncles sometimes branching further at that level. Foliage tending to darken on drying.

Flowers usually three or more. Sepals 4–6, broadly oblong to elliptic, the lateral margins sometimes inrolled distally, 1.3–2 cm long, 0.8–1.5 cm wide, white or greenish, tomentose dorsally. Anthers 0.7–1.2 mm long. Carpels very many in a compact, ovoid to cylindrical head.

Achenes many in a stoutly cylindrical, pale brownish spike 1.2–3 cm long by 7–11 mm or more thick. Achene body 2.5–3 mm long, largely concealed in the brownish wool. Stylar beak about 1.5 mm long, upcurving with a recurving tip, partly short-hairy except for the glabrous dorsal edge. 2n = 16.

Anemone riparia is found from British Columbia eastward to Newfoundland, and southward to New York and Illinois, on calcareous rocks and river gravels. It is rare in British Columbia; and known only from three records scattered along the Fraser River, and one from the Peace River basin.

Anemone drummondii Watson   **Alpine Anemone**

Low hirsute perennial with a branching caudex, and stems 10–20 (rarely to 25) cm tall: the plants sometimes forming broad, dense clumps.

Basal leaves numerous, petioled; the blades 2–6 cm wide, trifoliolate, the terminal leaflet stalked. Leaflets dissected into linear to lanceolate lobes 1–2.5 mm wide. Involucral leaves 3 or 4, similar to the basal leaves but on broad flat petiolar bases a centimetre or less long.

Flower solitary and terminal, or occasionally 2. Sepals 5–9, usually oblong-lanceolate, acute to rounded apically, 8–20 mm long, white within, pubescent and commonly bluish without. Stamens many. Carpels many. Receptacle ellipsoidal, densely tomentose.

Achenes in a globose or ovoid head; the achene body 2–4 mm long, dark and thinly pubescent dorsally, surrounded by long (3–5 mm) silky hairs that arise on its lower half, the filiform stylar beak 1½–4 mm long, ascending, straight or nearly so. 2n = 16, 32, 48.

KEY TO VARIETIES

1. Leaf lobes 1–1.5 (–2) mm wide, ± hirsute. Styles 2–4 mm long. 2n = 32 **var. *drummondii***
   Leaf lobes 1.5–2.5 mm wide, almost or quite glabrous. Styles 1.5–3 mm long. 2n = 48,
   (16) .... **var. *lithophila***

Anemone drummondii ranges from Kamtchatka through Alaska and the western Cordilleran region to California. It occupies alpine and subalpine habitats, often on open, eroded slopes. Within its overall range, this species comprises a number of regional varieties.

Variety *drummondii* occurs from southwestern British Columbia along the Cascade and Coast Ranges to California; and a more stoutly styled form is found from Alaska to Northwest Territories. Boraiah and Heimburger (1964) point out a peculiarity of the seed germination in the Californian population of this variety in which the petiolar bases of the cotyledons are joined (though the blades are distinct), and the shoot arises beside what appears to be a branch-like hypocotyl. The northern form has distinct cotyledons on very long petioles.

Variety *lithophila* (Rydberg) C. L. Hitchcock occurs across central and southern British Columbia to Alberta, and along the Rocky Mountains southward to Wyoming. Boraiah and Heimburger (1964) point out that the different chromosomal complements of these varieties indicate a definite genetic barrier between these taxa, justifying the treatment of *A. lithophila* Rydberg as a separate species. Because of the difficulty in separating them unambiguously on gross morphological criteria, it is considered here to be more practical to maintain them as varieties within one inclusive species, following Hitchcock and Cronquist (1964).

An added complexity has been identified in the discovery of diploid (2n = 16) populations of var. *lithophila* in Kamtchatka and in the unglaciated parts of Alaska and Yukon. These cytological differences are not associated with distinctions in gross morphological characters (Heimburger; personal communication).

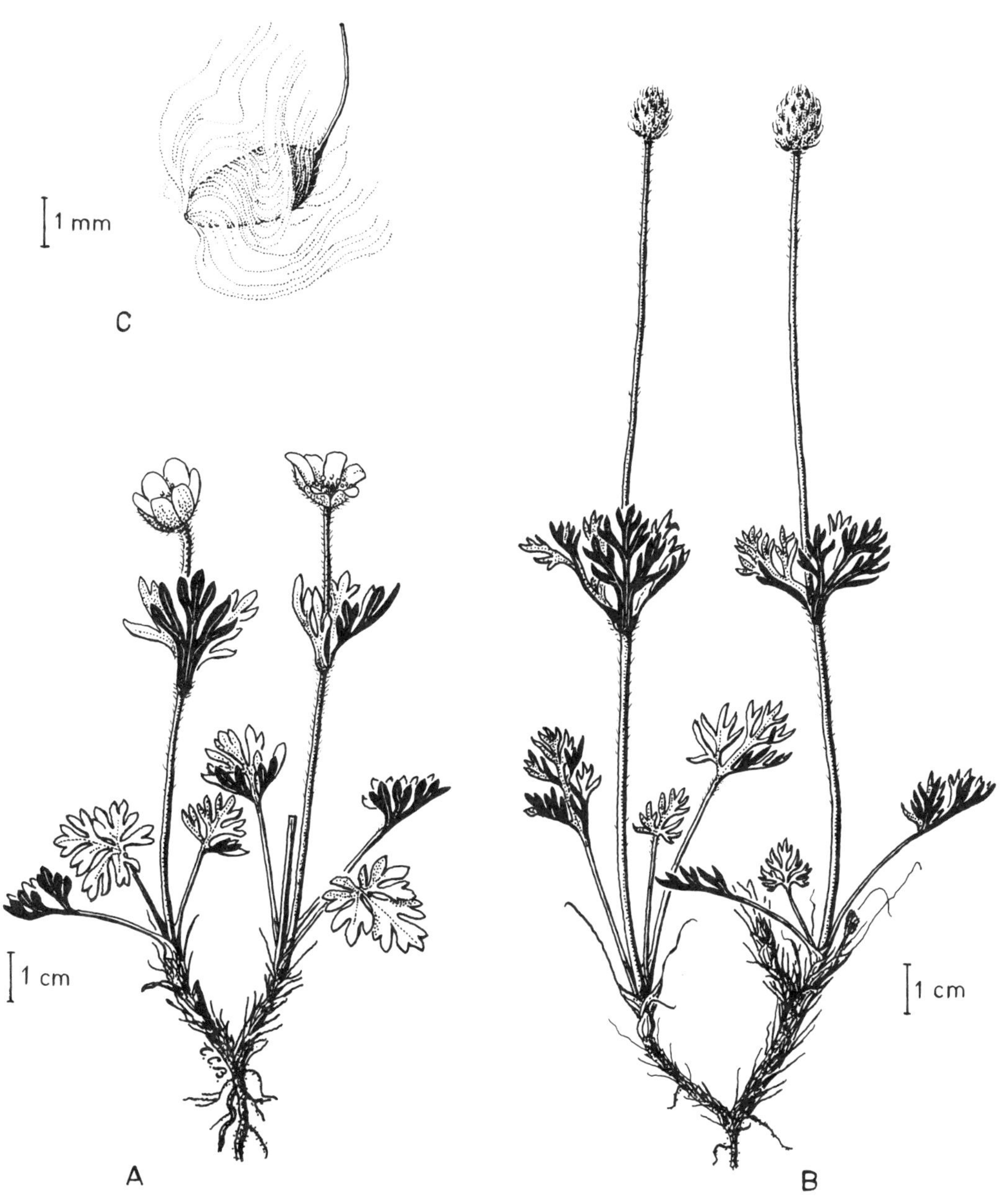

Anemone multifida **Poiret**

Cut-Leaf Anemone
Wind Flower

Variable hirsute perennial with an erect, often branching caudex giving rise to single or clustered stems 1–6 dm tall that may branch at the involucre.

Basal leaves clustered around the base of the flowering stem, long-petioled; the blades 3–8 cm wide, palmately divided into 3 (occasionally 5) divisions that are deeply dissected into narrow, linear-lanceolate, acute lobes 1.5–5 mm wide. Involucral leaves petioled, commonly 3 in the main involucre; or 2 and less dissected in the involucels, where they resemble the central divisions of the basal leaves.

Flowers usually 2 or 3, or up to 7; the oldest, terminal flower being followed by peduncles bearing involucels and flowers, arising from the axils of the involucral leaves. Sepals 5–9, but commonly 5, pubescent dorsally, 7–20 mm long, ovate or oblong, variously coloured; white, ochre-yellow, red, or purple, or bicoloured with purple outside. Stamens many. Carpels many, pubescent, with short slender styles 1–1½ mm long. Flowering from May to July.

Achenes in an ovoid to globose head, on a slenderly ellipsoid, woolly receptacle. The achene body mostly 3–4 mm long, almost concealed in the dense wool of wavy hairs up to 8 mm long; the longest hairs arising from near the base of the achene; the upper part of the achene with progressively shorter hairs. The stylar beak slender but tapering at the short-hairy base, usually straight, 1– 1.6 mm long. 2n = 32.

Flowering plants may be assigned to varieties and forms according to the following key:

KEY TO VARIETIES AND FORMS

1. Sepals bicoloured: pale cream to yellowish within; purplish to reddish without .... **var. *saxicola* 2**

   Sepals unicoloured: white, yellowish or purplish **3**

2. Plant grey-hirsute with dense, long (2–3 mm long) hairs .... **forma *hirsuta***

   Plant green and moderately hirsute with hairs 2 mm or less long. .... **forma *saxicola***

3. Sepals 7–10 mm long. .... **var. *multifida* 4**

   Sepals 11–17 mm long .... **var. *richardsiana* 5**

4. Sepals red to purplish .... **forma *sanguinea* (Pursh) Fernald**

   Sepals white to cream-coloured .... **forma *multifida***

5. Sepals red to purplish .... ***forma richardsiana***

   Sepals white to cream-coloured .... **forma *leucantha* Fernald**

Anemone multifida is widespread across North America, and occurs in parts of South America: in North America ranging from Alaska eastward to Hudson's Bay and Newfoundland, and southward to New Mexico and northern California. Variety *multifida* also occurs in Chile and Argentina. This species is typically found in dry, semi-open, coniferous woodland on coarse, well-drained calcareous soils; but may be found above the limits of forest locally.

Variety *multifida* [*A. multifida* var. *hudsoniana* DC., *A. hudsoniana* Richardson, *A. multifida* var. *globosa* Torrey & Gray, *A. globosa* (Torrey & Gray) Nuttall)] is found virtually throughout the range of the species. As the typical form of this species it was originally discovered on the Strait of Magellan in southern South America. However, extensive cultivation and breeding experiments by Dr. Heimburger and others (Boraiah & Heimburger, 1964; and Heimburger, personal communication) have demonstrated that there are no significant differences, either morphologically or genetically, between the South American plants and the North American plants defined as var. *hudsoniana*. Hitchcock combined them as var. *multifida* (Hitchcock and Cronquist, 1964).

Variety *richardsiana* Fernald (1917b) is weakly distinguished from var. *multifida*; the point of division, based on the size of the sepals, is rather arbitrarily set at 10 mm, since the range of variation

36

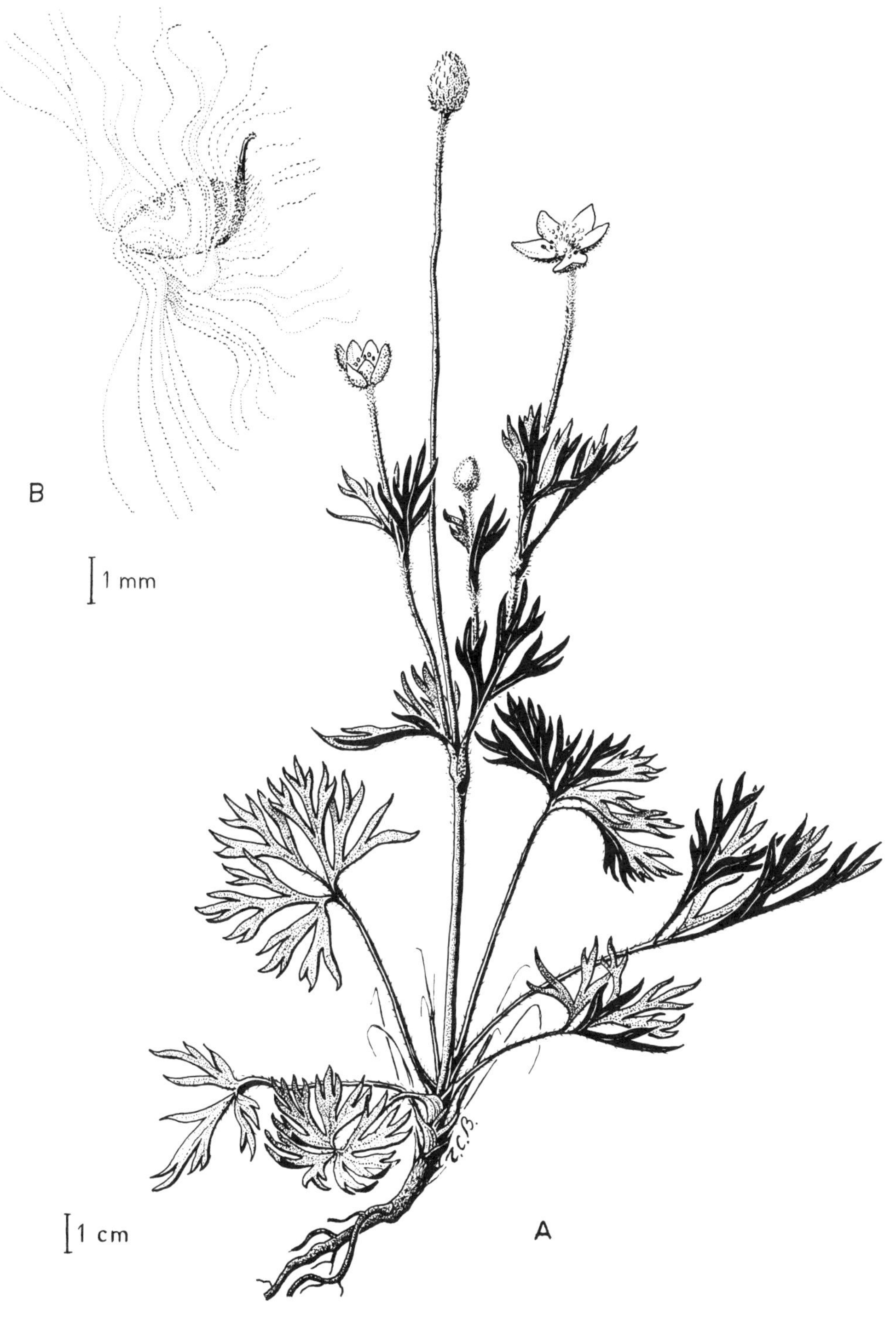

Figure 16. *Anemone multifida*

A. flowering and fruiting plant.   B. achene.

in sepal length is continuous between the extremes. Both pale-flowered and reddish-purple-flowered forms occur in both varieties. Variety *richardsiana* is scattered transcontinentally through the range of the species, at least in Canada.

Madahar and Heimburger (1969) found that while vars. *multifida* and *richardsiana* are fully interfertile, cross-breeding experiments between them and variety *saxicola* Boivin produced offspring that bore pollen of low viability; indicating the existence of partially effective genetic isolation of var. *saxicola*.

Variety *saxicola* Boivin occurs in the western Cordilleran and adjacent Great Plains regions of North America, often at high altitudes (Boraiah & Heimburger, 1964).

Anemone multifida Poiret var. *saxicola* B. Boivin forma *hirsuta* (C.L.Hitchcock) T.C. Brayshaw *stat. nov. (A. multifida* var. *hirsuta* C.L. Hitchcock in Hitchcock, C. Leo, and Arthur Cronquist, 1964; *Vascular plants of the Pacific Northwest;* Part 2: p. 327).

All specimens of this form seen at the University of Washington, including the type specimen (J.W. Thompson 8411, from Mount Angeles, Clallam County, Washington, June 18, 1932 (WTU)) are indistinguishable from var. *saxicola* on the basis of floral characters. The degree of hirsuteness varies among individuals, but is generally greater than the average for var. *saxicola*. It also varies seasonally; specimens collected in May or June being, on average, more densely hirsute than those collected later. This can be expected, since the hairs are dispersed progressively more widely as the underlying stems and leaves continue growth through the season, while some hairs are lost through normal wear and tear.

This author feels that, since var. *saxicola* and var. *hirsuta*, which are based on single and quite different characters, overlap completely, the latter one being completely contained within the former, it is preferable to treat *hirsuta* as a form of var. *saxicola*, which has priority of publication (1951), as well as being based on a relatively clear-cut floral character.

Plants of var. *saxicola* at the Royal British Columbia Museum that are assignable to forma *hirsuta* have been collected from Mt. Pavilion, a number of stations in Manning Provincial Park, and Mile 19, Flathead Road.

It has been postulated by Dr. M. Heimburger, in personal communication, that *Anemone multifida* originated as a hybrid between two diploid ancestors. Two morphologically distinguishable halves of its tetraploid chromosome complement can be matched: one by the chromosome complement in the diploid species *A. riparia* (with which *A. multifida* can form hybrids), or *A. cylindrica*, and one by that of the diploid form of *A. drummondii* var. *lithophila* found in Kamtchatka and the unglaciated parts of Alaska and Yukon.

Anemone parviflora Michaux   **Northern Anemone**

Colonial plant with a slender, elongate, horizontal rhizome, that gives rise to terminal ascending short caudices that bear leaves and flowering shoots. Stem pubescent, up to 20 cm tall at flowering time, and to 30 cm tall in fruit, though usually much less. Involucre variously placed on the stem.

Basal leaves clustered around, or adjacent to, the base of the flowering shoot; the blades shiny dark green and glabrous, 1.5–4 cm wide, seldom more, 3-parted into divisions that are distally lobed and toothed with rounded or blunt teeth; the lateral divisions commonly dissected about half way to base into unequal halves. Involucral leaves usually sessile, sometimes on petioles up to 1 cm long, cuneate, and divided into 3 narrow lobes that are often more acute-tipped than those of the basal leaves.

Flower solitary, on a pubescent pedicel that elongates significantly in fruit. Sepals 5 or 6, 9–20 mm long, broadly elliptic to nearly orbicular, white, the outer surface often bluish-tinged and pubescent. Stamens 70–80. Carpels 70–100, densely white-woolly, with linear styles 1–2.5 mm long. Receptacle cylindric to narrowly ellipsoid, finely puberulent.

Achenes in an ovoid head, densely woolly, the body of the achene nearly concealed among the especially long hairs arising from near its base; and with a slender stylar beak 1–2.5 mm long. $2n = 16, 32$.

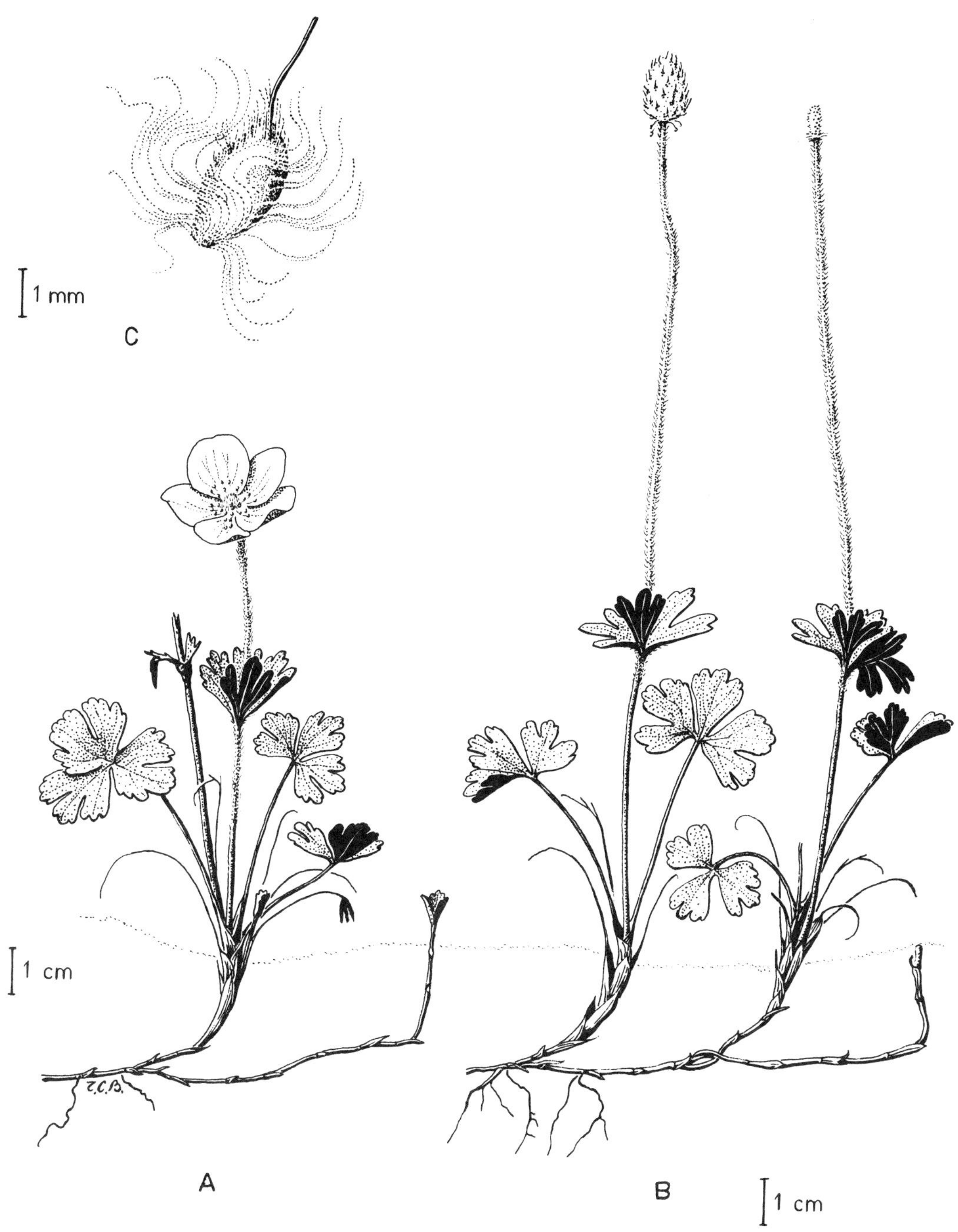

Figure 17. *Anemone parviflora*

A. flowering plant.   B. fruiting plant.   C. achene.

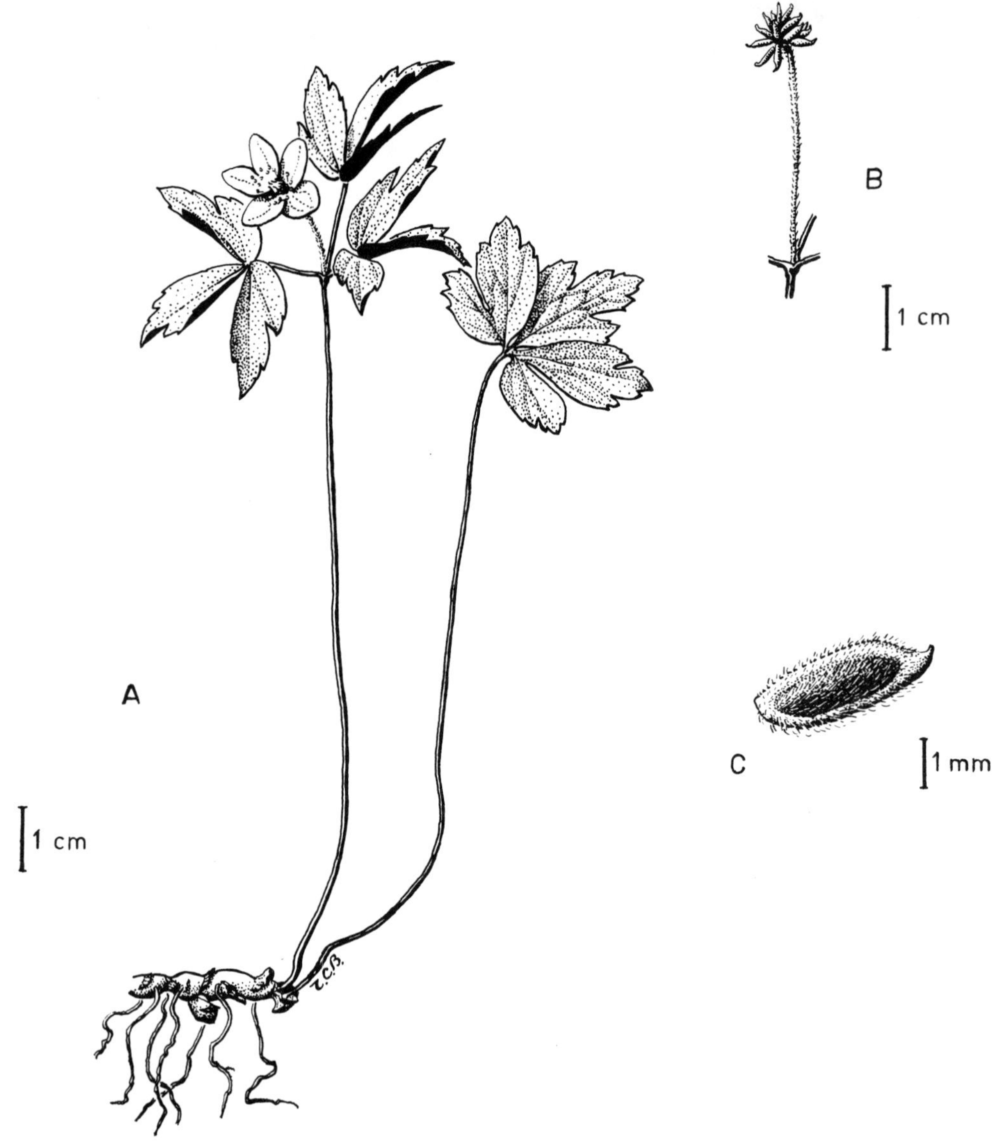

Figure 18. *Anemone lyallii*

A. flowering plant.   B. head of achenes.   C. achene.

Anemone parviflora is a mainly arctic species that is transcontinental in North America, and also found in eastern Siberia; and extends southward along the Rocky Mountains and other ranges to Washington, Oregon, and Colorado. It is found in arctic and alpine tundra, often around late-melting snow-patches, and occasionally it is found at lower altitudes, along stream banks and in woodland.

Anemone lyallii **Britton**   **Lyall's Anemone**
    A. quinquefolia **L.** var. *lyallii* **(Britton) Robinson**   **Wood Anemone**
    A. nemorosa **L.** ssp. *americana* **Ulbrich var.** *lyallii* **(Britton) Ulbrich** *in* **Engler**

Very slender perennial with a pale, scaly rhizome; the puberulent to glabrous stem 5–30 cm tall.

Basal leaves, if present, arising apart from the flowering stem, directly from the rhizome; the slender erect petiole supporting three rhombic, sessile or subsessile leaflets with the lateral ones sometimes deeply dissected into two lobes; the leaflets with lobed and toothed, ciliate margins, and very sparsely hispid surfaces. Involucral leaves (often the only ones seen) three, petioled; the blades similar to those of the basal leaves.

Flower terminal, solitary, erect, or sometimes nodding, on a puberulent pedicel that does not elongate significantly in fruit, rather small. Sepals 5–7, 4–8 mm long, elliptic, white or pale bluish or pinkish. Stamens 12–30. Carpels 12– 50, short-hispid, with a short glabrous style ½ mm long or less, and a minute stigma. Receptacle globose and glabrous.

Achenes in a globose, sometimes nodding, head, slenderly ellipsoidal to lanceolate, compressed, 3–4 mm long, thinly short-hispid, tapering to conical, glabrous stylar beaks about ½ mm long. 2n = 16.

Anemone lyallii ranges from western British Columbia to California along the coast and in the Cascade Range. It is found in damp, shady coniferous forest and adjacent clearings in the Coast Forest region, extending from sea level up to around 1300 metres in altitude in the Coastal Subalpine forest.

Anemone piperi **Britton** *ex* **Rydberg**   **Piper's Anemone**

Slender perennial closely resembling *A. lyallii*; with a deep, branching rhizome; the aerial shoots arising from ascending, dark brown rhizome branches.

Basal leaves, if present, trifoliolate, doubly serrate and ciliate-margined; the lateral leaflets often cut half way or more toward the base into two unequal main lobes. Involucral leaves three, petioled, similar to the basal ones but often larger, strigose-hairy on the surface.

Flower solitary, similar to that of *A. lyallii* but larger. Sepals 5–7, 8–16 mm long, elliptic to obovate, white to pink. Stamens 30–60. Carpels 10–20, shortly stipitate, the oblong body shortly hispid; the glabrous style slightly longer (½–¾ mm long) and more slenderly tapering than in *A. lyallii*, and with recurved stigmas. Flowering from April or May to July.

Achenes 3–5 mm long, on very short stipitate bases, hirsute, tapering to rather conical glabrous beaks ½–1 mm long, with small recurved stigmas. 2n = 32.

Anemone piperi inhabits moist sites in coniferous woods, mainly at altitudes between 1000 and 2000 metres, in the Columbia and Interior Subalpine Forest regions. It is known from Montana to eastern Washington State, and southward to eastern Oregon. It has recently (1983) been identified in southeastern British Columbia, where it has been found growing on Buchanan Mountain, near Kaslo, at an altitude of about 1900 metres. It may be looked for at higher, moister levels in the southern Selkirk and Purcell Mountain Ranges.

Anemone canadensis **L.**

Colonial perennial herb that reproduces itself vegetatively principally by means of buds on its roots (Raju *et al.* 1966): these buds giving rise to short ascending rhizomes or caudices, or to aerial shoots directly. Stems pubescent, 2–7 dm tall, commonly bearing axillary peduncles with secondary involucres (involucels) in the axils of the primary involucral leaves.

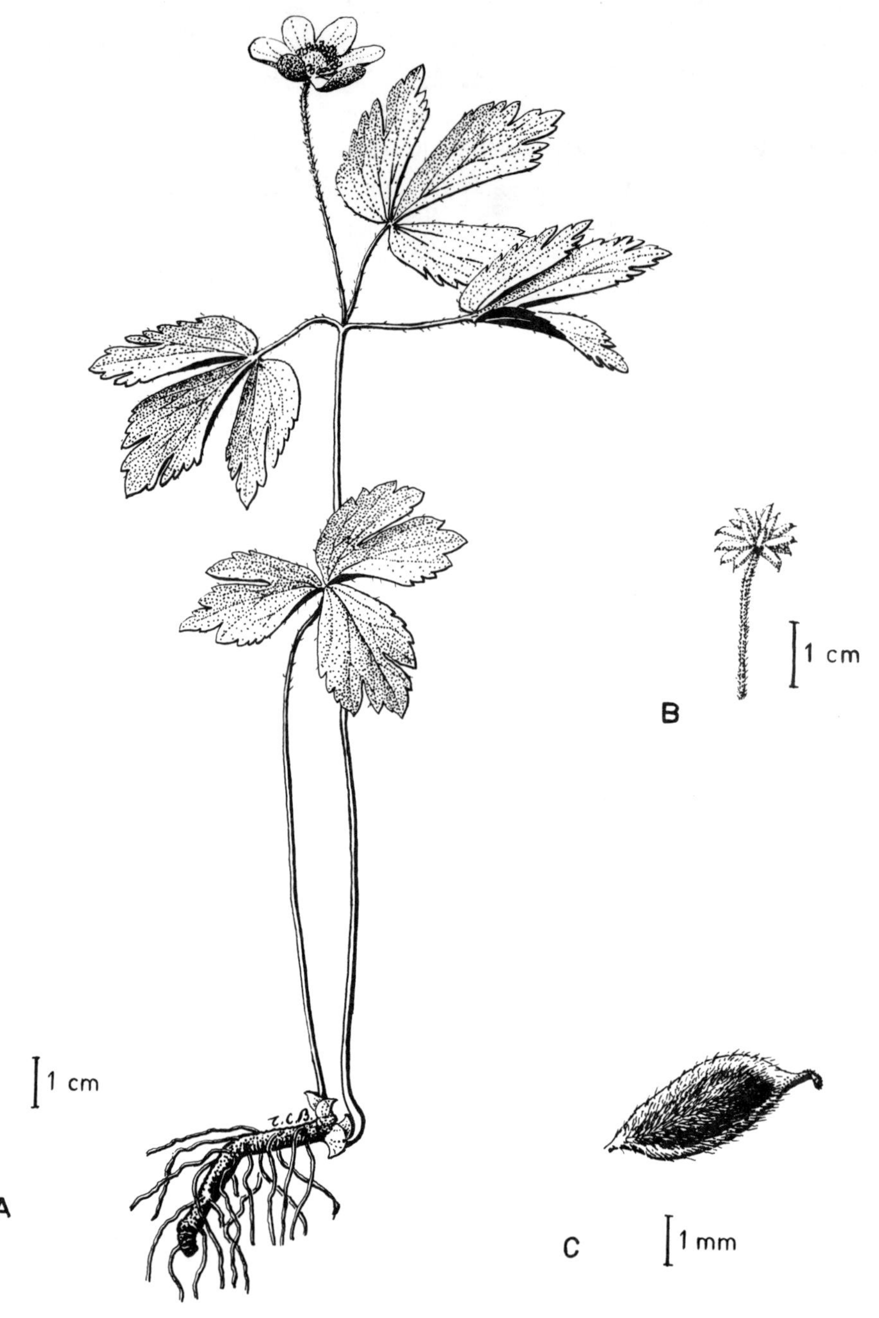

Figure 19. *Anemone piperi*

A. flowering plant.　　　　B. head of achenes.　　　　C. achene.

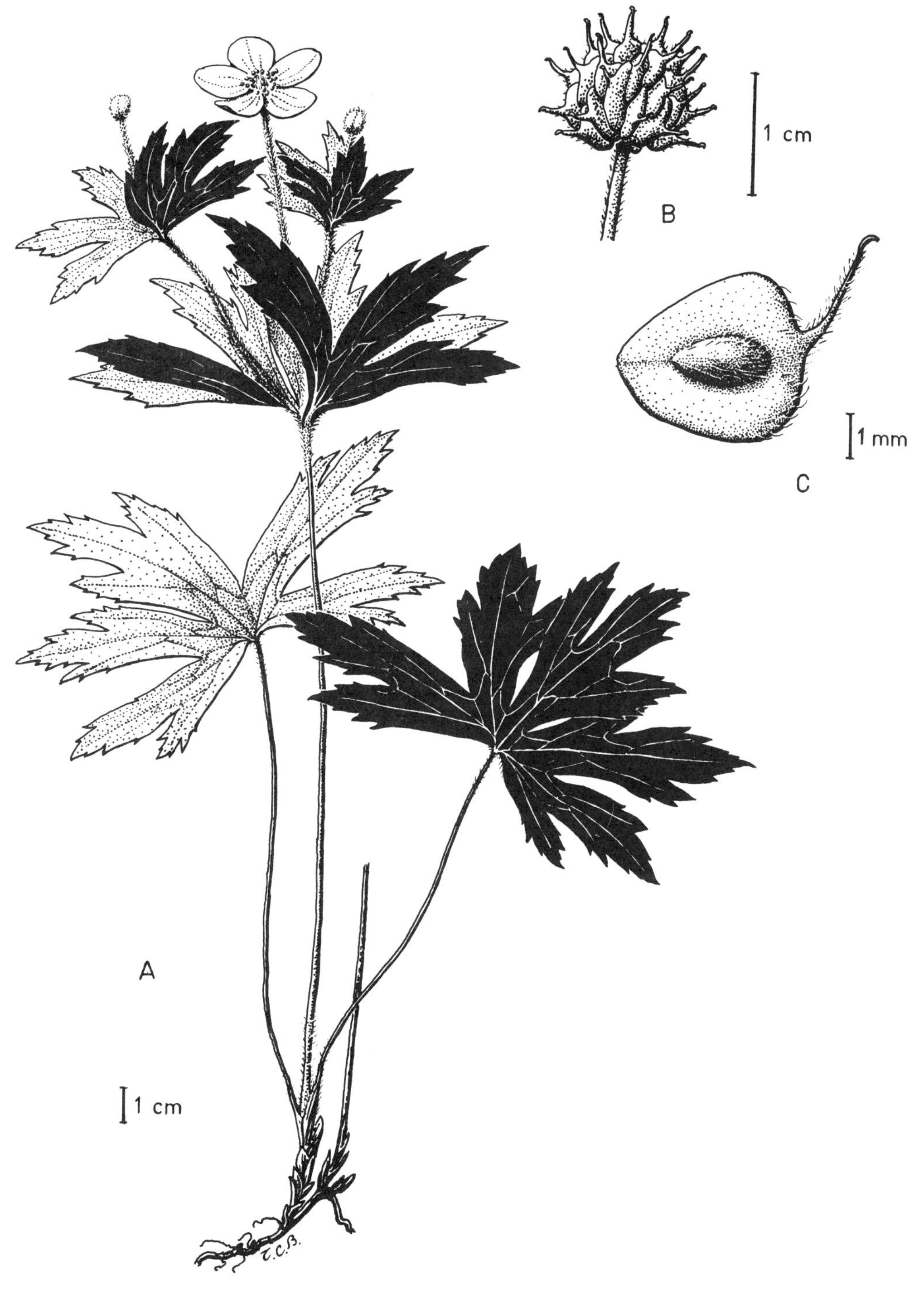

Figure 20. *Anemone canadensis*

A. flowering plant.   B. head of achenes.   C. achene.

Basal leaves clustered around the stems, long-petioled, pubescent; the blades 5–15 cm wide, dissected nearly to base into 3–5 rather broad, cuneate divisions that are again coarsely and doubly toothed. Involucral leaves similar but sessile; the primary involucre of three leaves; the involucels of two, rather smaller leaves.

Flowers commonly three or more, seldom solitary and terminal in depauperate plants. The terminal and oldest flower, on a pedicel without an involucel, being followed by later flowers on the secondary pedicels or peduncles. Tertiary pedicels and flowers may sometimes also be produced. Sepals normally 5, broadly elliptic to obovate, 1–2.5 cm long, white, strigose without near base. Stamens and carpels many; the carpels strigose.

Achenes in a globose head. The achene body flattened and broadly wing-margined, 3–6 mm long; with a slender ascending beak 2–3 or more mm long; both the body and beak thinly strigose. $2n = 14$.

Anemone canadensis occurs from eastern British Columbia and southwestern Northwest Territories eastward to the Atlantic coast, and southward to New Mexico. In British Columbia it has been found at a few widely scattered localities at moderate altitudes in the Rocky Mountains. It is typically found in meadows, and moist, especially deciduous woods, particularly where the soil is calcareous; and often forms large patches.

Anemone narcissiflora L.

Highly variable perennial with a stout, oblique or erect, fibrous-coated caudex, and one or more villous to glabrous stems 1–5 dm tall.

Basal leaves clustered around or adjacent to the base of the flowering stem; the long petioles sheathing at the bases; the blades pentagonal to orbicular in outline, 2–10 cm wide, palmately divided into 3–5 leaflets that in turn are dissected into narrow, linear to cuneate, acute-tipped lobes. Involucral leaves mostly above the middle of the stem, even in fruit, sessile, dissected into linear or lanceolate, acute lobes, often with axillary flower buds.

Flowers 1–5, but most often solitary, on pedicels that elongate in fruit. Sepals 5–8, 9–20 mm long, elliptic, obovate or rhombic, white, sometimes bluish dorsally, glabrous or almost so. Stamens many, on short filaments. Carpels 20–30, glabrous, dark green to purplish, with short styles.

Achenes in a compact head; the broadly obovate bodies flattened and marginally keeled, 6–10 mm long; with stylar beaks up to 1 mm long, with an upward turn. $2n = 14$.

KEY TO SUBSPECIES

1. Basal leaves divided into 3 leaflets. Flowers normally single, sometimes 2 or 3 .... **ssp. *interior***

    Basal leaves divided into 5 deeply dissected leaflets. Flowers usually 2–5 .... **ssp. *alaskana***

This species varies widely and continuously across its extensive range, that spans Europe and Asia, and extends eastward across Alaska and Yukon to the Mackenzie District of the Northwest Territories, and southward into northern British Columbia. The subspecies are not sharply distinct, and many specimens cannot be decisively assigned to one or another. In general, ssp. *alaskana* Hulten occurs along the Pacific coast and adjacent mountains, merging with ssp. *villosissima* (DC.) Hulten in the Aleutian Islands; while ssp. *interior* Hulten is prevalent inland, and appears more closely related to the Siberian and western Alaskan ssp. *sibirica* (L.) Hulten (*A. sibirica* L.).

Anemone narcissiflora is found in alpine and arctic tundra, and extends downward into the upper, more open communities of the subalpine forest.

Anemone richardsonii Hooker   **Yellow Anemone**

Slender perennial; spreading by pale, horizontal, filiform rhizomes that terminate in erect flowering shoots; prolongation of the rhizomes proceeding from axillary buds. Stems thinly pubescent, up to 1.5 dm tall in flower, and up to 2.5 dm tall in fruit.

Basal leaves arising at intervals from the rhizome; with erect slender petioles, and 3-lobed and toothed blades that are sparsely pubescent and ciliate, bright green above and rather glaucous

44

Figure 21. *Anemone narcissiflora* ssp. *interior*

A. flowering plant.   B. fruiting plant.   C. achene.

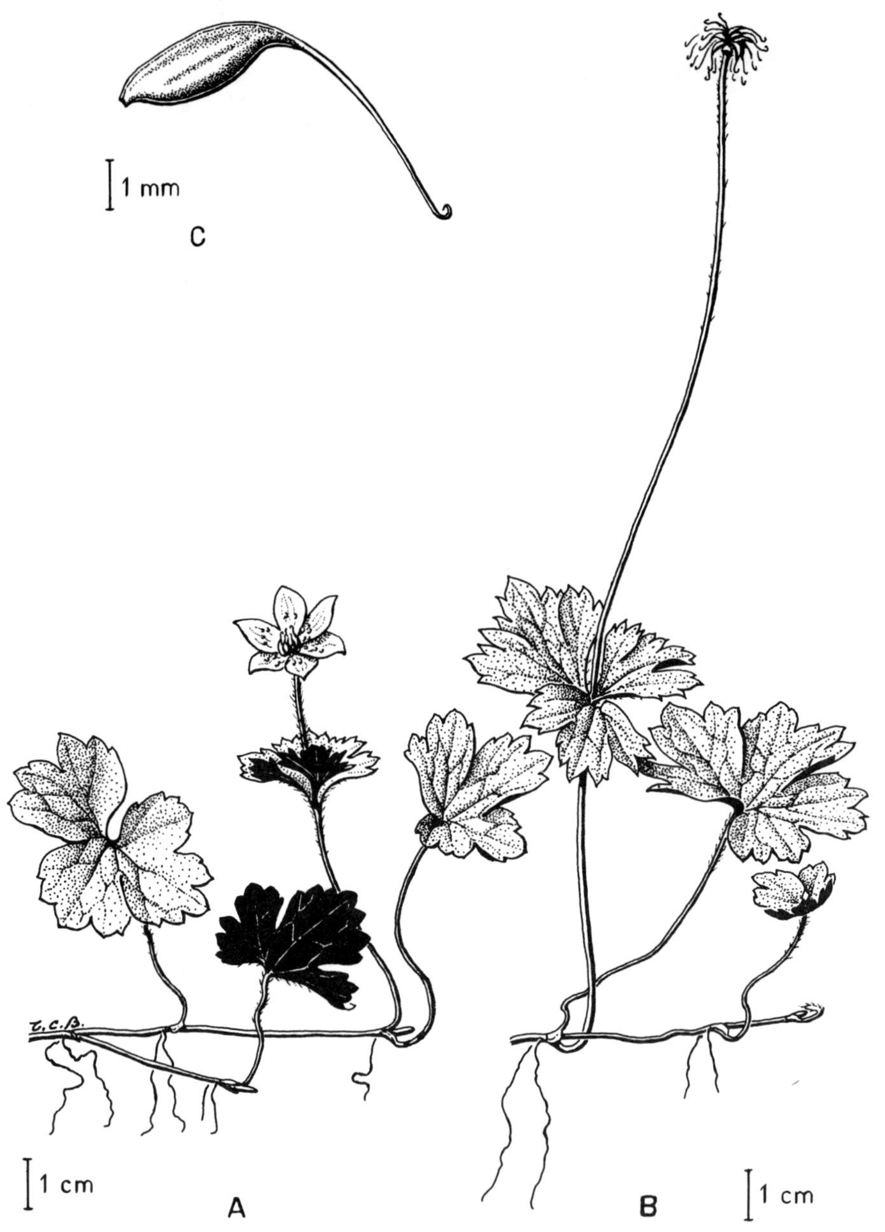

Figure 22. *Anemone richardsonii*

A. flowering plant.   B. fruiting plant.   C. achene.

beneath. Involucral leaves 2 or 3, sessile, cuneate-based, and narrower, less deeply lobed, and often more sharply toothed than the basal ones.

Flower solitary on a pubescent pedicel that elongates considerably as the fruit develops. Sepals 5–8, commonly 6, 10–15 mm long, narrowly elliptic, yellow, finely silky-pubescent dorsally, especially near the bases. Stamens 25–55. Carpels 25–55, glabrous, with long styles hooked at the tips. Receptacle globose, glabrous.

Achenes in a depressed-globose head. The achene body 3–4 mm long; tapering into a slender beak 5–6 mm long, that curves downward to an upwardly tightly hooked tip. 2n = 14.

Anemone richardsonii is a species ranging from eastern Siberia, across Alaska and northern Canada to Labrador and Greenland. In western Canada it ranges southward to about 53°N. latitude. It is found in rich Subalpine and Boreal Forest stands and open arctic and alpine tundra, at altitudes from about 600 metres up to 2200 metres in the southern parts of its range.

Anemone occidentalis **S. Watson**   **Western Anemone**
  Pulsatilla occidentalis **(S. Watson) Freyn**   **Mountain Pasque Flower**
  Tow-Headed Baby

Stout, densely hirsute perennial with a thick, short, tap-rooted caudex. Stem 1–3 dm tall at flowering time, elongating to 3–6 dm in fruit.

Basal leaves tufted, arising from an axillary bud adjacent to the base of the stem, and scarcely evident at flowering time; the blade ternately and then pinnately to bipinnately compound. Involucral leaves sessile or subsessile; the first division of the rachis occurring at the top of the sheath; the blades similar in character to those of the basal leaves.

Flower solitary and terminal, on a pedicel that usually raises it above the involucre. Sepals usually 6, elliptic, white, occasionally bluish near the bases outside, 2–3 cm long. Stamens many, all fertile. Carpels very many, densely hirsute, with elongate styles. Flowering starts before the expansion of the leaves.

Fruiting pedicel much elongated. Fruit a spike of achenes, often apparently in rough whorls, on an elongate, cylindrical, short-hispid receptacle. The achenes spreading, hirsute; their hirsute styles, 3–5 cm long, drooping. 2n = 16.

A plant of alpine meadows, from British Columbia and Alberta southward to Idaho and California. Showy in flower, this species is particularly conspicuous in fruit.

Plants of the Flathead Valley, in the extreme southeastern corner of British Columbia, and also in the adjacent Waterton Lakes National Park, Alberta, as represented in the herbarium at the Royal British Columbia Museum, differ somewhat from the typical form of the species in flowering late, when the basal leaves are well grown (late July and August), and in having smaller flowers and fruiting spikes.

Anemone patens **L. var.** *multifida* **Pritzel**   **Wind-Flower**
  A. ludoviciana **Nuttall** *nomen illegitimum*   **Pasque Flower**
  A. nuttalliana **DC.**   **Prairie-Crocus**
  Pulsatilla nuttalliana **(DC.) Sprengel**
  A. patens **var.** *wolfgangiana* **(Besser) Koch**
  Pulsatilla ludoviciana **(Nuttall) Heller**
  A. multifida **(Pritzel) Juzepczuk** *non* **Poiret**

Low (8–30 cm), scapose, hirsute, long-lived perennial herb with a branching subterranean caudex arising from a taproot.

The active caudex segment terminates in a scape, surrounded by old dead leaves, and bearing a pair of opposite or subopposite, multifid, hirsute, sessile bracts, and a single large terminal flower. Leaves of the season arise from a bud in the axil of the uppermost bud scale at the base of the scape. In the following year, this lateral shoot will produce a flower and further axillary buds. Small buds in the upper leaf axils may remain dormant, but have the potentiality for producing branches (Wildeman & Steeves, 1982).

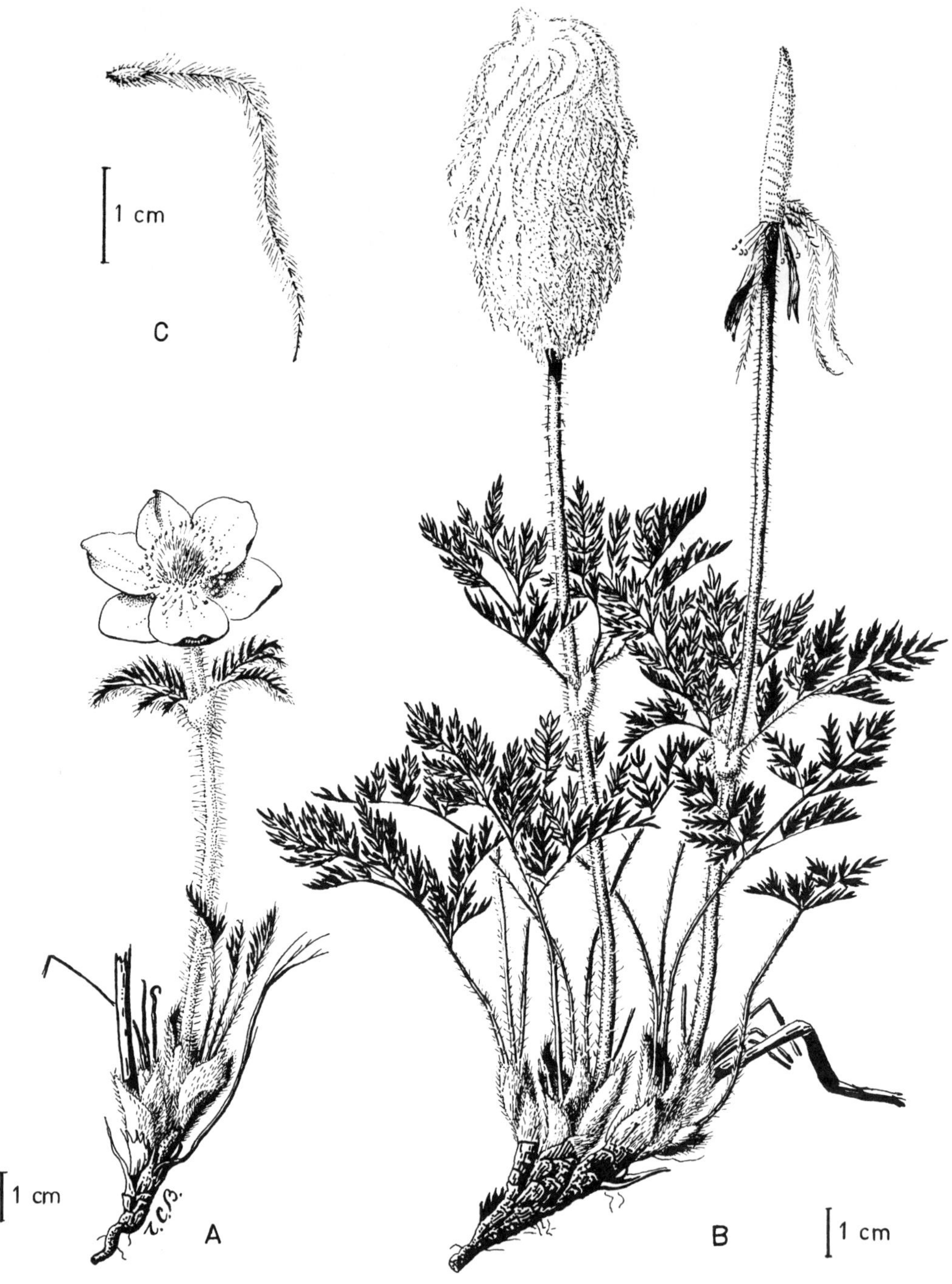

Figure 23. *Anemone occidentalis*

A. flowering plant. B. fruiting plant. C. achene.

Figure 24. *Anemone patens* var. *multifida*

A. flowering plant.   B. fruiting plant.   C. achene.

Basal leaves are long-petioled, with biternately compound blades; the terminal leaflet petioluled, the laterals sessile; the outer ones pinnately or ternately dissected into linear lobes. Involucral leaves sessile, deeply divided, their bases clothed in long silky hairs.

Flower single, terminal, erect or inclined, on a pedicel shorter than the involucral leaves, large, with 2 whorls of 3 sepals each. Sepals 3–4 cm long, pale violet to blue, and hirsute dorsally. Stamens many, the outermost ones reduced to minute sterile appendages. Carpels many, with long hirsute styles and single ovules. Flowering starts before the expansion of the leaves.

Fruiting pedicel elongating. Fruit a head of many achenes on a globose receptacle; each achene with a long, plumose style 3–4 cm long. $2n = 16$.

Anemone patens is a circumboreal species with a number of subspecies and varieties in Eurasia. Variety *multifida* occurs in Siberia and across North America from Alaska to the Great Lakes, and southward to Texas.

In British Columbia it is found in the eastern and northern regions. It typically inhabits open grassy prairies and balds, commonly on well-drained gravelly soils; but occasionally it is found in dry open woods.

AQUILEGIA L.

Columbine

Perennial herbs with biternate or triternate leaves and radially symmetric, often pendent flowers. Sepals 5, large, petaloid and coloured, with distinctly clawed bases. Petals extending back past the floral base in prolonged tubular spurs containing nectar glands within the apices, often with a very abbreviated lamina (blade). Stamens numerous, in several whorls of 5, with one or two inner whorls of broad membranous staminodes crimped together by their edges to form a sheath around the ovaries. Carpels 5–12 (5 in ours) separate, multi-ovuled, with terminal styles and linear stigmas, fruit a group of erect to spreading follicles. $x = 7$.

About 70 species of the Northern hemisphere including Africa.

Aquilegia provides an exception to the general rule of incompatibility between species. Many, if not all species of this genus possess equal, and apparently compatible, basic chromosome sets. Fertile hybrids can readily be produced horticulturally, and are occasionally found in nature. The maintenance of separate species here appears to depend less on intrinsic sterility barriers than on differences in flowering time, or on external mechanisms, such as different geographical ranges, or selection of the habitat or of the pollinating agent.

KEY TO SPECIES

1. Sepals and spurs bluish to purplish, spur hooked **2**
   Sepals and spurs yellow or red. Spur as long as petal lamina or longer **3**

2. Petal lamina longer than spur, whitish. Style and stigma 2–5 mm long. Sepal lanceolate .... *A. brevistyla*
   Petal lamina shorter than spur, purplish or blue. Sepal ovate above the claw. Style and stigma 7 mm long .... *A. vulgaris*

3. Petal lamina half or more as long as usually curved spur. Sepals usually yellow .... *A. flavescens*
   Petal lamina less than half as long as usually straight spur. Sepals usually red *A. formosa*

Aquilegia brevistyla Hooker   **Blue Columbine**

Taprooted perennial, similar to *A. formosa* in habit but averaging rather smaller throughout, puberulent and glandular.

Flowers smaller than in our other species: 2–3.5 cm wide, inclined to pendent. Sepals pale blue, lanceolate; petal with whitish blade longer than the blue, hooked spur, and rounded to truncate or shallowly notched at the apex. Stamens not, or barely, exserted. Staminodes lanceolate, nearly as long as stamens. Carpels 5, densely glandular-pubescent in flower, becoming less so in fruit; styles

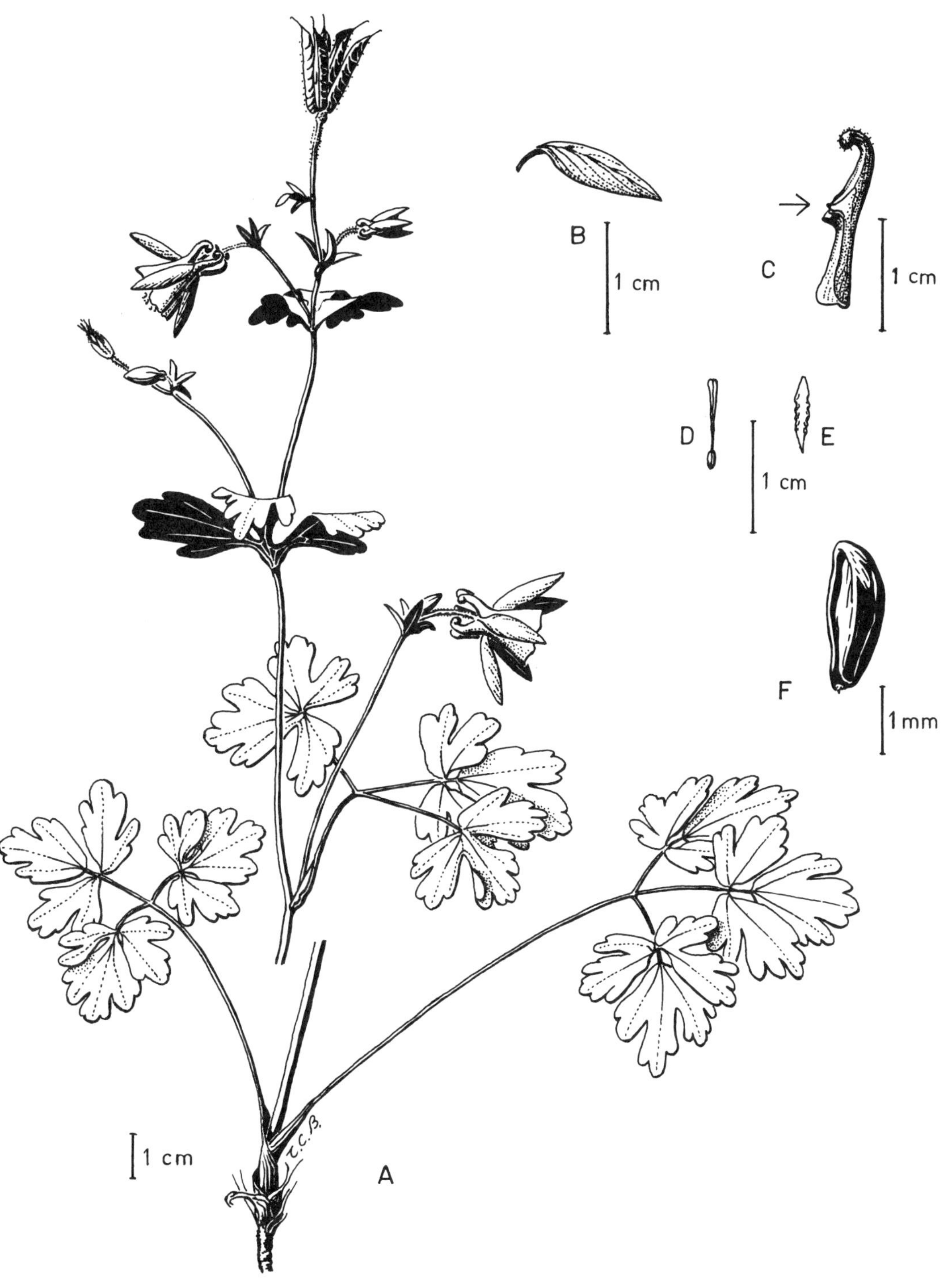

and stigmas 2–5 mm long. Follicles 2–2.5 cm long, with short stylar beaks commonly 3–4 mm long, slightly diverging. Seed ± ellipsoid, black and shiny. 2n = 14.

A species primarily of the Boreal Forest region from Alaska to James Bay, it extends southward into the northern prairie region to South Dakota, and to the northern Montane and Subalpine Forest regions in British Columbia and Alberta; with an isolated area in the Cariboo region.

Aquilegia flavescens S. Watson   Yellow Columbine

Taprooted perennial very similar, vegetatively, to *A. formosa*.

Flowers about as large as those of *A. formosa*, and similarly disposed. Sepals spreading to somewhat reflexed, typically yellow, but sometimes pale pinkish to red. Petals with spurs usually curved or hooked at the tips, and yellow, and a lamina half as long to fully as long as the spur. Stamens numerous; at least the anthers exserted. Staminodes as in *A. formosa*. Carpels normally 5, glandular-pubescent, tipped with styles 7–10 mm long.

Follicles 2–3 cm long, erect with spreading tips, and stylar appendages up to a centimetre long; the numerous seeds ellipsoidal, black, wrinkled and pebbled, like those of *A. formosa*.

A species ranging from south central British Columbia to Idaho, Montana and Colorado.

In British Columbia, *A. flavescens* is found in the Columbia and Interior Subalpine Forests, and extends upward into alpine meadows. A record from the McConnell Range, near latitude 57° North, shown on the map, may be in error, and should be re-examined. This species tends to occupy moister forest habitats than *A. formosa*; even to areas that may be seasonally flooded, and at higher elevations; but the ranges of these species overlap.

Where these species occur together, intermediate forms of presumed hybrid origin are commonly found. Typically these hybrids have the reflexed sepals and prominent petal blades of *A. flavescens* but the sepals are pink or red as in *A. formosa*, and the petal spurs also reddish and commonly curved. These plants are called *A. flavescens* var. *miniana* Nelson & Macbride. In this connection it is interesting to note that Gregory (1941) remarked on the exceptional degree of interfertility among species of the genus *Aquilegia* worldwide.

Aquilegia formosa Fischer   Red Columbine

A deeply taprooted perennial; the taproot supporting one or a few caudices, each giving rise to a leafy stem up to a metre tall, glabrous to glandular-puberulent. Leaves biternate, the leaflets palmately divided into rounded lobes.

Flowers scattered in a diffuse cyme, pendent, opening one or two at a time, 3–5 cm wide. Sepals spreading to 90° or rather less, commonly brick-red, but occasionally yellow. Petal with a very short, almost truncate, yellow lamina, less than half as long as the spur, which is usually straight and red, rarely yellow. Stamens about 20, in about 4 whorls, exserted. Staminodes 5–10, white, together ensheathing the ovaries; and persisting for a time after the fall of the outer floral parts. Carpels 5, with exserted styles and stigmas 8–11 mm long.

Follicles glandular-puberulent, erect, with spreading tips and widely spreading stylar beaks; with many ellipsoidal seeds with a black, pebbled and wrinkled seed coat. 2n = 14.

Ranging from Alaska and Yukon southward in the Cordilleran regions of western North America, including all but southeastern British Columbia. Growing mainly in the forests, but extending into Alpine meadows: by far our commonest Columbine. Pollinated by hummingbirds.

Aquilegia vulgaris L.   European Columbine

Erect perennial, vegetatively similar to *A. formosa*. Leaves biternate or triternate; the lobes rather more broadly rounded than in *A. formosa*.

Flowers blue, violet or purple; the sepals with ovate laminae. Petals with stout, funnel-shaped, hooked spurs, and coloured broad blades nearly as long as the sepals, and half to almost as long as the spurs. Stamens exserted. Staminodes about ten, obtuse and mucronate apically. Carpels usually 5, puberulent and glandular; with styles and stigmas about 7 mm long. Flower bud plumper than in *A. formosa*.

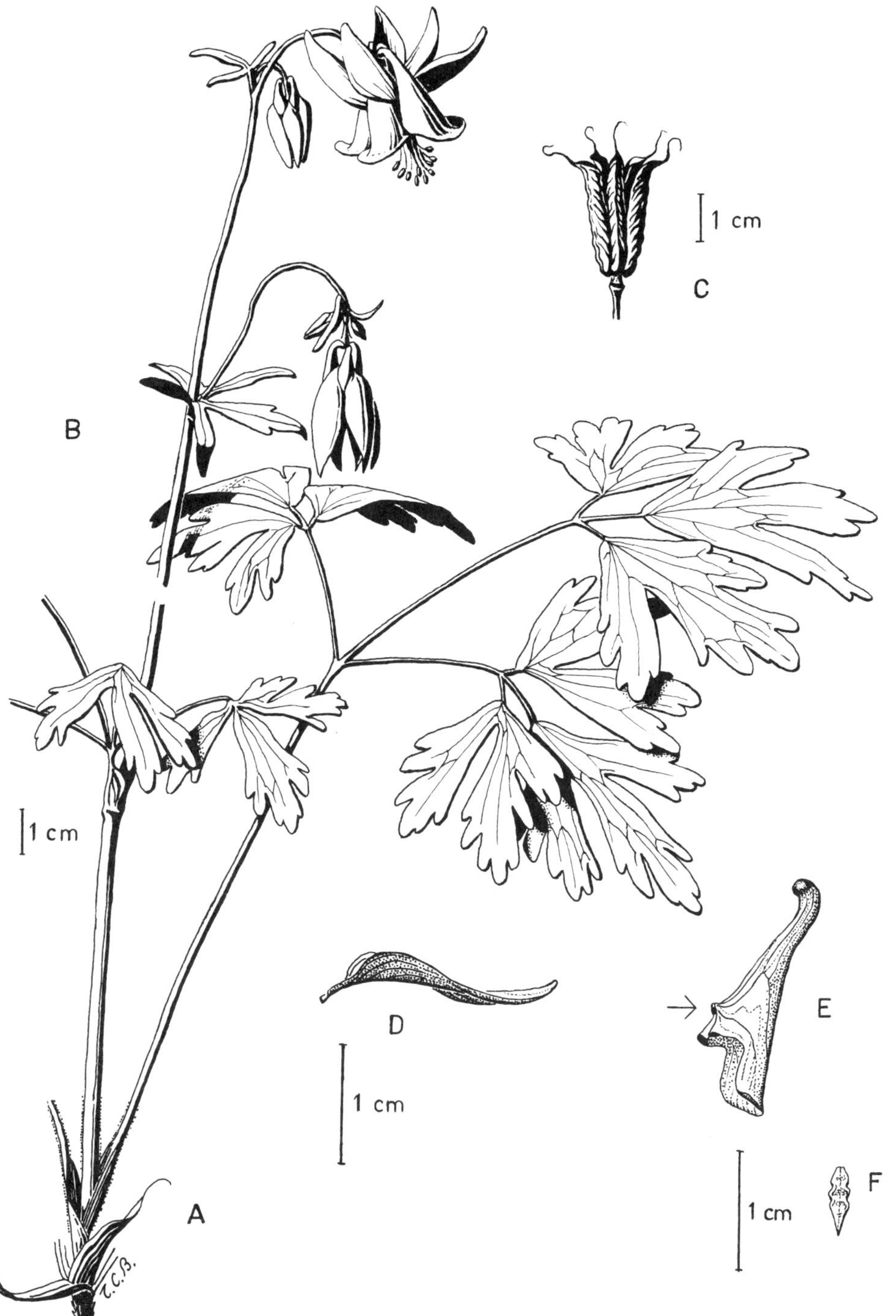

Figure 26. *Aquilegia flavescens*

A. basal leaf.
B. flowering branch.
C. cluster of follicles.
D. sepal.
E. petal: arrow indicates point of attachment.
F. staminode.

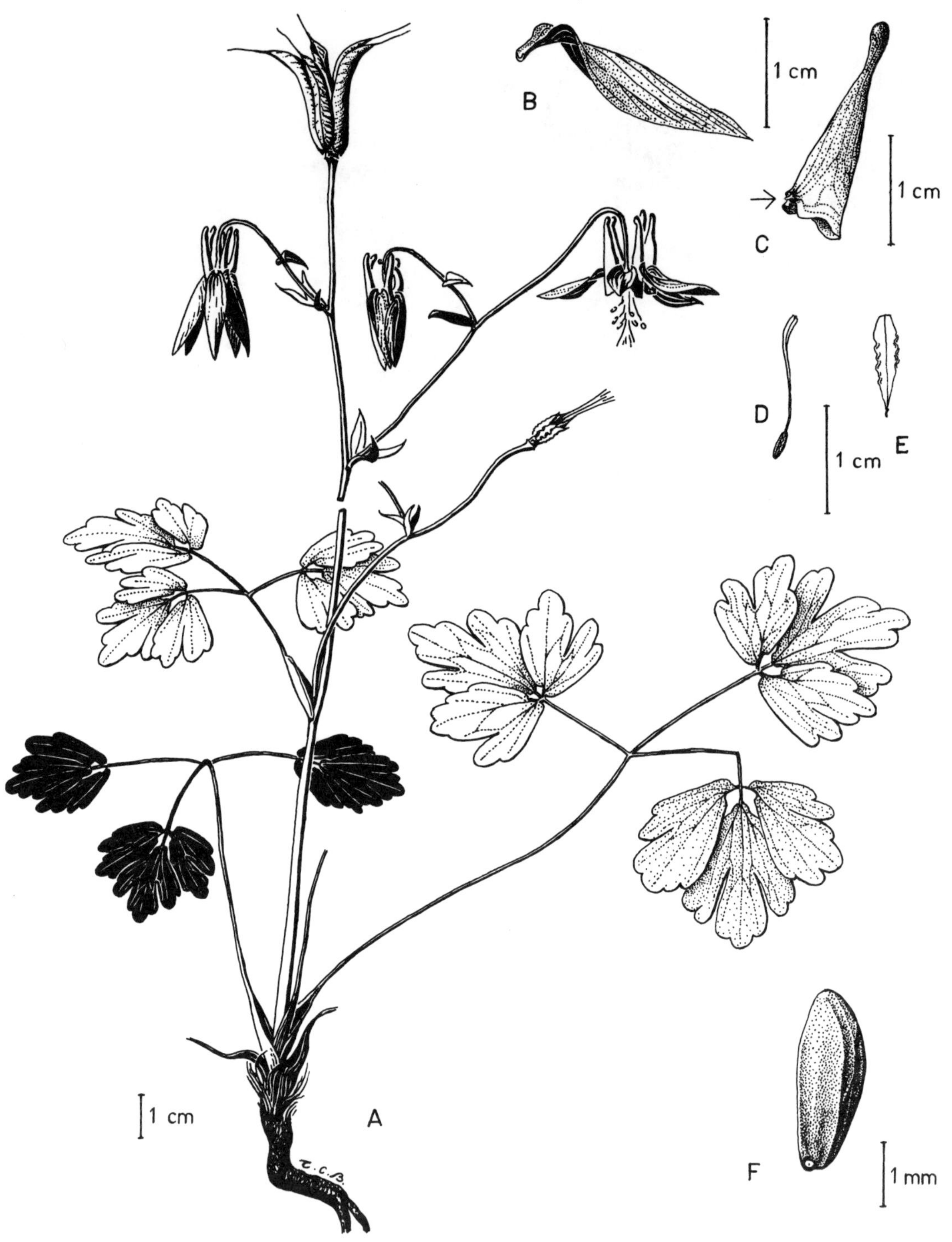

Figure 27. *Aquilegia formosa*

A. plant.
B. sepal.
C. petal: arrow indicates point of attachment.
D. stamen.
E. staminode.
F. seed.

Follicles 2–2.5 cm long, excluding the 7 mm long stylar beaks. 2n = 14.

Native of Europe and long-cultivated in gardens in various horticultural forms, this species has been reported once as an escape in a ditch near Rosedale, in the lower Fraser Valley.

CALTHA L.

Marsh-Marigold

Glabrous, herbaceous perennials with rather fleshy stems and leaves, simple leaves and terminal, solitary or loosely clustered, radially symmetric flowers. Sepals 5–12, large, coloured, and showy; petals none; stamens several to many, carpels several to many, multi-ovulate, developing into a head of many-seeded, sessile or stipitate follicles.

A genus of about 20 species, mostly of north temperate latitudes, but a few extending to south temperate regions. Plants usually of wet, swampy habitats.

KEY TO SPECIES

1. Plant stemless or with only one stem leaf. Sepals white within **2**
    Plant with leafy stems **3**

2. Leaf blades longer than wide, oblong. Flowers commonly appearing solitary on leafless scapes .... *C. leptosepala*
    Leaf blades at least as wide as long, orbicular to reniform. Flowers often 2 on a stem with one leaf .... *C. biflora*

3. Stems stout; erect or arching and prostrate. Mature leaves more than 5 cm wide, ascending, aerial. Sepals 1 cm or more long, yellow .... *C. palustris*
    Stems more slender, creeping in mud or water. Mature leaves less than 5 cm wide, floating. Sepals less than 1 cm long, white .... *C. natans*

Caltha biflora **DC.**	**Broad-Leafed Marsh-Marigold**
C. leptosepala? **var.** *biflora* **(DC.) Lawson**	**Two-Flowered Marsh-Marigold**

Plants usually less than 30 cm tall with a buried caudex or short rhizome. Leaves mostly basal, 4–10 cm long and as wide or wider, orbicular to reniform, deeply cordate, the margin regularly crenate to dentate.

Flowers solitary on a scape, or more commonly in pairs with one terminating the stem and one on a pedicel in the axil of a short-petioled leaf. Sepals 6–12, narrowly obovate to oblanceolate, rounded at apex, white above and beneath or sometimes faintly bluish beneath. Stamens many, carpels many.

Fruit a conical head of stipitate follicles with short stout erect stylar beaks. 2n = 48.

Found mainly in the Coast Forest region, in wet seepage areas at low or moderate to subalpine levels, usually below 1000 metres, along the coast and the Coast and Cascade Ranges from southeastern Alaska to northwestern Washington State. It has also been reported from Idaho and eastern Oregon to Colorado; but the identity of the plants in this latter area is questioned (Morris, 1973).

Caltha biflora is obviously closely related to *C. leptosepala*, and there is some justification for combining these species, as Lawson did. However, they appear to be clearly separable as a rule, and few intermediates have been found in the collection at the Royal British Columbia Museum examined by the author.

Caltha leptosepala **DC.**	**Mountain Marsh-Marigold**
	Elkslip

Similar in general aspect to *C. biflora* but often shorter; the leaf blades broadly obovate to oblong, up to twice as long as wide, variably cordate at base and rounded above; the margin crenate

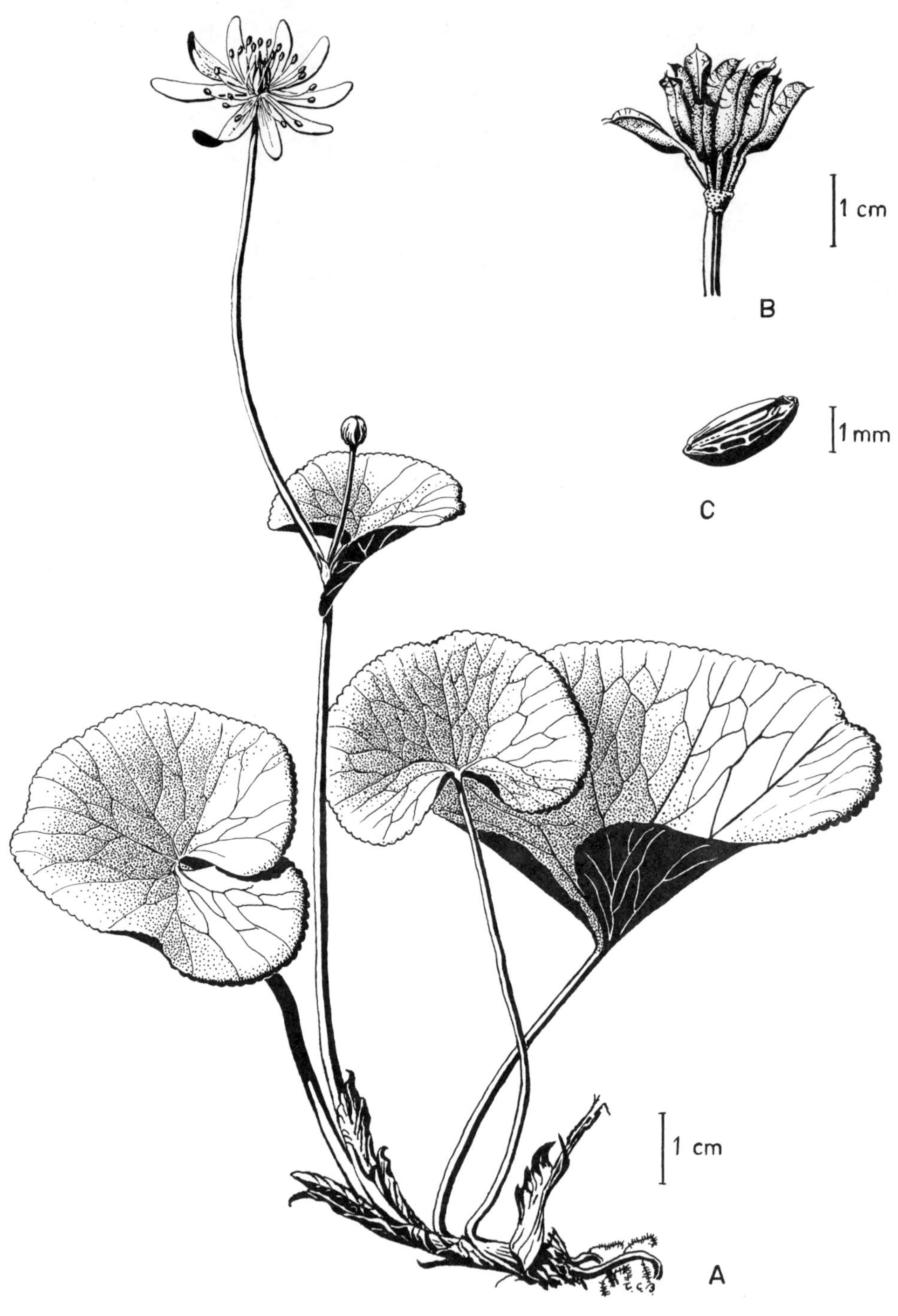

Figure 28. *Caltha biflora*

A. flowering plant.   B. cluster of follicles.   C. seed.

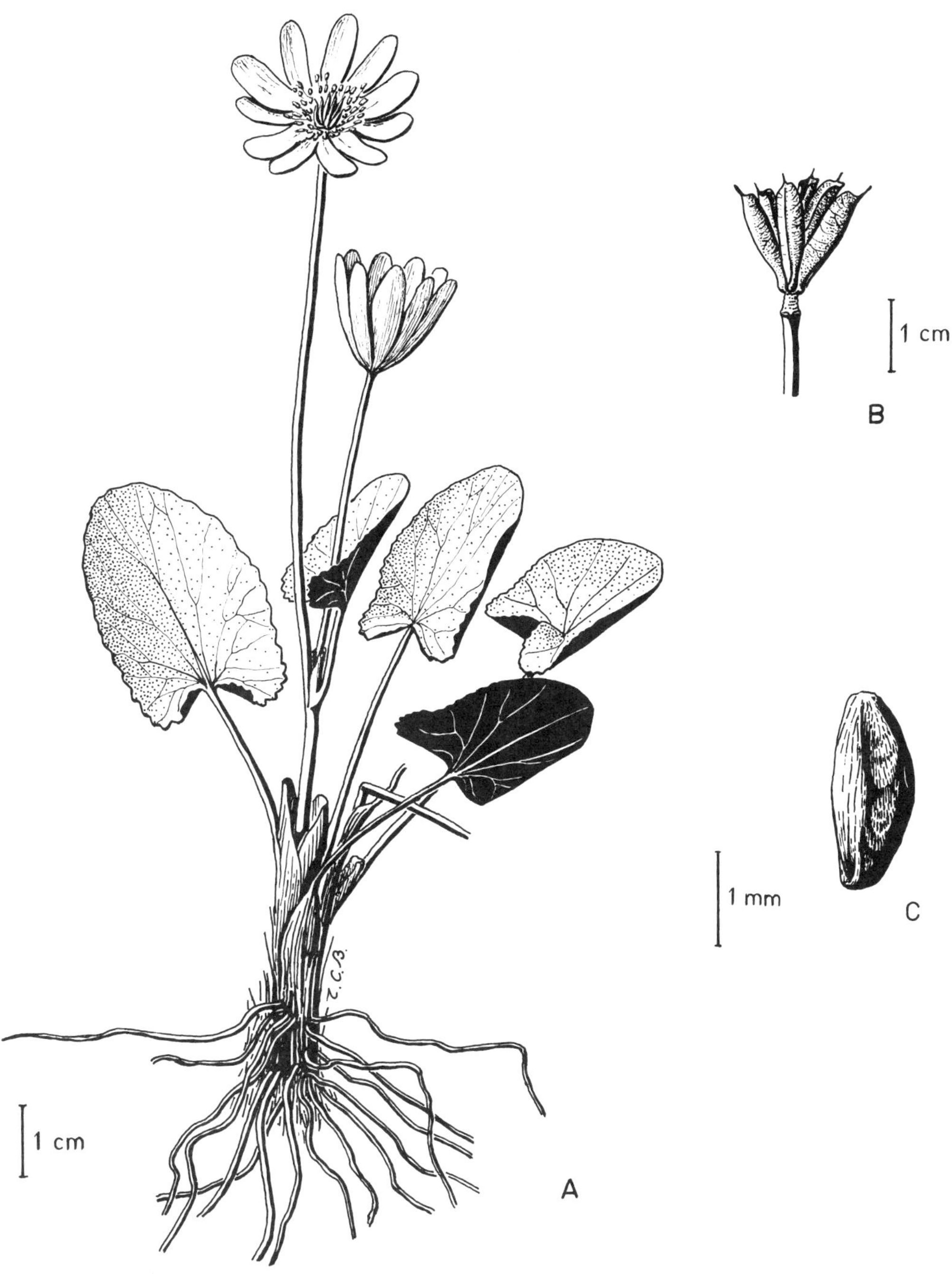

Figure 29. *Caltha leptosepala*

A. flowering plant.   B. cluster of follicles.   C. seed.

or dentate, sometimes deeply so near the base, and shallowly and broadly crenate to sinuate or almost entire toward the apex. Stipular sheaths projecting as ligules.

Flowers solitary on leafless scapes, or more often in pairs as in *C. biflora*, on once-branched stems, the lowest internode of which may be so short as to be concealed among the basal leaves, thus giving an impression of a pair of scapes. Sepals 6–11, 10–20 mm long, lanceolate to oblanceolate, the apex rounded, white above, commonly bluish beneath. Stamens many (50 or more). Carpels (3) 5–10(–32).

Fruit a conical head of sessile or subsessile (rarely stipitate) follicles with short, more or less erect stylar beaks. Seeds black or dark brown, longitudinally striate.

Caltha leptosepala ranges from the Alaska Peninsula to Oregon and Colorado; and in British Columbia from the coast to the Rocky Mountains. It is found usually in wet seepage areas of the Subalpine Forest region, and in alpine meadows, at moderate to high elevations, seldom below 1000 metres.

Caltha leptosepala DC. var. *rotundifolia* Huth, (*C. biflora* var. *rotundifolia* (Huth) C. L. Hitchcock, 1964), distinguished by being more than 30 cm tall and having rotund leaf blades more than 5 cm long, described originally from the Rocky Mountains, occurs sporadically throughout the populations of *C. leptosepala* in the Rocky Mountain region, intergrades with other populations of this species, and appears to be little more than an adjustment of the plant to favourable growing conditions. Morris (1973) reports that plants growing at lower elevations and in shaded habitats are generally taller than those growing in more exposed sites at higher elevations; a very normal type of response. In all cases the leaf blades are longer than wide. In view of this, var. *rotundifolia* does not appear to this author to merit formal recognition.

Caltha natans Pallas   Floating Marsh-Marigold

Creeping aquatic plant with stems floating, or creeping on mud and rooting at the nodes.

Leaves commonly floating, with ± free stipular sheaths forming ligules, slender weak petioles, and reniform blades 2–4 cm wide, with shallowly dentate margins, sometimes purplish beneath.

Flowers solitary and terminal or in loosely organized cymose terminal shoots with reduced bract-leaves. Sepals 5–6, 4–7 mm long, obovate, white, occasionally purplish-tinged. Stamens about 20, shorter than the numerous (40 or more) sessile carpels. Carpels lacking distinct styles, with minute terminal stigmas; each carpel containing 20 or more ovules.

Fruit a globose head of sessile, more or less beakless, many-seeded follicles 4–6 mm long. Seeds brown, fusiform, 0.5–0.8 mm long. 2n = 32.

Found in shallow ponds and ditches in the Boreal Forest region. Ranging across northern North America and Asia, from Lake Superior to the Ural Mountains: in British Columbia extending southward east of the Rocky Mountains to the Peace River basin.

Caltha palustris L.   Yellow Marsh-Marigold

Including *C. palustris* L. var. *asarifolia* (DC.) Huth (*C. asarifolia* DC.)

Stout-stemmed perennial with an erect caudex, the thick rather soft stem erect in var. *palustris* to arching or prostrate in var. *asarifolia* and in var. *arctica* (R. Brown) Huth.

Basal leaves long-petioled from conspicuously sheathing bases, the apices of the stipular sheaths projecting as ligules; the reniform to orbicular blades 8–15 cm wide with regularly crenate or shallowly dentate margins. Stem leaves with short petioles and smaller blades.

Flowers one or few to several, cymosely arranged in leafy bracted clusters on the ends of stems and branches. Sepals 5–6, deep yellow, 1–3 cm long, broadly obovate to somewhat rhombic. Stamens many. Carpels 5–25, sessile or subsessile, with stout, diverging terminal styles and stigmas.

Fruit an obconical aggregate of follicles with diverging beaks. 2n = 32, 44, 56, 64.

KEY TO VARIETIES

1. Stem permanently erect .... **var. *palustris***
   Stem tending to sprawl with age and sometimes to root at the nodes .... **var. *asarifolia***

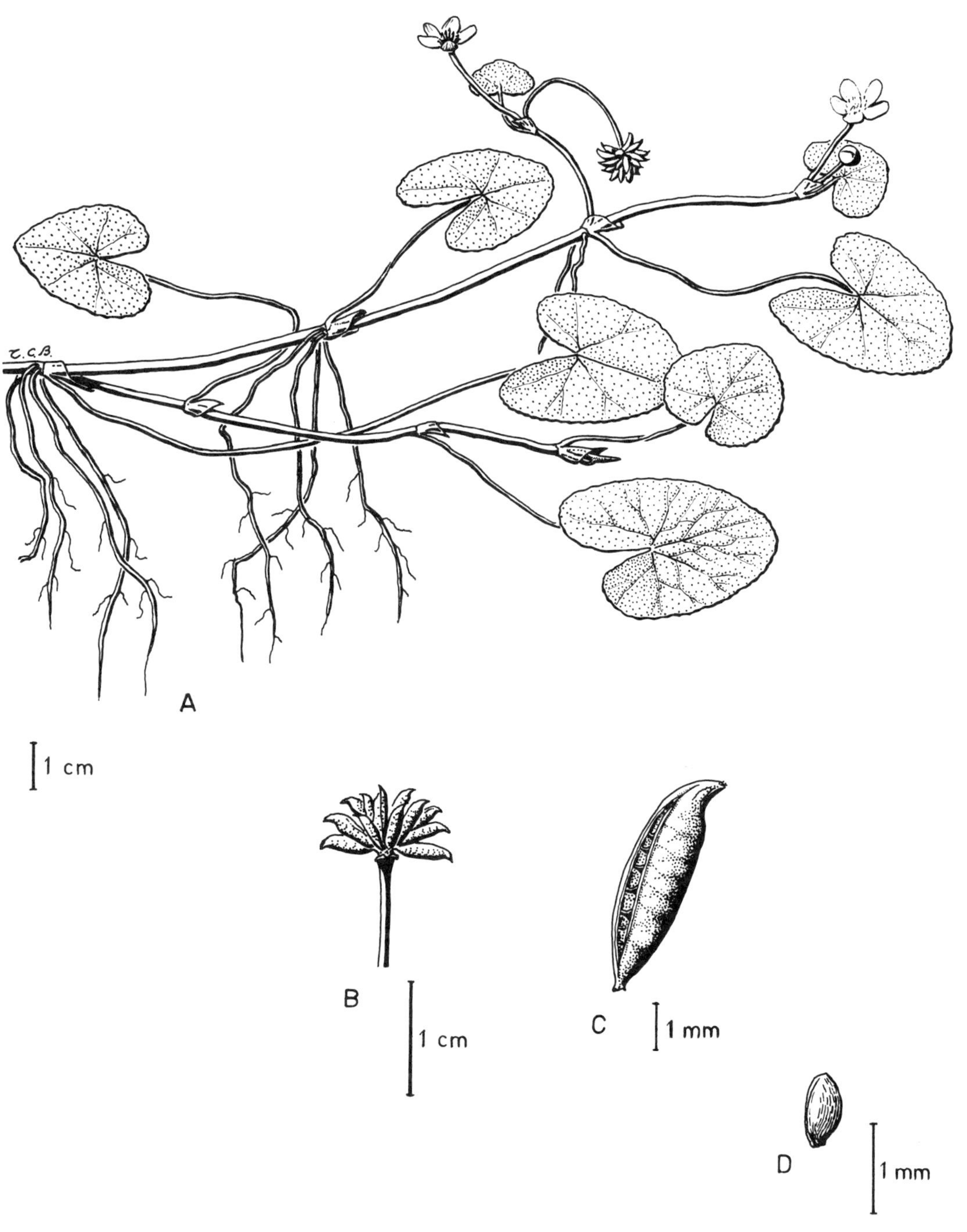

Figure 30. *Caltha natans*

A. plant in flower and fruit.   B. cluster of follicles.   C. follicle.   D. seed.

Figure 31. *Caltha palustris* var. *asarifolia*

A. flowering plant. B. cluster of follicles. C. seed.

A plant typically of swamps. The species is circumboreal; and represented in British Columbia by var. *asarifolia* (DC.) Huth (*C. asarifolia* DC.) according to present records. Var. *asarifolia* is coastal, from the Aleutian Islands, Alaska, to Oregon, but is scarce in our area. A specimen in the Royal British Columbia Museum, from Sturgeon Lake, south of Valleyview, Alberta, which is probably var. *palustris*, suggests that this variety may be looked for in the Peace River basin and in other Boreal Forest areas of British Columbia.

CIMICIFUGA L.

Bugbane

Coarse herbaceous perennials with large compound leaves and small flowers in terminal compound racemes or panicles. Sepals 5, white to pinkish, deciduous early. Petals small and 2-lobed, or absent. Stamens many, white. Carpels 1–8, forming follicles tipped by stylar beaks, and containing several finely corrugated or wrinkled seeds.

Eight species of the Northern Hemisphere; only one of them in British Columbia.

Cimicifuga elata Nuttall *in* Torrey & Gray
Actaea elata (Nuttall) Prantl

Tall (1–2 m) finely puberulent perennial from a woody rhizome; the solid stem commonly branching above.

Leaves, except the uppermost, large (up to 80 cm long), biternate; the leaflets 5–20 cm long, ovate to orbicular, palmately veined and lobed, and coarsely doubly serrate, somewhat glaucous and finely puberulent beneath, and ciliate on the margins.

Flowers in a terminal, compound, crowded, bracted raceme, with racemes terminating the upper branches; the whole inflorescence finely glandular-puberulent. Flowers open from the base toward the apex of each raceme, except that the terminal flower may open slightly ahead of its immediate neighbours; the terminal raceme flowering first. The minutely bracted pedicels are shorter than to about equal to the flowers in length. Sepals obovate, glabrous, 3–4 mm long, white, or sometimes faintly pinkish-tinged, deciduous as the flower opens. Petals none. Stamens many, white, their orbicular anthers on filaments 3–4 mm long. Carpels 1–3, or rarely 4, separate, with short, stout styles and recurved stigmas. Flowering May to August, fruiting June to September.

Follicles (4–) 9–12 mm long by 2–5 mm thick when mature, glandular, very shortly stipitate, with stipes 1 mm long or less, and with stout, somewhat recurving stylar beaks. Seeds ellipsoidal, 2.5–3 mm long by about half as thick; the surface appearing strongly wrinkled.

Cimicifuga elata ranges from the lower Fraser River, in southern British Columbia, southward, west of the crest of the Cascade Mountains, to northwestern Oregon. It is apparently rare in British Columbia. It is found in shady woods at elevations up to about 1000 metres.

CLEMATIS L. (including *Atragene* L.)

Clematis

Vines, shrubs or herbs, with opposite, usually compound leaves; the climbing species clasping their supports by means of twining petioles and petiolules.

Flowers terminating main stems or branches, solitary or in cymose panicles, and perfect or dioicous. Sepals 4 (in ours) to 8, meeting edge to edge (valvate) in bud, variously coloured. Petals none. Stamens many. Carpels many, separate, containing one or sometimes more ovules, and tipped by long hairy styles and linear stigmas.

Fruit a head of achenes; each achene containing one seed and bearing a long feathery 'tail' formed of the elongated style and stigma; and thus adapted to dispersal by wind. x = 8, at least for our species.

A genus of about 300 species, mostly in, but not confined to, north temperate latitudes. 2 species native in British Columbia.

Many exotic species of *Clematis* are cultivated for ornamental purposes; and a few of these may be found established as wild-growing plants in disturbed ground.

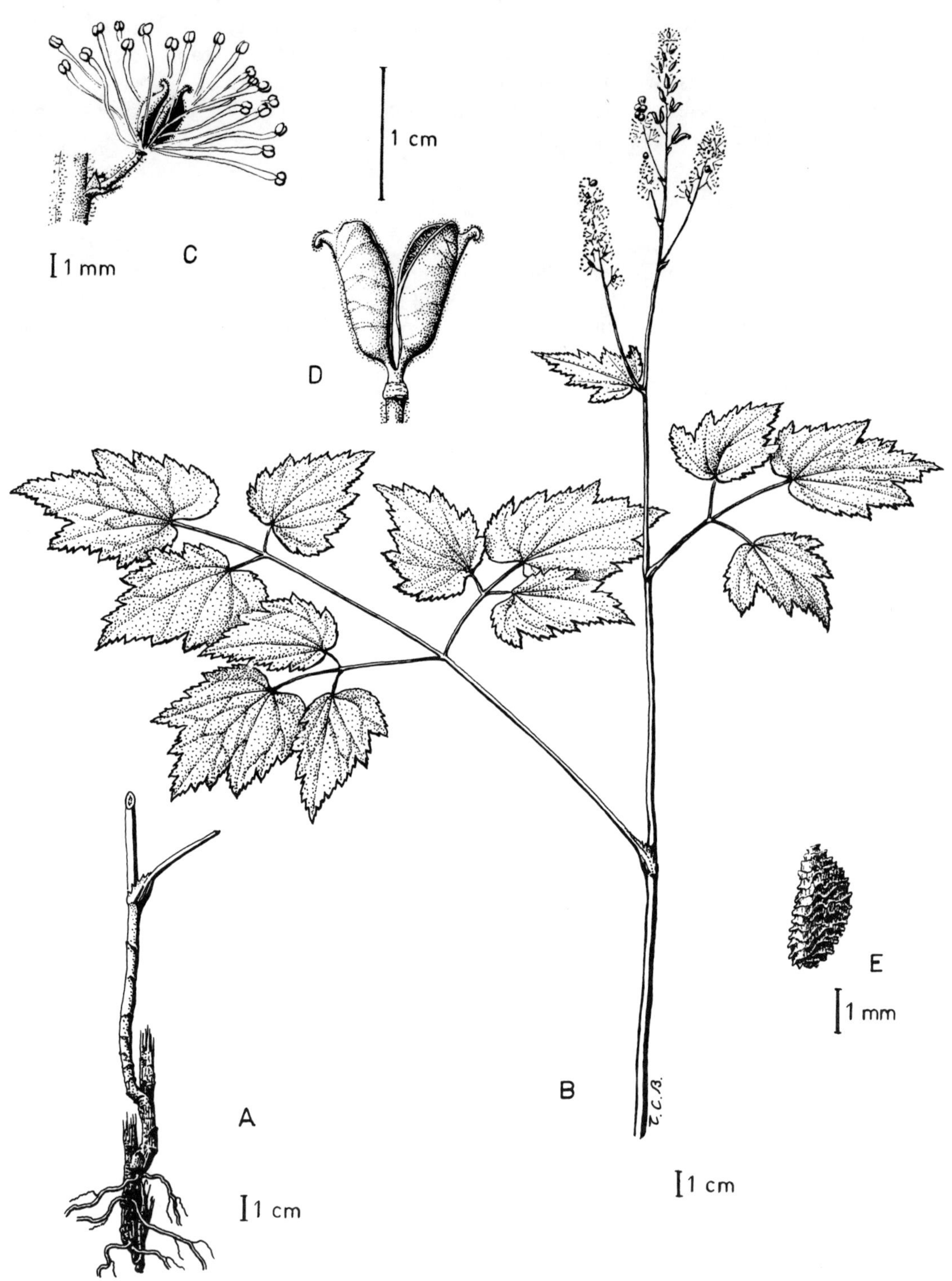

Figure 32. *Cimicifuga elata*

A. base of plant.
B. top of plant.
C. flower.
D. fruit: a pair of follicles.
E. seed.

KEY TO SPECIES

1. Flowers in cymose panicles, white. (subgenus *Clematis*) **2**

   Flowers solitary in leaf axils **3**

2. Flowers dioicous .... *C. ligusticifolia*

   Flowers perfect .... *C. vitalba*

3. Flowers yellow. Leaves often bipinnate, with small leaflets on long petioles, (subgenus *Clematis*) .... *C. tangutica*

   Flowers blue (subgenus *Atragene*) **4**

4. Leaves biternate. True staminodes present, short, spatulate .... *C. alpina*

   Leaves usually ternate (trifoliolate). True staminodes not present, but outermost stamens with oblanceolate filaments and reduced anthers .... *C. occidentalis*

Clematis ligusticifolia Nuttall *in* Torrey & Gray — White Virgin's Bower / Old Man's Beard

Robust, high-climbing dioicous vine with ridged stems and shreddy, light yellowish brown bark; reaching up to 20 m tall.

Leaves pinnately compound with 5–7 deciduous leaflets with coarsely toothed or lobed to entire margins, the persistent petioles and petiolules serving as tendrils.

Flowers dioicous, in few- to many-flowered panicles terminating branchlets, or axillary on the current year's growth. Sepals 4 or more, sometimes partly coherent, white, densely and finely sericeous outside, villous within, and ciliate on the margins. Staminate flowers with 40 or more glabrous white stamens and no carpels; pistillate flowers with a few staminodes and 20–40 or more sericeous carpels with prominent sericeous styles and linear stigmas.

Fruit a head of pale to dark brown, pilose achenes, each with a feathery 'tail' 2.5–5 cm long. $2n = 16$.

A vigorous climber, this species grows best in alluvial sites where it may ascend to the tops of the trees, covering them with dense canopies of foliage. It flowers and fruits profusely, in the latter stage being known locally as "Old Man's Beard". It may also be seen in drier sites covering roadsides, fences, etc.

Distributed from southern British Columbia to California and east to the Dakotas; in British Columbia only east of the Coast and Cascade Ranges, in the climatically dry areas of the Montane Forest region. One record from Hope, where it was found on an abandoned railway grade, probably owes its introduction there to the railway. Another, from Port Moody, is regarded as suspect with respect to its origin.

Clematis occidentalis (Hornemann) DC. — Blue Clematis

    Atragene americana Sims *non A. americana* Miller
    Atragene occidentalis Hornemann
    Clematis verticillaris DC.
    C. columbiana of many authors, not (Nutt.) Torrey & Gray
    C. occidentalis var. *grosseserrata* (Rydberg) Pringle
    C. verticillaris var. *columbiana sensu* A. Gray *non Atragene columbiana* Nuttall

Woody vines with slender stems with shredding bark; climbing to 2 m or trailing on the ground.

Leaves opposite, trifoliolate, with strong, often coiling, petioles and petiolules that serve as tendrils, and persist after the leaflets have been shed. Leaflets ovate, acuminate, usually sparsely and coarsely toothed, or entire, or occasionally, in var. *dissecta* (C. L. Hitchcock) Pringle, deeply lobed to even divided into two or three leaflets; pilose when young, often becoming glabrescent.

Flowers solitary, nodding or inclined on erect pedicels terminating short branchlets from axillary buds on stems of the previous year, rarely terminal on long shoots of the current year's growth; perfect; the 4 sepals meeting at their edges in bud, not overlapping, lanceolate, acuminate,

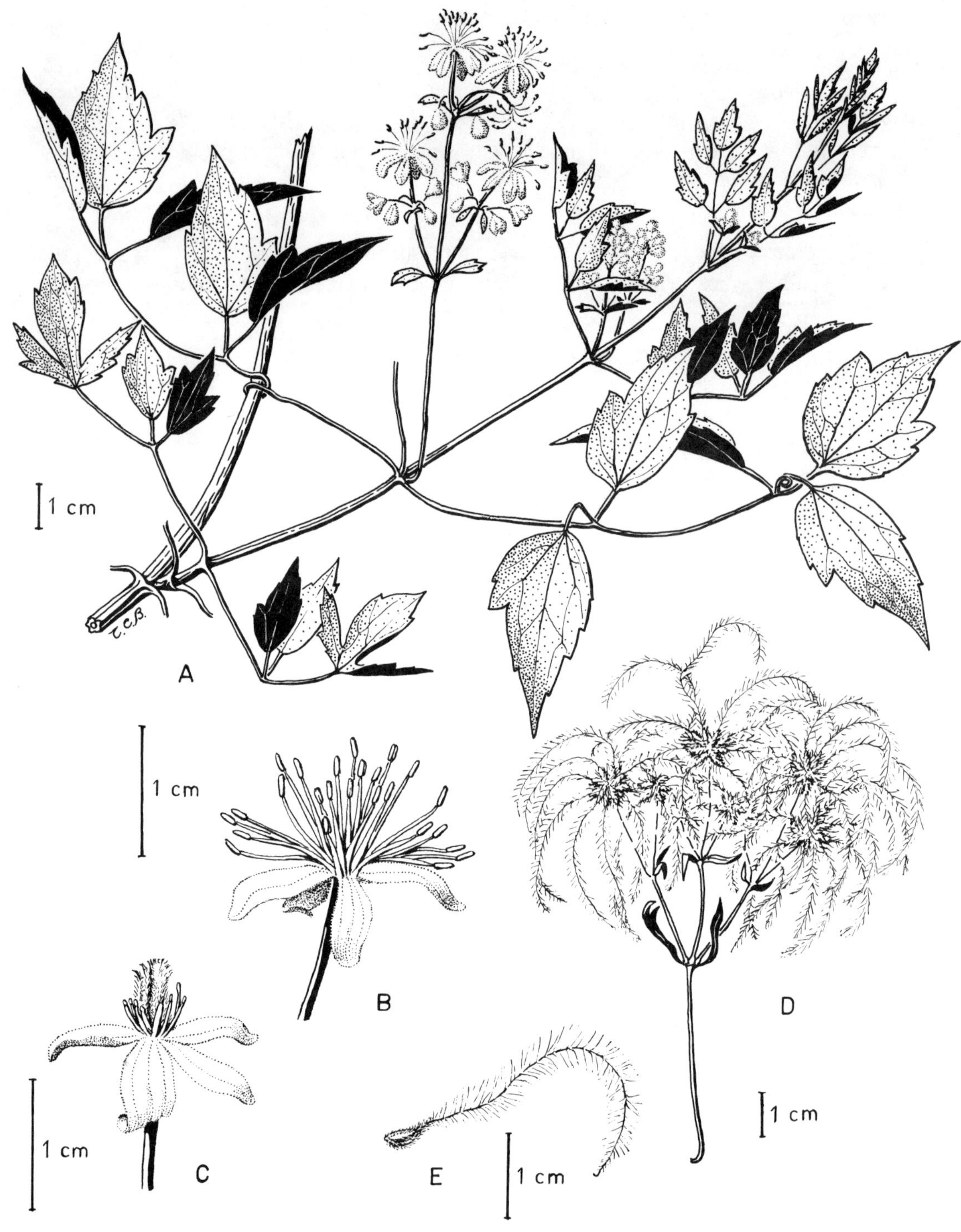

Figure 33. *Clematis ligusticifolia*

A. staminate flowering branch.
B. staminate flower.
C. pistillate flower.
D. fruiting inflorescence.
E. achene.

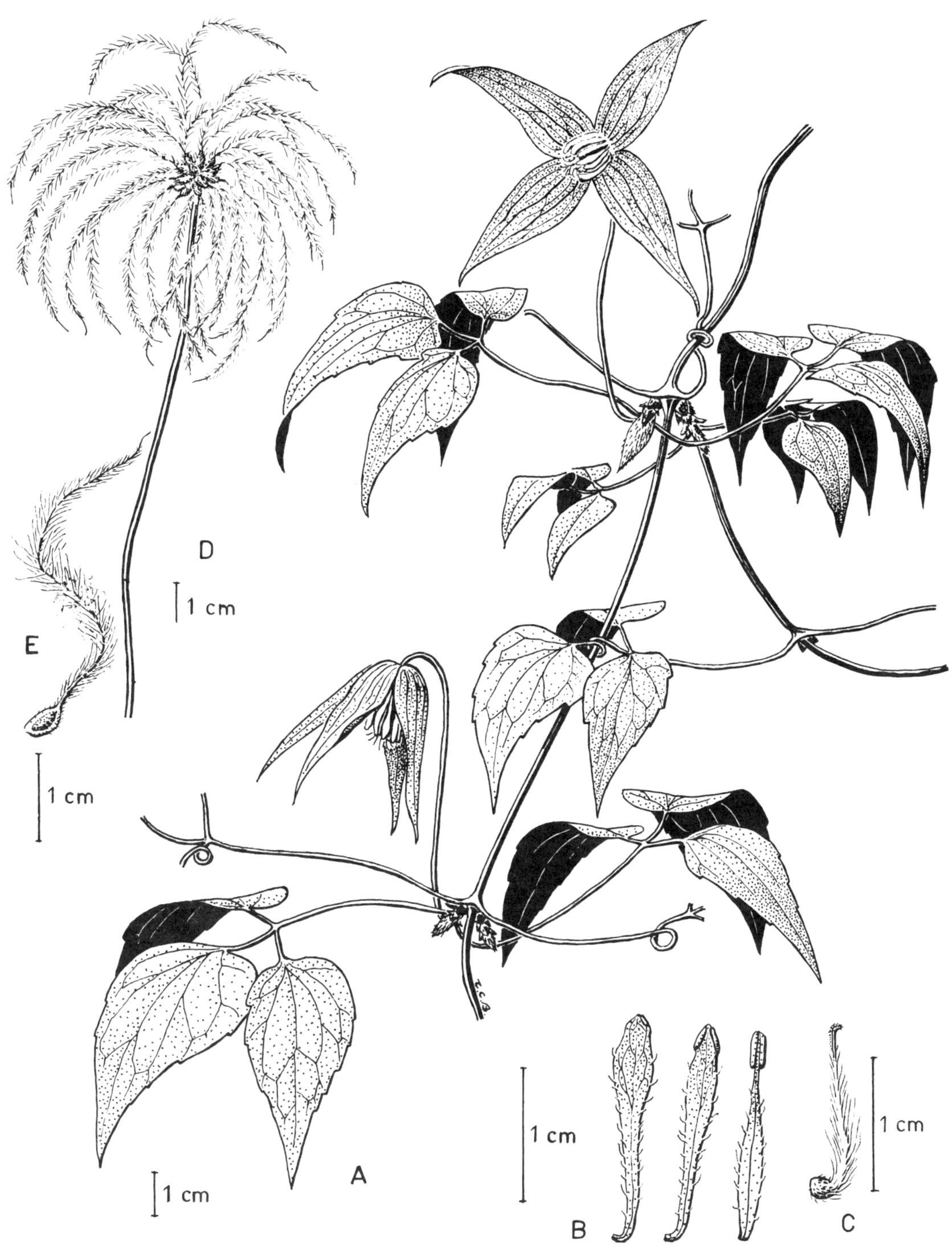

Figure 34. *Clematis occidentalis* var. *grosseserrata*

A. flowering branch.
B. outer and inner stamens.
C. carpel from flower.
D. head of achenes.
E. achene.

violet blue, moderately divergent or sometimes widely spreading with age, pilose. Stamens 20–50, the filaments flattened and pilose, the outer stamens with broadly oblanceolate filaments and reduced anthers, sometimes sterile. Carpels 40–175, pubescent, tipped by styles that elongate to several centimetres in fruit. 3–5 ovules may be produced in each carpel, but only one functions, the fruit formed being an achene with a long feathery style. $2n = 16$.

Our variety in British Columbia is var. *grosseserrata* (Rydberg) Pringle.

Growing in stony or rocky soils in generally rather open wooded environments, in the Cordilleran region of western North America, from the Peace River basin in British Columbia to Colorado. A widely disjunct record from Dease Lake, British Columbia, looks questionable; but similar occurrences of other more southern Interior species are also known there.

On a local scale, the distribution of this species suggests an adaptation to basic rocks (limestone, dolomite, basalt) more than to acidic rocks such as granites or gneisses (Pringle 1971).

Variety *occidentalis* (*C. verticillaris* DC.), with mauve-pink, less divergent sepals, occurs in eastern North America, from the Mississippi valley eastward.

Variety *dissecta* (C.L. Hitchcock) Pringle, a local variety of the Cascade Mountains of Washington State, has not been found in British Columbia, but, if present, may be looked for in the unglaciated mountain area south of the Similkameen River. It is distinguished from var. *grosseserrata* by a shorter, more compact, lower growth habit, more deeply dissected and toothed leaflets, and reddish violet sepals.

The existence of this variety may account for the report of *C. alpina* (L.) Miller, a European species, by Taylor & MacBryde (1977) from British Columbia. That report is based on a specimen, now at DAO, collected by J.A. Calder, J.A. Parmelee and R.L. Taylor, No. 19560, on 1 August, 1956, from the "trail to Ashnola Range, c. 49°06′N, 120°11′W. Locally common in coniferous woods at about 6600 ft.; all sterile except one plant." It was annotated as *C. alpina* by B. Boivin in 1963. That specimen has one head of achenes but no flowers. The leaves are biternate; the leaflets rather attenuate-tipped with rounded teeth. The location is not near human habitation or cultivation, where one might expect a European species to be introduced.

Clematis alpina is a vine up to 2 metres tall, with normally biternate leaves. The blue or sometimes white flowers have antherless staminodes about half as long as the stamens, with a spatula-like dilation of the upper staminode filament. This species is occasionally cultivated.

Some extremely dissected-leafed specimens of *C. occidentalis* var. *dissectum* (C.L. Hitchcock) Pringle, from the Wenatchee Mountains, part of the Cascade Range on the border of Chelan and Kittitas Counties, Washington State, resemble the above specimen of "*C. alpina*", at least in vegetative features. Noteworthy in this respect is J. William Thompson No. 8320 (WTU), collected 29 May, 1932 on Tronson Ridge, Chelan County, at 3000 feet (ca. 1000 m.). The latter plant bears flowers typical of *C. occidentalis*, and biternate leaves. No material of this variety has been collected from the 200 km. gap between the locations of these two collections; but similar material may be looked for, especially in unglaciated areas on the eastern spurs of the Cascade Range.

Of other North American species, *C. columbiana* (Nuttall) Torrey and Gray, in the sense of Pringle (1971), a species of the Rocky Mountain States from southern Montana southward to Texas, resembles the above specimen of "*C. alpina*" in having biternate leaves. Our *C. occidentalis* has commonly been confused with that species. The Ashnola Range (Okanagan Range), a part of the Cascade Range, is far outside the range of *C. columbiana*.

Flowering material of this plant of the Ashnola Range will be needed in order to arrive at a definite conclusion as to its identity.

Clematis tangutica (Maximowicz) Korshinsky   Golden Clematis
C. orientalis L. var. *tangutica* Maxim.

Vine trailing or climbing to 3 m. Stems thinly villous when young. Leaves pinnate or bipinnate, the leaflets small, lanceolate, with diverging sharp teeth.

Flowers usually solitary on axillary short branchlets, or terminal. Sepals 4, ovate-lanceolate, yellow, sometimes 'fading' to purplish. 2–4 cm long. Stamens many, with pubescent filaments. Achenes with plumose stylar beaks up to 7 cm long. $2n = 16$.

Figure 35. *Clematis tangutica*

A. flowering branch.   B. head of achenes.

Introduced from central Asia, this hardy vine has escaped at a few widely scattered localities, mainly in the Interior, in British Columbia.

Clematis vitalba L.

Travellers Joy
Old Man's Beard

Woody vine similar to *C. ligusticifolia*, climbing up to 10 metres; with fluted, thinly pubescent branches. Leaves pinnately compound with usually five rather distant leaflets on long petiolules, the basal leaflets sometimes ternate; the leaflets coarsely toothed or entire, sparsely tomentose, becoming glabrous.

Inflorescence an axillary or terminal cymose panicle with few to many flowers, small bracts, and puberulent peduncle and pedicels. Flowers perfect, (1–) 2 cm wide when fully open, sweet-scented. Sepals 4 (–6), whitish, closely tomentose both outside and inside. Stamens many, white, with flattened filaments. Carpels many, long-styled, sericeous.

Achenes with elongate plumose styles, in large clusters. $2n = 16$.

Clematis vitalba is native of Europe and North Africa. It has been introduced into North America as a garden plant; and has escaped along roadsides and in semi-open habitats. In British Columbia it is found growing wild locally along the coast; on southern Vancouver Island and the adjacent Gulf Islands; where it is sometimes mistaken for the related *C. ligusticifolia*. The latter species does not grow at the coast; and is distinguished from this species by its normally unisexual flowers.

COPTIS Salisbury

Goldthread

Low, rhizomatous, evergreen, mostly glabrous, perennial herbs with ternately or more divided basal leaves. One to few flowers on a scape, perfect to dioicous, radially symmetric, inconspicuous. Sepals petaloid, deciduous early. Petals small and slender, with a stipe-like base and a linear-lanceolate blade or almost no blade; bearing a thickened or pocket-like nectary at the base of the blade on the ventral side. Stamens 12–25. Carpels 5–12, stipitate; the stipes elongating as the fruits mature. The fruit is an umbel-like cluster of stipitate follicles, each of which contains several seeds and opens along the apical half of the ventral suture. $x = 9$.

A circumboreal genus of about ten species; usually in wet wooded or boggy habitats.

KEY TO SPECIES

1. Leaf biternate or ternately pinnate with several leaflets. Petals with linear blades.. *C. aspleniifolia*

    Leaf ternate, with three leaflets... **2**

2. Leaflets unlobed or merely shallowly notched. Petal with little or no blade beyond the nectary.. *C. trifolia*

    Leaflets distinctly lobed, with sinuses reaching halfway to base. Petal with extended blade.. *C. occidentalis*

Coptis aspleniifolia Salisbury

Fern-Leafed Goldthread
Cut-Leafed Goldthread

Leaves fern-like; ternate with the pinnae ternate or pinnately divided into several lobed and toothed, dark green, shiny leaflets.

Scape about as tall as the leaves (around 8–12 cm) in flower; elongating to around 25 cm in fruit, when well overtopping the leaves; bearing 2 or 3 flowers, the lateral ones in the axils of minute bracts. Flowers nodding, perfect or staminate. Sepals 5–6, tapering-linear, 6–15 mm long, reflexed and ascending, pale greenish white. Petals shorter than the sepals, with broad, swollen nectaries at the bases of the recurved blades. Stamens several, with thickly ellipsoidal to almost globose anthers and slender filaments. Carpels up to 11. Each carpel tapers into a beak with stigmatic lines running

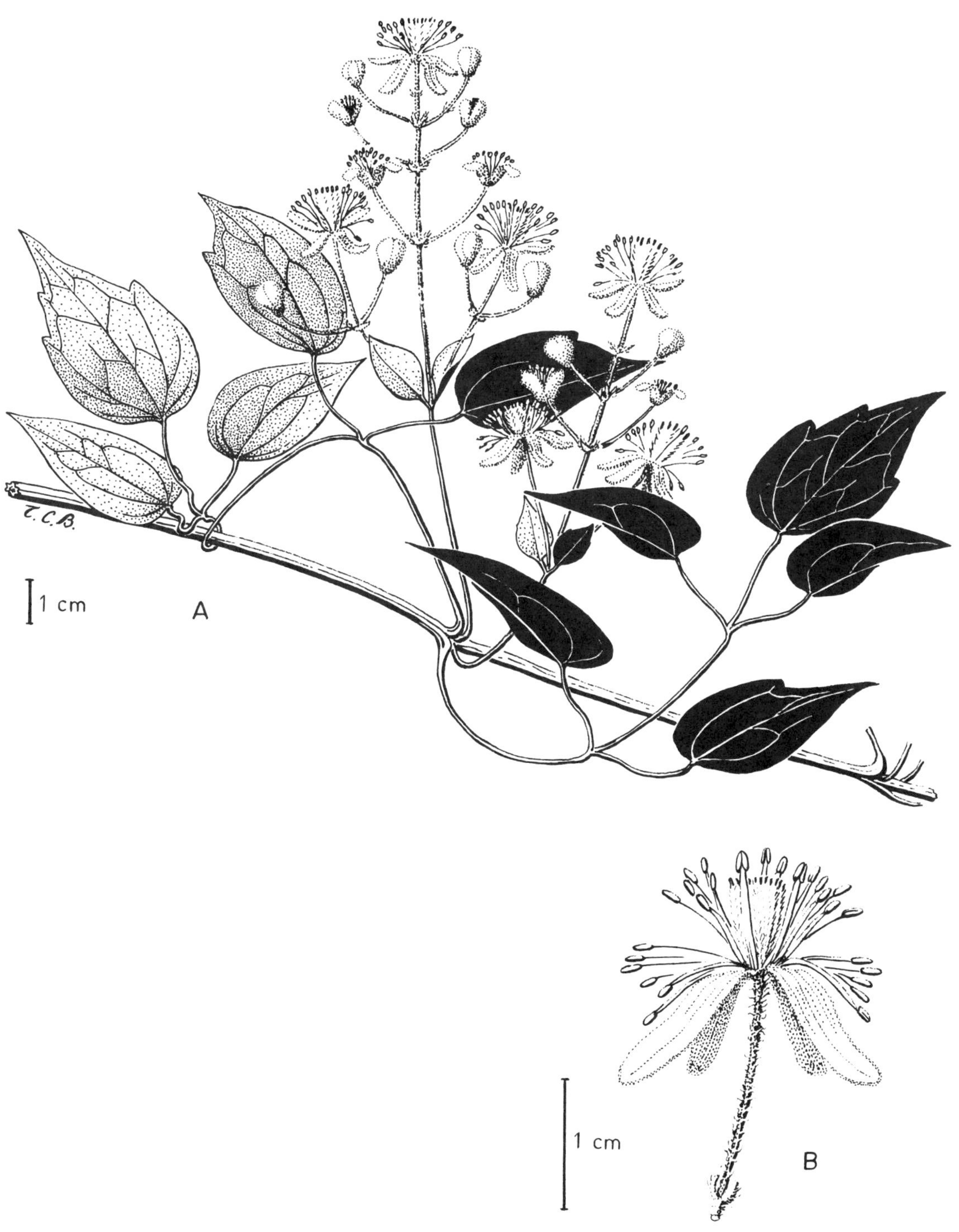

Figure 36. *Clematis vitalba*

A. part of flowering branch.   B. flower.

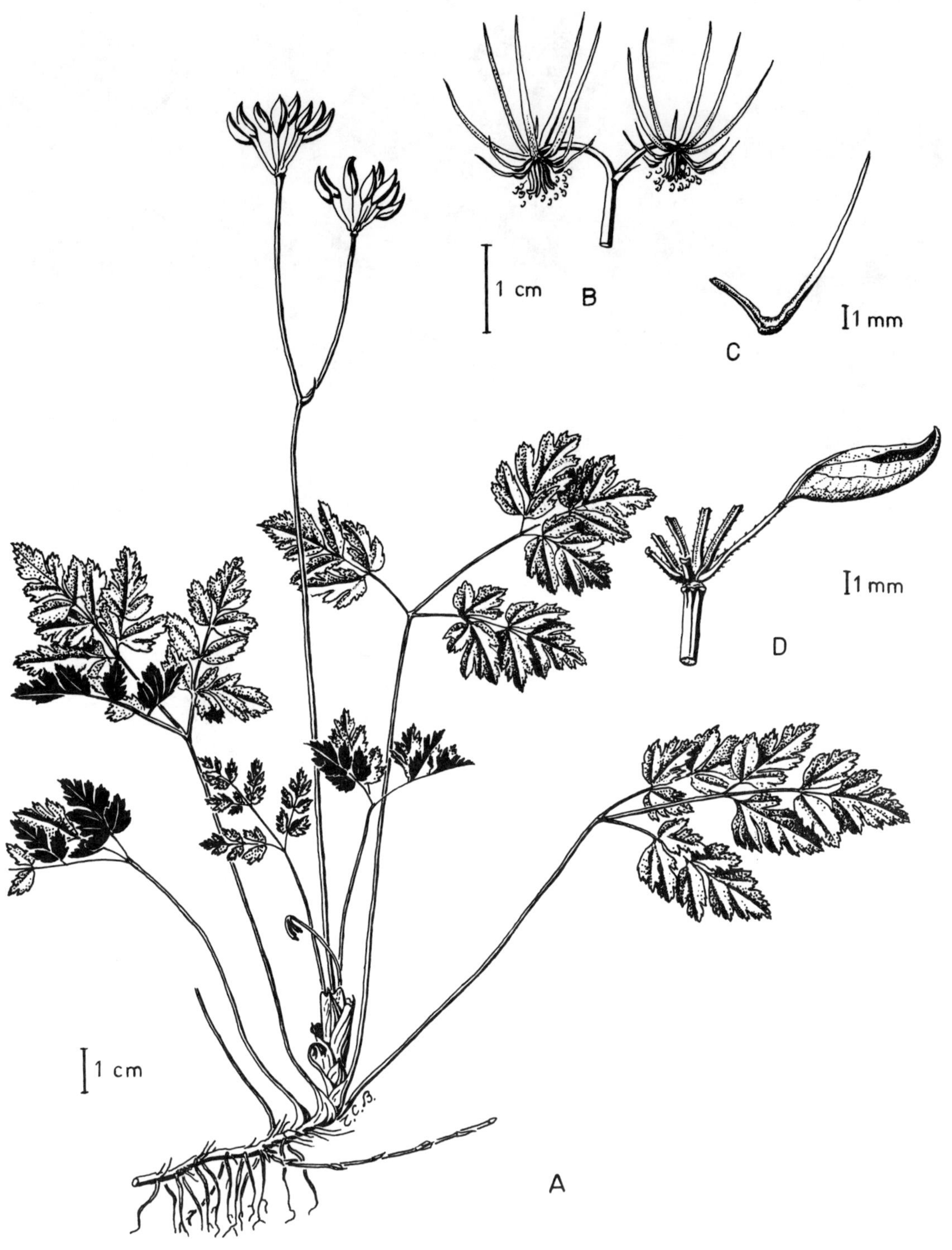

Figure 37. *Coptis aspleniifolia*

A. fruiting plant.   B. flowers.   C. petal.   D. follicle.

down the upper carpel margins, which appear to be pressed together but not joined above, though joined below, where they do not separate.

Follicles 7–9 mm long on stipes about as long; opening by spreading apart of the distal carpel margins, which may start to separate before the follicle is fully grown. Each follicle contains 5–10 seeds. 2n = 18.

Found in moist forest, seeps, and bogs, along the coast from Alaska to Vancouver Island.

Coptis trifolia (L.) Salisbury   **Gold-Thread**

Low perennial herb with a yellow, thread-like rhizome, trifoliolate leaves with toothed but scarcely lobed leaflets 1–2 cm long, and scapes up to 12 cm tall, minutely puberulent above.

Flower solitary on each scape. Sepals usually 5, oblanceolate to narrowly elliptic, white, often with a purplish flush dorsally. Petals about half as long as the sepals, each with a yellow, fleshy, pocket-like nectary and only a blunt rudiment of a blade beyond it. Carpels 3–7, distinctly stipitate, with a recurved stigma; the stipes minutely puberulent.

Fruiting follicles 5–10 mm long, on stipes about as long or often longer; the style and stigma forming a distinct, ascending beak 2–3 mm long and terminally hooked; the follicle containing 5–12 seeds. 2n = 18.

Found in moist woods, mossy bogs and tundra, from Greenland and eastern North America [as var. *groenlandica* (Oeder) Fassett] westward across North America, mainly in the Boreal Forest region, and in the Aleutian Islands, Japan and eastern Siberia. In British Columbia, *C. trifolia*, as var. *trifolia* , has been found mainly on or near the coast, with a few records in the east central Interior. More occurrences of this species inland, including east of the Rocky Mountains, may be looked for. The flowers are pollinated by hover-flies (Syrphidae) (Douglas, 1983).

Coptis occidentalis (Nuttall) Torrey & Gray

Scapose herb with a slender, yellow to pale brown rhizome terminating in erect aerial shoots, and branching from buds in the axils of scale-like leaves.

Leaves 1–3, basal, trifoliolate; the leaflets 2–6 cm long, more deeply dissected than those of *C. trifolia*; cut about half way or more to the base into lobes that may be again lobed; the margin sharply or apiculately toothed.

Flowers 2–5, usually 3, on a scape that is shorter than the leaves at flowering time, but elongates more by fruiting time. Pedicels glabrous or finely puberulent. Flower similar to that of *C. aspleniifolia*, but larger, and held erect. Sepals commonly 5, linear, spreading to reflexed, longitudinally 3-veined, whitish. Petals 6–10, slenderly lanceolate, ⅔ to ¾ as long as the sepals; with a linear ascending claw topped by a swollen nectar gland, the nearly linear blade being outwardly flexed at the gland. Stamens 15–20. Carpels 5–10 or more, on glabrous or finely puberulent stipes; each carpel containing 5–10, commonly 8, ovules, and tapering into a hollow stylar beak ending in a reflexed stigma. Flowering in April and May, fruiting from May to August.

Follicles on conspicuous stipes shorter than the follicle bodies, which taper into upcurved, blunt beaks, and open along the distal third of the ventral sutures.

Coptis occidentalis is found in moist coniferous woods, in northeastern Washington State, Idaho, and northwestern Montana. It has been reported by Hitchcock and Cronquist (1964) to occur in southeastern British Columbia, and is listed for this province by Taylor and MacBryde (1977). It is on the basis of this statement that this species is included here; however, to this date (January, 1987) no British Columbian specimens have been seen by this author.

DELPHINIUM L.

Delphinium, Larkspur

Erect perennial or annual herbs with alternate, palmately dissected leaves, and flowers in terminal, and sometimes axillary, racemes.

Flowers bilaterally symmetric, blue in ours. Sepals (the most conspicuous parts of the flower) 5, the uppermost one prolonged back from the base into a hollow spur. Petals 4 or 2, smaller than the

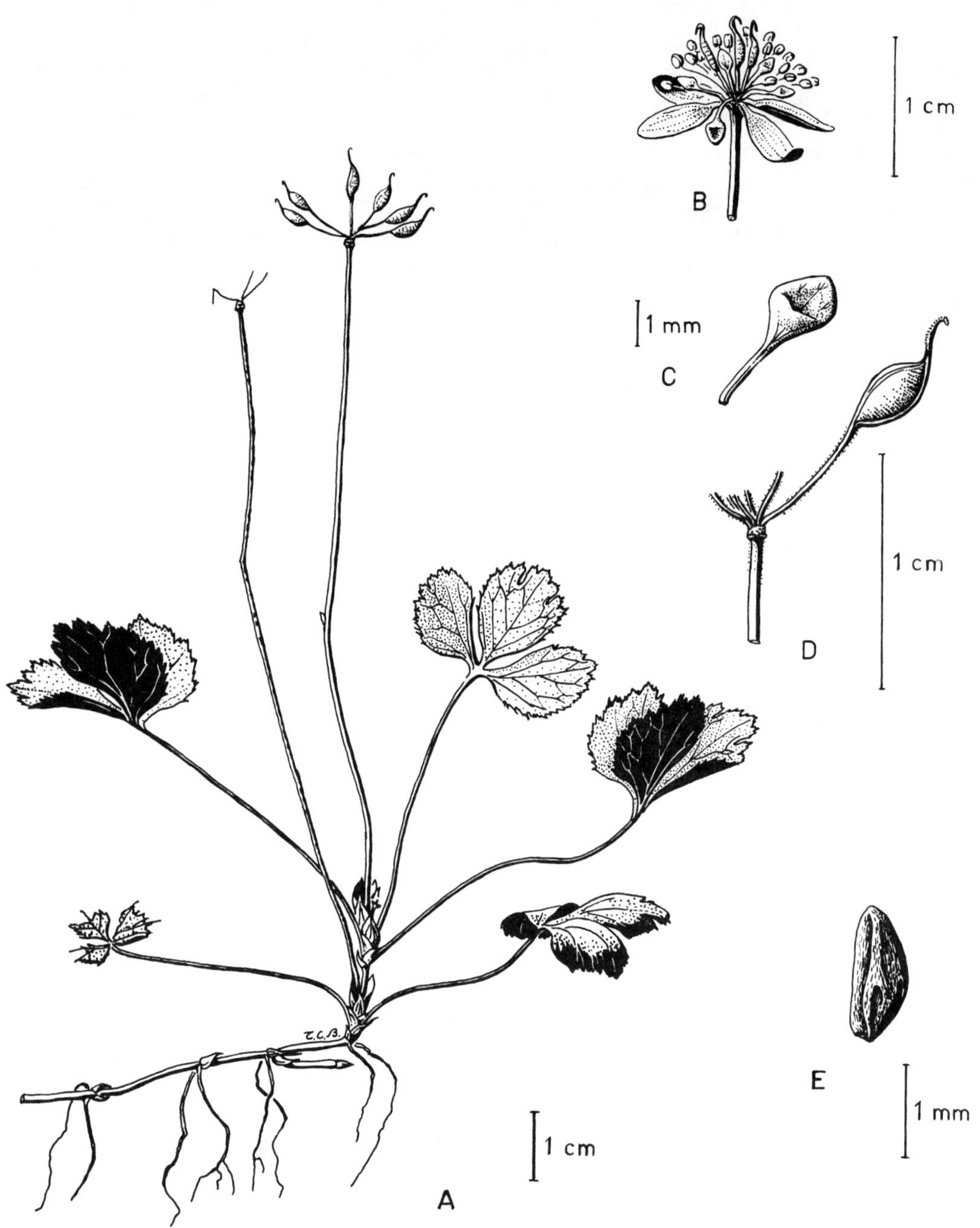

Figure 38. *Coptis trifolia*

A. fruiting plant.
B. flower.
C. petal.
D. follicle.
E. seed.

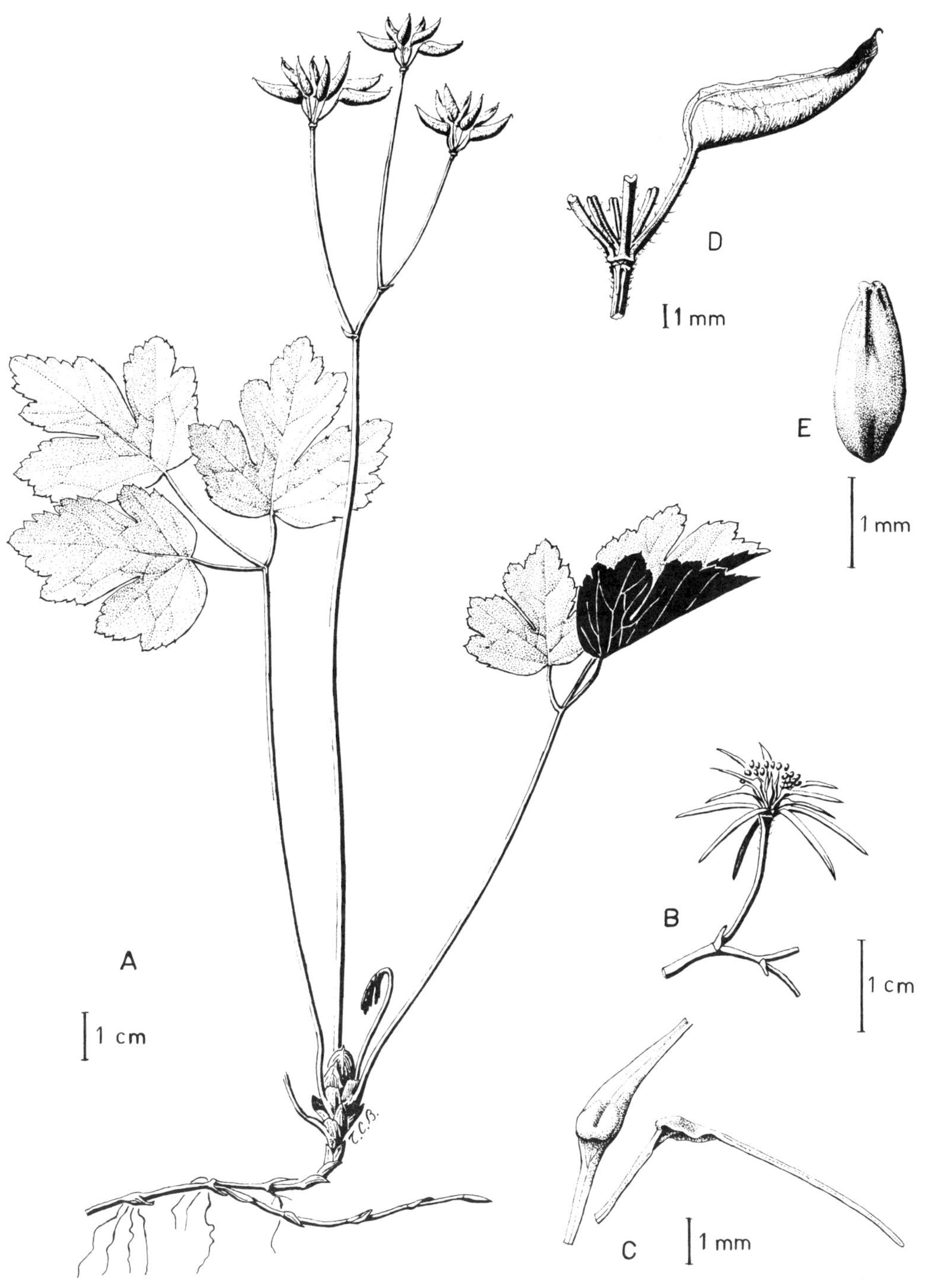

Figure 39. *Coptis occidentalis*

A. fruiting plant.
B. flower.
C. petals.
D. follicle.
E. seed.

sepals, the upper two (sometimes the only ones) bearing nectariferous spurs that are enclosed in that of the upper sepal. Stamens many, in spiral series, the filaments with wide, flattened bases. Carpels 1–5, normally 3 in our species; containing many ovules, and tipped by styles and short, bilobed, terminal stigmas.

Fruit 1–5 follicles per flower (3 in ours), tipped by the persistent styles. Seeds many, with loose, wrinkled seed-coats (testa). x = 8.

Delphinium is a genus of about 200 species, mainly of the north temperate latitudes, with a few in Africa. Five species are native to British Columbia. They are associated with a variety of habitats: from moist shady woodland to dry open steppe. Pollination of our species is by bees and bumblebees. All species are poisonous.

Several exotic species are cultivated as ornamental garden plants. Notable among these here are *D. elatum* L., the common tall (to 2 metres) garden delphinium, originally native of Siberia, and the annual larkspur, *D. ambiguum* L., a native of Europe, which occasionally escapes on Vancouver Island.

KEY TO SPECIES

1. Plant a taprooted annual with finely dissected leaves. Upper petals united, lower petals absent. Follicle one. Introduced; occasional garden escape **Section *Consolida; D. ambiguum***

    Plant perennial with fibrous or tuberous roots. Flower with 4 separate petals. Follicles normally 3. Native .... **Section *Delphinastrum* 2**

2. Flowers mostly cup-shaped: not opening widely. Sepal blades 6–12 mm long. Stem often thick and hollow **3**

    Flowers opening widely. Sepal blades 12–25 mm long. Stem slender. Leaves dissected into linear, forked divisions **4**

3. Plant tall (6–30 dm tall) with notably hollow (fistulose) stem, and fibrous roots. Leaf segments broad and wedge-shaped. Plants of Rocky Mountains and northern B.C. .... ***D. glaucum***

    Plant 2–7 dm tall, with somewhat hollow stem, and thick fleshy roots. Basal leaves broadly segmented, the upper linearly segmented. Southeastern Interior .... ***D. burkei***

4. Lower petal blade white to pale blue, with darker veins; and notched at least ¼ of its length .... ***D. nuttallianum***

    Lower petal blade light to dark blue; notched for ⅕ or less of its length **5**

5. Plant glandular above. Stem firmly attached to fleshy-fibrous root. Southeastern Interior .... ***D. bicolor***

    Plant not glandular above. Stem easily detached from tuberous root. Coastal ***D. menziesii***

***Delphinium ambiguum* L.**   **European Larkspur**
    ***D. ajacis* L.**
    ***Consolida ambigua* (L.) P.W. Ball & Heywood**

Annual, taprooted, finely puberulent herb up to a metre tall but usually less; the slender stem simple or branched.

Leaves many times dissected. Basal and lowest cauline leaves long-petioled, the blades appearing to be trifoliolate with oblong or cuneate leaflets, the middle and upper cauline leaves sessile or subsessile, and multifid into fine linear divisions; often giving a tufted appearance to the nodes.

Inflorescence a simple or branched raceme (rarely a panicle), the lower flowers subtended by divided bracts, the upper ones by simple bracts. Pedicels 12–30 mm long, each with a pair of minute bractlets.

Flowers with broader sepals than in our other, native, Delphiniums; purplish blue, purple, pink, or occasionally white. The upper two petals united to form a corolla with an ascending bifid lobe,

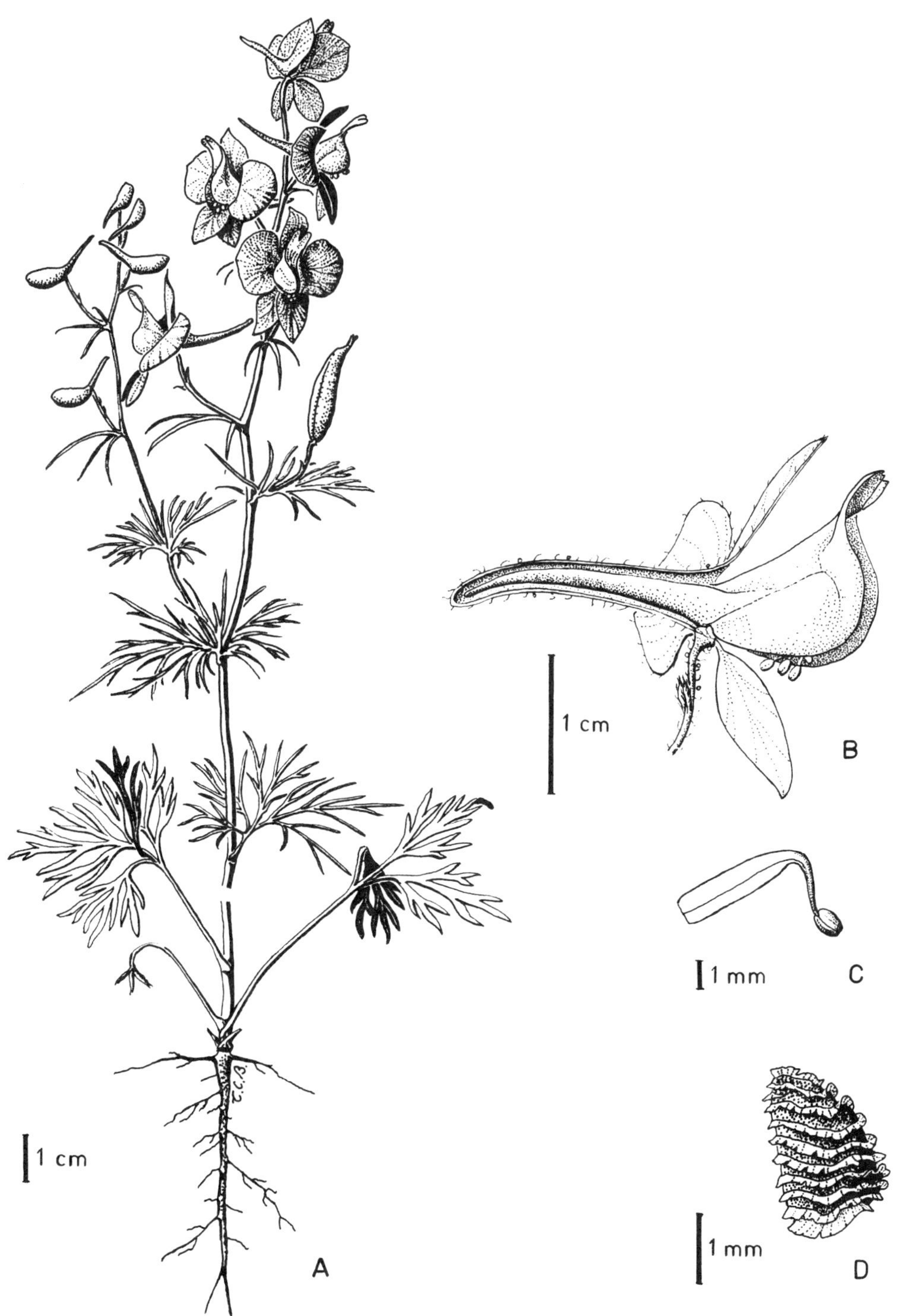

Figure 40. *Delphinium ambiguum*

A. base and top of flowering and fruiting plant.
B. flower: section, with near sepals and near half of upper sepal removed; showing spurred corolla and a few of the
anthers.
C. stamen.
D. seed.

two rounded lateral lobes, and a hollow, nectar-bearing spur. Lower petals absent. Stamens about 15. Carpel one, pubescent.

Follicle erect, tapering abruptly to the short stout stylar beak; thinly pubescent; containing many small black or dark brown seeds, with narrow, paler, transverse fringes. 2n = 16.

Delphinium ambiguum is native to the Mediterranean countries. It is sometimes cultivated in this country as an ornamental plant, and occasionally escapes locally along roadsides on this coast.

Delphinium bicolor Nuttall

Rather low (10–70 cm tall) erect, stiff perennial, the stem usually single, and usually velutinous, not fistulose, firmly attached to the fibrous and sometimes rather fleshy, extensive root system.

Leaves few, mainly basal or near-basal, on short to long petioles, the blade palmately divided into bipinnately dissected divisions, the ultimate lobes linear. The upper bract-leaves progressively reduced upward.

Inflorescence usually a short simple raceme. Pedicels longer than the flowers, velutinous and sometimes glandular; each commonly bearing one or two minute bractlets below the flower.

Flower deep blue to purplish, rather larger on the average than those of our other native Delphiniums. Upper sepal with a blade shorter than the other sepals; its spur longer than the blade and downwardly curved at the tip. The other sepals broadly ovate, and sometimes apiculate. Upper petals white or pale bluish, with purple-lined veins near the apex; their obliquely hollowed spurs enclosed in the spur of the upper sepal. Lower petals with broadly ovate, outwardly twisted, dark blue blades centrally bearded with white hairs on the outwardly facing (morphologically inner) surfaces; the margin erose to nearly entire; the notch, if present, less than ⅕ the length of the blade. Stamens many, with filaments widened toward the bases. Carpels 3, velutinous, and sometimes glandular.

Follicles somewhat divergent, sparsely velutinous to occasionally glabrous, tipped by stylar beaks about 3 mm long. Seeds about 2 mm long, the dark brownish body bordered by pale, flange-like, extended margins.

Delphinium bicolor is a plant of the Rocky Mountains and adjacent ranges and plains, from Alberta to eastern Washington, Wyoming and South Dakota. In British Columbia it is rare, and found only in the extreme southeastern part of the province, as in the Crowsnest Pass area, in open park-like and grassy areas at subalpine levels. One specimen in the Royal British Columbia Museum from Crawford Bay, on Kootenay Lake, shows some characteristics of *D. nuttallianum*.

Delphinium burkei Greene
D. simplex Douglas *ex* Hooker *non* Salisbury

Erect perennial up to a metre tall, with rather fleshy, thick roots; the stem hollow and puberulent with curled hairs.

Leaves of two forms: The basal leaves with long petioles and blades parted nearly to base into 3–5 cuneate-based divisions with obtuse to rounded tips, often withered by flowering time. Lowest stem leaves similar to basal leaves or transitional to the upper stem leaves, which are short-petioled and commonly crowded and overlapping on the lower half of the stem, with blades divided to base into ascending, linear, or linearly lobed divisions.

Flowers in simple or compound racemes; the bractletted pedicels crisp-puberulent and shorter than the spurs of the sepals. Flowers rather cupped, with forward-turned sepals. Sepals blue to purplish, puberulent on the backs, the blades 5–11 mm long; the spurs 11–17 mm long, projecting back well past the axis of the raceme. Upper petals white or nearly so, often bluish and shallowly notched at the tips. Lower petals blue; the blades notched more or less half way to the tops of the claws, and bearing a central patch of sparse long white hairs, and marginal cilia. Carpels 3, finely and densely puberulent with yellowish hairs, and commonly glandular. Flowering from late May to August.

Follicles erect, 6–15 mm long, distinctly puberulent and glandular. Seeds about 1.0–1.5 mm long, with relatively wide pale flanges on the angles.

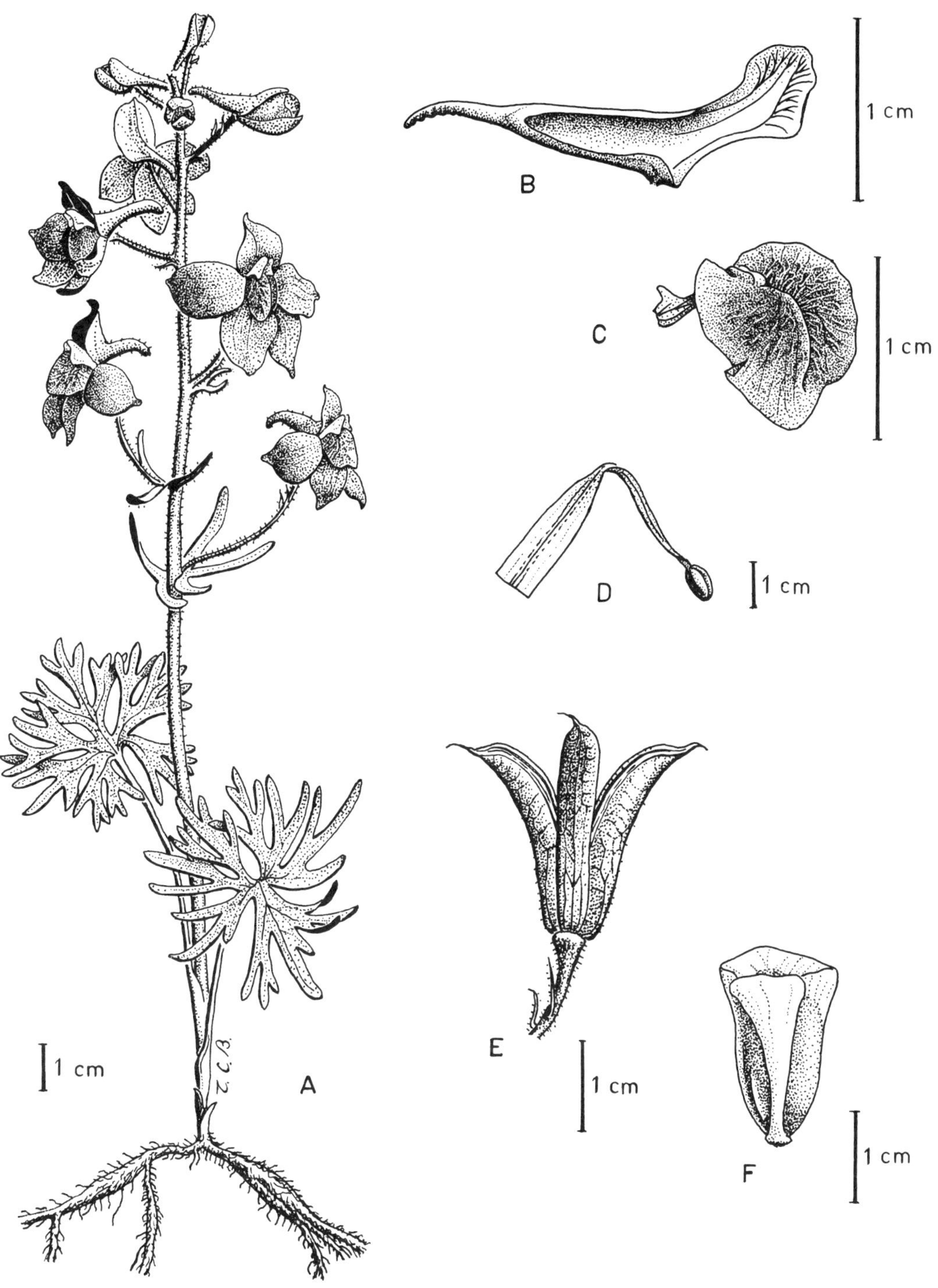

Figure 41. *Delphinium bicolor*

A. flowering plant.
B. upper petal (seen from inner side).
C. lower petal (from outer side at base).

D. stamen.
E. follicles.
F. seed.

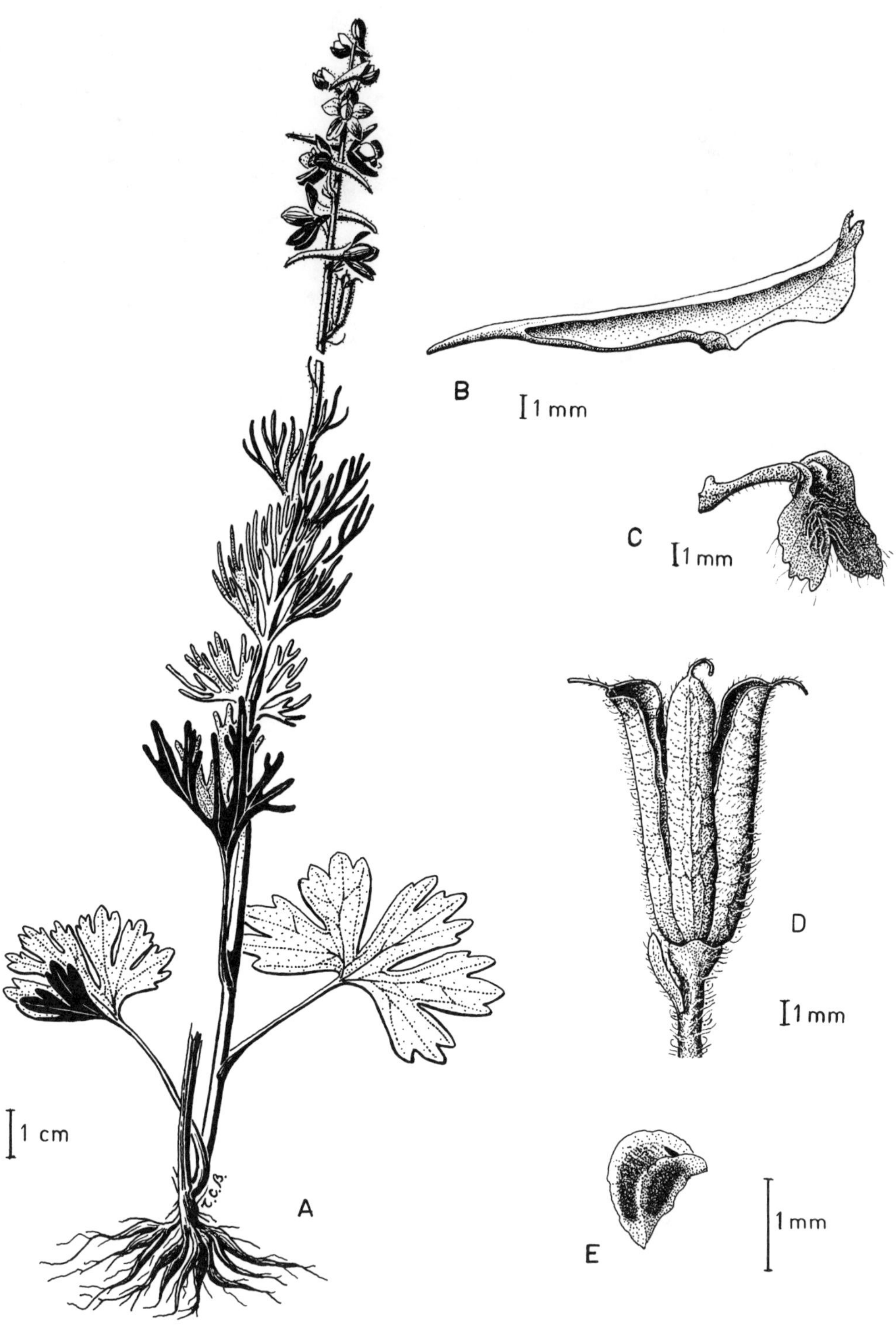

Figure 42. *Delphinium burkei*

A. flowering plant.
B. upper petal.
C. lower petal.

D. follicles.
E. seed.

Delphinium burkei is found from the southern interior of British Columbia south to central Oregon and eastward to Idaho. In British Columbia it has been found in the Columbia Valley near Trail and in the Kettle Valley. It inhabits seasonally spring-moist low ground in the dry forest and steppe zones.

Delphinium glaucum S. Watson   Tall Delphinium

**   *Delphinium scopulorum* Gray var. *glaucum* (S. Watson) Gray**
**   *Delphinium exaltatum* Aiton var. *glaucum* (S. Watson) Huth**
**   *Delphinastrum glaucum* (S. Watson) Nieuwland**
**   *Delphinium brownii* Rydberg with respect to Canadian material.**

Tall (1–2 metres or more) perennial herb, often multi-stemmed from a caudex; the stem coarse, simple below the inflorescence, hollow, and glabrous except in the inflorescence; the plant resembling *Aconitum columbianum* in general aspect.

Leaves without stipules, the petioles from longer than to shorter than the blades; the blades deeply palmately dissected into 3 (rarely 5) cuneate-based main lobes, which in turn are dissected into usually acute lobes and teeth; glabrous to sparsely puberulent above, often pubescent beneath. Leaf blades generally alike up to the inflorescence where intergrading with the bracts. The bracts leaf-like though subsessile below, but reduced upward through linear-lanceolate to minute and linear, and adnate to the pedicels.

Inflorescence a dense, almost spike-like terminal raceme of crowded flowers, extending and opening out as the fruit matures. Often, lateral racemes arise from the axils of the lower, leaf-like bracts, and flower later than the terminal raceme. Peduncles and pedicels often finely puberulent.

Flower rather cup-like, not opening widely; greyish blue to dark blue. Sepal blades 8–17 (mostly 10–15) mm long, ovate, pubescent dorsally along their mid-lines; the spur about as long as or somewhat longer than the blades, and wrinkled. Petals scarcely evident: the upper pair spurred, the blades glabrous, pale within, streaked bluish without, with wide white margins, and each with a double tooth at the apex: the lower pair pale to dark blue, the expanded upper part ciliate, and pubescent on the inner surface (which is turned outward by a twist at mid-length of the petal); the hairs white or pale yellowish; the expanded upper part of the blade notched up to half of its length, the notch 1–3 mm deep.

Stamens many (around 20); the filaments broadly flattened below, becoming recurved above; the anthers blue. Carpels 3, glabrous or sometimes finely crisp-puberulent, with styles 2–3 mm long, and containing 12–14 ovules each. Carpel margins not or scarcely united at flowering time.

Follicles 3, erect, 12–14 mm long in ours (14–18 mm in *D. glaucum* in the strict sense), normally glabrous, but sometimes puberulent; tipped by filiform stylar beaks about 3 mm long, and containing around 12 seeds. Seeds 2–3 mm long (up to 3.5 mm long in *D. glaucum* in the strict sense), pale to dark brown with projecting, wide, paler margins.

Delphinium glaucum is distributed from Alaska to Alberta, and from western Washington State (Cascade and Olympic Mountains) to California, mainly in the Subalpine Forest regions.

These two geographically separated populations are sometimes treated as distinct species (e.g. by Ewan, 1945); the population in the Pacific states being then treated as *D. glaucum* in the strict sense. That in Alaska and Canada is then called *D. brownii*. However, the morphological differences between them in follicle and seed lengths, and in the darker seeds in *D. glaucum* in the strict sense, seem insufficient for specific distinction. The floral dimensions overlap completely. Ewan (1945), in his keys, distinguished them only by their different geographic ranges. This species appears to be most closely related to *D. exaltatum* Aiton, of the eastern United States.

Delpinium menziesii DC.   Menzies' Larkspur

**   *Delphinastrum menziesii* (DC.) Nieuwland**

Short (up to 50 cm tall) single-stemmed perennial; the slender stems weakly attached to the globose to fusiform tuberous roots, and simple or branched above, glabrous, pubescent, or sometimes glandular; the hairs curved or wavy.

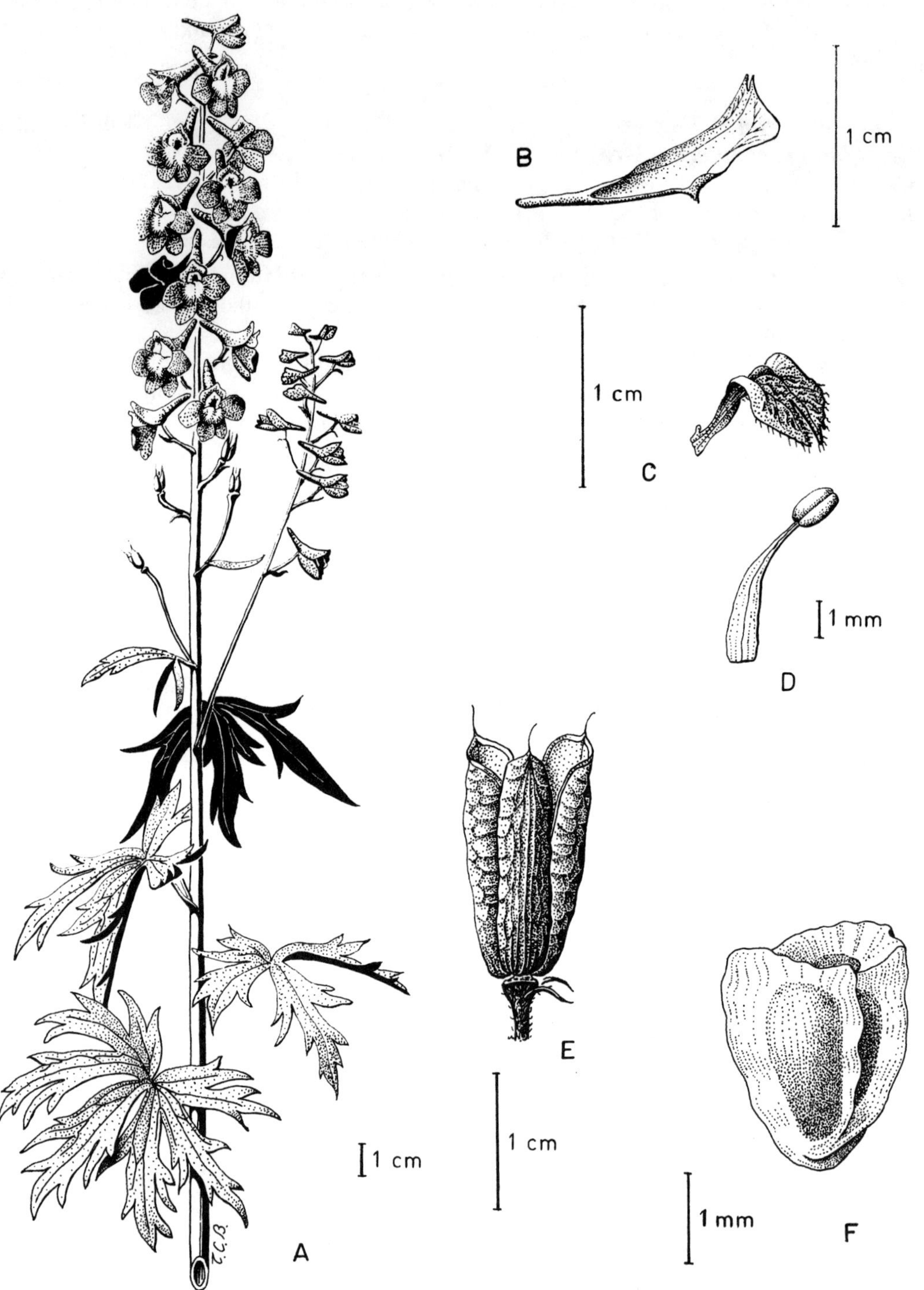

Figure 43. *Delphinium glaucum*

A. top of flowering plant.
B. upper petal: inner side.
C. lower petal: outer side at base.
D. stamen.
E. follicles.
F. seed.

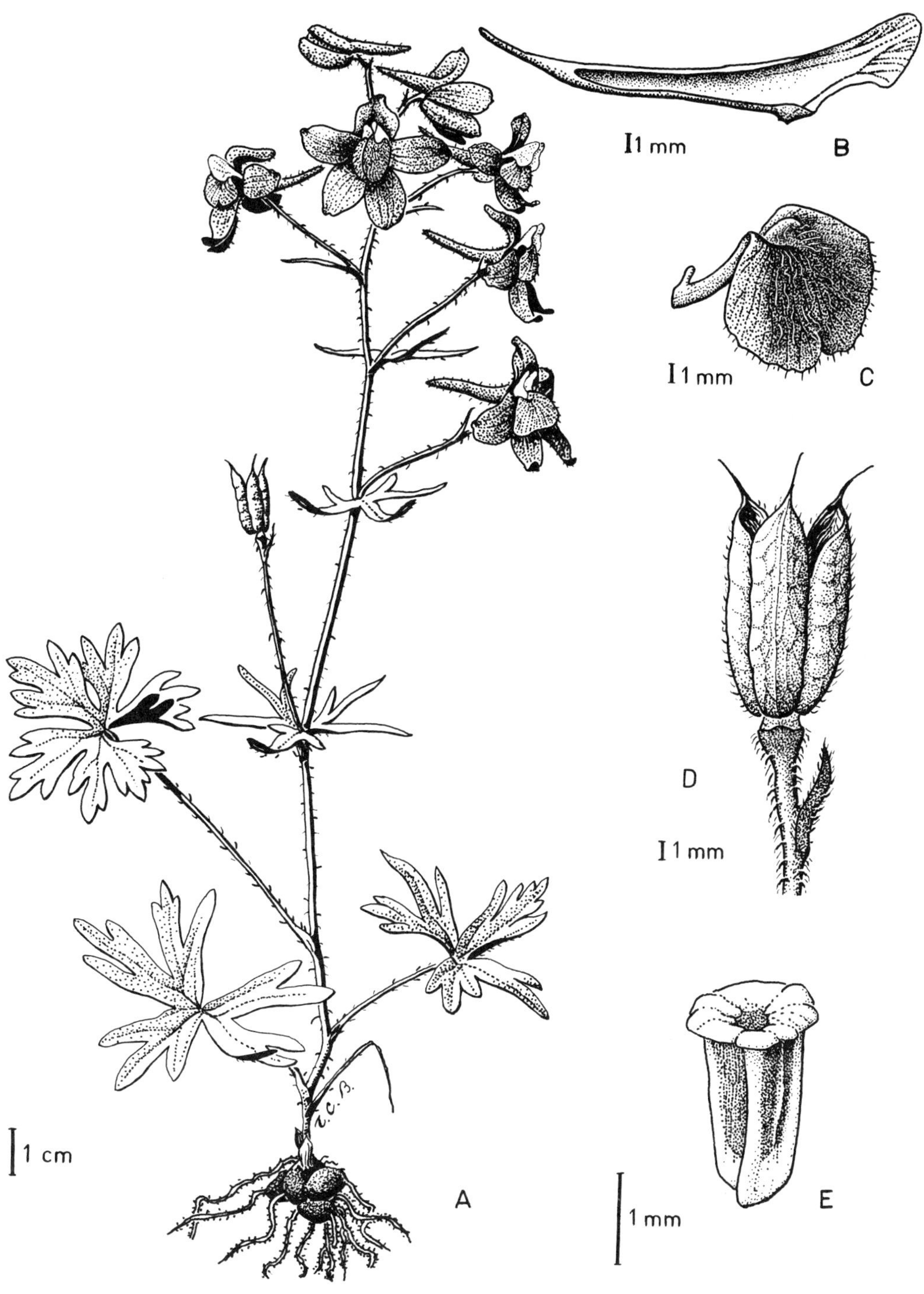

Figure 44. *Delphinium menziesii*

A. flowering plant.
B. upper petal: ventral (inner) view.
C. lower petal: dorsal view at base.

D. follicles.
E. seed.

Leaves mainly cauline and petioled; the few basal ones long-petioled; the upper stem leaves reduced and sessile. Main leaf blades palmately dissected into linear or cuneate-lanceolate divisions.

Inflorescence a simple or compound raceme; the pedicels equalling to much exceeding the flowers in length, pubescent; the hairs often curved downward. Sepals 12–18 mm long, dark blue to purplish blue; the spur of the upper sepal up to 18 mm long, about 1½ times as long as its blade. Upper petals pale blue to white. Lower petals blue, entire, sinuately toothed, or with a shallow notch ¼ or less as long as the expanded part of the blade. Carpels 3, usually thickly puberulent with ascending short hairs, not glandular. Flowering from late April to June, or to July at higher altitudes.

Follicles erect or spreading, 9–16 mm long. Seeds about 1½ mm long, narrowly flanged on the angles.

Delphinium menziesii is found on open bluffs and coastal prairies; occasionally up to 1000 metres in elevation; from coastal British Columbia to northern California, west of the Cascade and Coast Ranges. A single record from Sorrento, in the interior (see map) may be an error. The specimen, at UBC, has entire petals, but could be an aberrant *D. nuttallianum*.

Delphinium nuttallianum Pritzel   **Nuttall's Larkspur**

Highly variable, erect, slender, usually single-stemmed perennial herb 1–8 dm tall. The root system varying from tuberous and compact to fibrous and often fleshy, at least partly in response to soil conditions.

Leaves basal and cauline, the lower ones long-petioled, without stipules; the blade palmately divided into slender divisions that are again divided pinnately into linear divisions. Stem and leaves puberulent to pubescent or glabrous, occasionally glandular.

Flowers in terminal, and sometimes also lateral, bracted racemes. Peduncle and pedicels often glandular-pubescent; the pedicels from shorter to longer than the flowers. Sepals deep, or rarely pale, blue, ciliate, the upper one with a spur from shorter than to much longer than the blade; the lateral sepal blades the widest, but of medium length; the lowest pair the longest. Petals 4, with blades about half the length of the sepal blades. The upper petals with almost white, lanceolate, ascending, glabrous blades with blue veins, and basally projecting nectariferous spurs enclosed in that of the upper sepal. The lower petals unspurred, with downwardly and outwardly reflexed blades notched for ¼–⅔ their length; generally solid blue, often pale, sometimes pale brownish tinged, and with white bearding on the outer (morphologically inner) surface beyond the point of flexure, and with ciliate margins. Stamens 10 to many. Carpels 3, glandular-pubescent, multi-ovuled.

Follicles erect to somewhat diverging, beaked by the persistent styles and terminal stigmas. Seeds 1½–2 mm long, dark grey-brown with paler, flanged edges. 2n = 16.

KEY TO VARIETIES

1. Petal blades yellowish or brownish-tinged. Inflorescence and carpels crisply puberulent but not glandular. Southeastern B.C.. **var. *fulvum* C.L. Hitchcock**

   Petal blades without yellowish or brownish tinge. Inflorescence glandular as well as puberulent. Widespread in the Interior... **var. *nuttallianum***

Delphinium nuttallianum is found in dry open woodland, Ponderosa Pine parkland and open steppe communities, on well-drained, medium-textured to sandy soils, at low to medium altitudes.

It is found from British Columbia, east of the Coast and Cascade Ranges, to Alberta, southward to California and Arizona, and eastward to Nebraska. In British Columbia it is abundant in a well defined range in the southern interior; but a single record from Dease Lake, at 58½°N. latitude, is hard to account for. The specimen, at the National Museum of Natural Science, at Ottawa, in spite of rather shallowly notched lower petals, appears to be this species. Before suspecting anyone of mixing up their labels, one may recall that *Clematis occidentalis* and *Juniperus scopulorum*, which have similar distribution patterns in this province, also have far outlying records in the Dease Lake-Stikine Valley area, where dry open bluffs would appear to offer congenial conditions. It is not

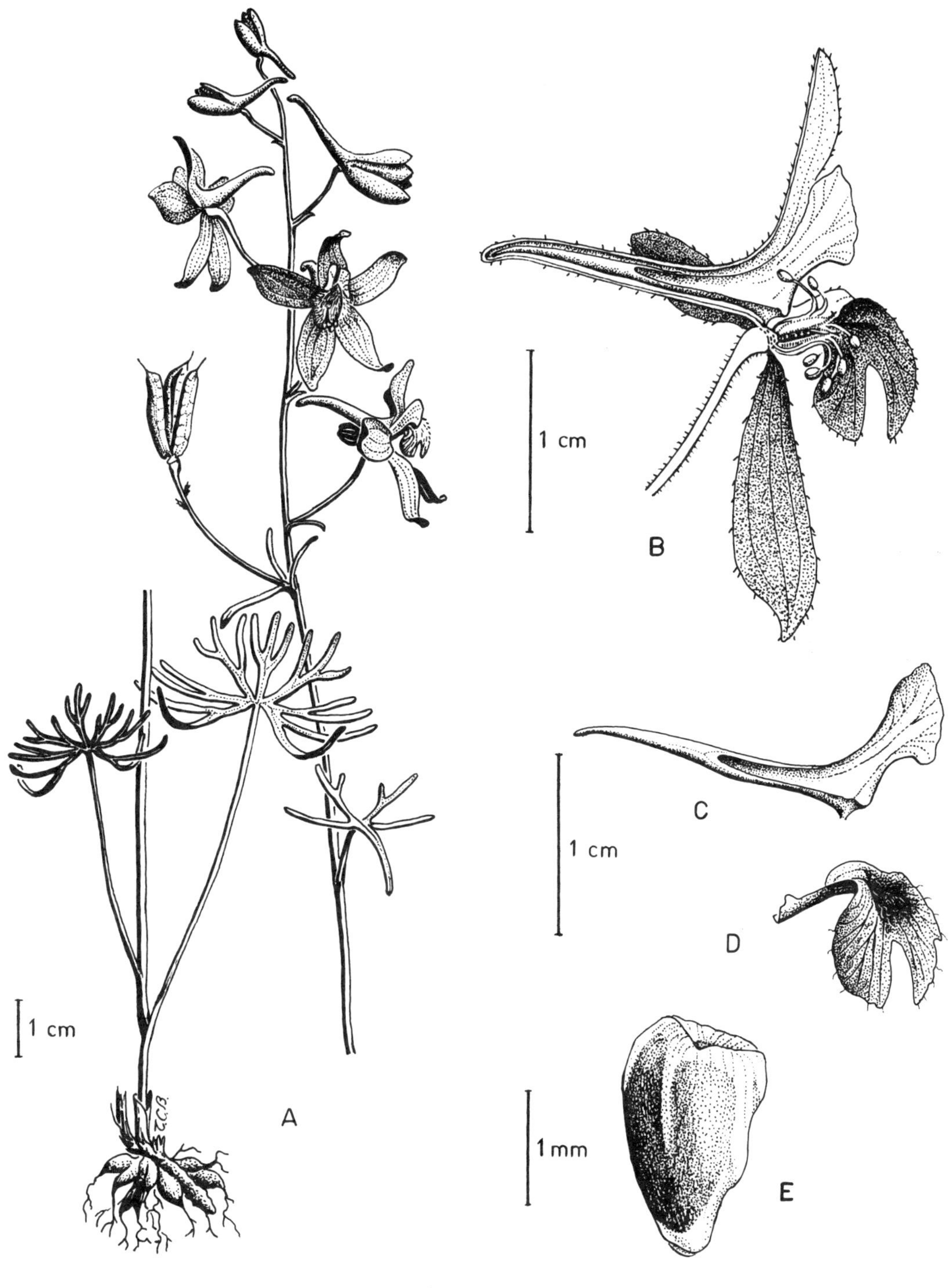

Figure 45. *Delphinium nuttallianum*

A. flowering and fruiting plant.
B. longitudinal section of flower.
C. and D. upper and lower petals, respectively.
E. seed.

inconceivable that, with more intensive and more informed collecting in the appropriate habitats, all of these species may yet turn up in intervening localities; as well as in the area reported, where the Juniper is definitely present, and not uncommon.

ISOPYRUM L.
Rue-Anemone

Rhizomatous perennial herbs. Flowers solitary or clustered, terminal, and sometimes also axillary. Sepals conspicuous, petaloid, deciduous. Petals absent (in ours) or reduced to minute nectaries. Stamens numerous. Carpels 2–20, free, sessile, each with few to several ovules. Fruit a group of free, sessile follicles, each containing 2–several seeds. x = 7.

About 25 species of north temperate latitudes in both the Old and New Worlds.

Isopyrum savilei **Calder & Taylor**
Enemion savilei **(Calder & Taylor) Keener**

Delicate, glabrous, rhizomatous perennial, 10–35 cm tall; with basal and cauline leaves, though not always with both kinds together.

Basal leaves biternately compound, the main divisions stalked, on long slender petioles. The usually two stem leaves with shorter petioles and smaller blades that may be only once ternate. The leaflets glaucous beneath, deeply to shallowly lobed, the lobes broadly rounded and commonly shallowly emarginate at the tips, the veins ending in minute 'glands' (hydathodes?) in the notches.

Flower solitary at the top of the stem, though a flower bud, that normally does not develop, occurs in the axil of the upper stem leaf. Sepals 5, obovate, white, occasionally pink-tinged at tips, 9–16 mm long, readily deciduous. Petals absent. Stamens 40–60, 3–8 mm long, the filaments slender and somewhat dilated above. Carpels 2–8, sessile, tapering into short styles and stigmas, each containing 2–8 ovules.

Fruit a group of sessile, divergent follicles, 11–16 mm long, with short tapering beaks, and containing few to several seeds. 2n = 14.

Discovered in 1957 on the Queen Charlotte Islands, this species was thought to be endemic to those islands until its further discovery twenty years later on the Brooks Peninsula, on the west coast of Vancouver Island. More recently (1987) it has also been found on Mt. Egeria, on Porcher Island, south of Prince Rupert. It typically inhabits wet rocky cliffs and runnels, ranging from alpine and subalpine levels, as in the Queen Charlotte Range, down, locally, nearly to sea level.

MYOSURUS L.
Mouse-Tail

Slender, scapose, glabrous annuals with linear basal leaves and fibrous roots. Flowers small, solitary on axillary and terminal scapes. Sepals normally 5, sometimes more, spurred at base, commonly longer than the 5 or fewer inconspicuous (sometimes lacking) whitish, nectary-bearing petals. Stamens 5–10. Carpels 10– 500, on a receptacle that elongates greatly in fruit. Fruit a linear or lanceolate spike of rhombic flattened achenes with dorsal keels projecting into beaks. x = 8.

A widespread but mainly temperate genus of 7 species.

KEY TO SPECIES

1. Achene beak 0.8–1.5 mm long. Mature spike 5–10 mm long, with 20–50 achenes.. *M. aristatus*
    Achene beak 0.25–0.7 mm long. Mature spike 5–50 mm long, with 70 or usually more achenes.. *M. minimus*

Myosurus aristatus **Bentham *ex* Hooker**

Low scapose annual 2 up to 10 cm tall, the leaves filiform to sometimes narrowly spatulate. Scapes shorter than the leaves at flowering time but elongating to exceed the leaves as the fruit matures.

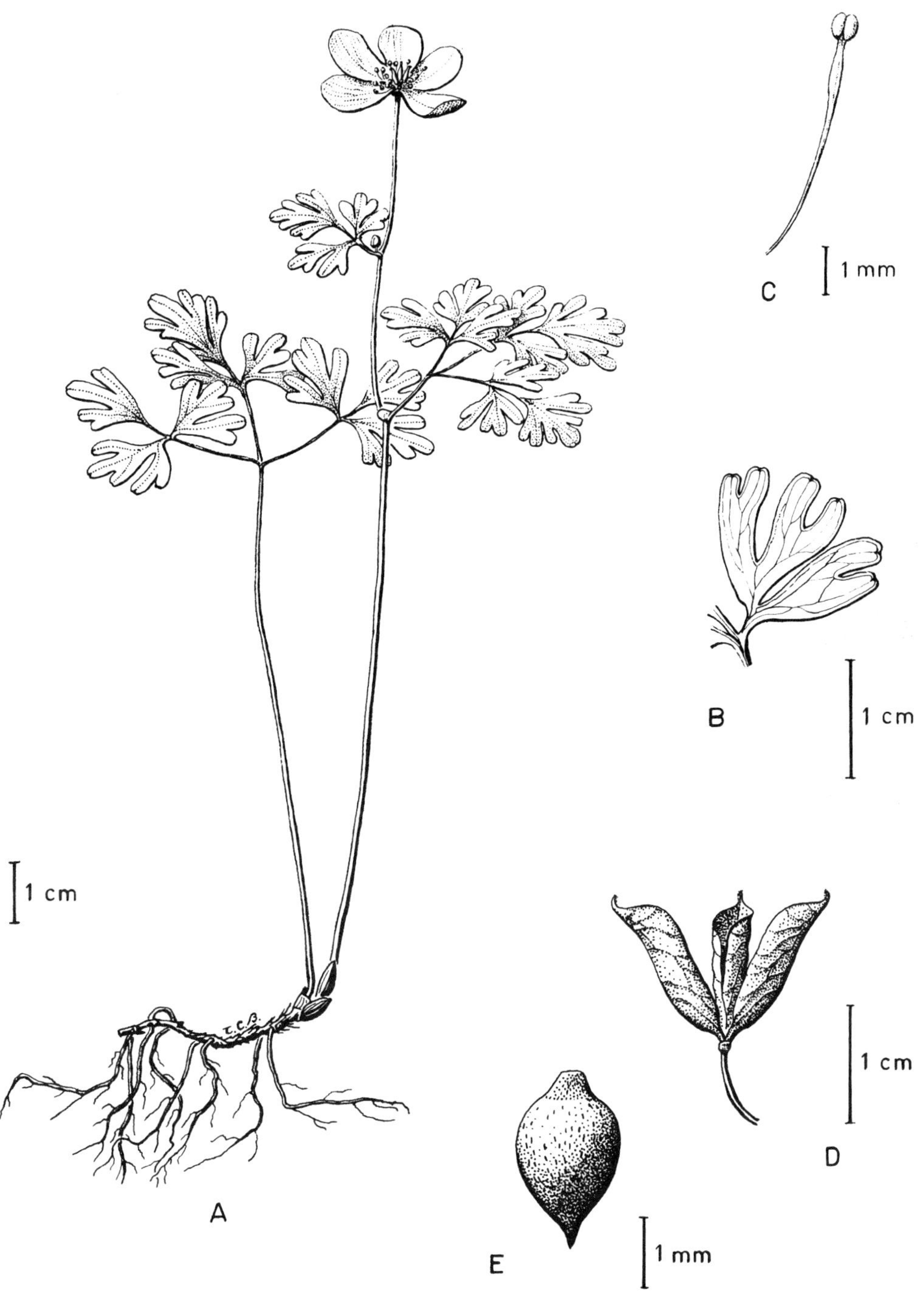

Figure 46. *Isopyrum savilei*

A. flowering plant.
B. leaflet.
C. stamen.
D. follicles.
E. seed.

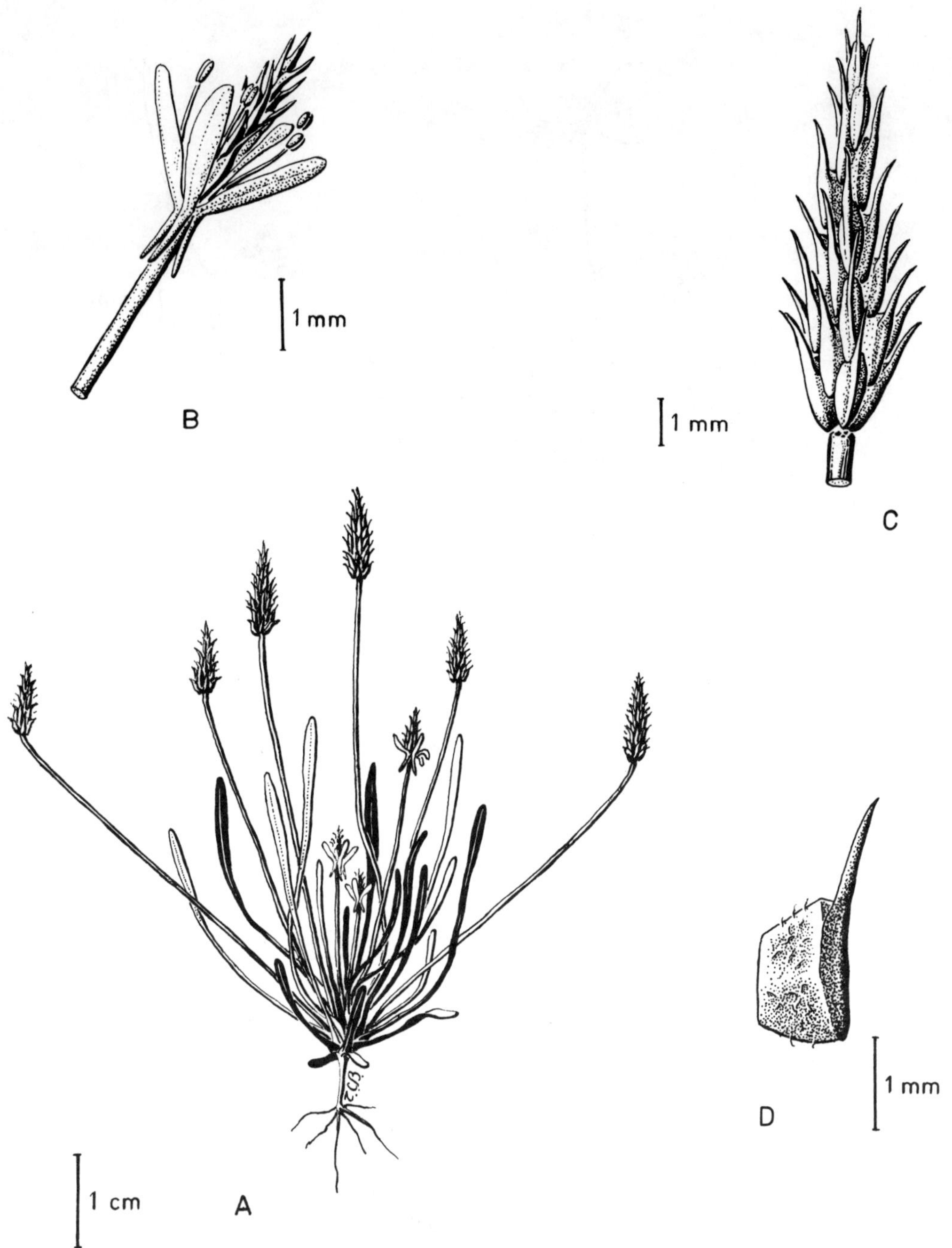

Figure 47. *Myosurus aristatus*

A. flowering and fruiting plant.
B. flower.
C. spike of achenes.
D. achene.

Sepals with blades slightly longer than the spurs, usually single-veined. Petals, when present, linear, equalling the sepals, but usually absent in our material. Stamens 5. Carpels up to 50 in a conical spike that grows up to 1 cm long in fruit. Each achene with a projecting beak 0.8–1.5 mm long, and puberulent on the lateral surfaces.

Found in vernally moist but later dry sites in the Cordilleran region from British Columbia to California, and east to Wyoming. In British Columbia it is recorded from the upper Fraser Canyon southeastward: much less common than *M. minimus*.

Myosurus minimus **L.**   **Mouse-Tail**
    M. apetalus **Gay**
    M. major **Greene**

Slender, tufted, taprooted, spring-flowering annual up to 15 cm tall, with few to many linear leaves 2–9 cm long by 0.5–1 mm wide, one to several scapes, often short (2–5 cm long) at flowering time and elongating as the fruit matures.

Sepals ascending, the blades 1–3 mm long, 3-veined, the spurs appressed to the peduncle, 1–2 mm long, greenish to creamy coloured. Petals linear to spatulate, about equalling the sepals, sometimes absent; the claw longer than the narrow blade, which has a minute, pocket-like nectary near its junction with the claw.

Achenes 70–100 or more in a spike that elongates to 0.5 –5 cm long as the fruit matures. Each achene rhombic and flattened, the keel projecting into an erect beak 0.25–0.7 mm long; the achenes at the base of the spike with longer beaks than those near the tip. The achene with an elongate attachment to the slender receptacle. 2n = 16.

Widespread in North America, and in the Mediterranean countries of Europe, *Myosurus minimus* is found in open sites that are vernally wet but that dry out by summer. This species is probably commoner than the records suggest; but tends to be overlooked because of its inconspicuous nature and early flowering time.

NIGELLA L.

Erect branching annuals with alternate, bipinnately to tripinnately dissected leaves with linear divisions. Flowers solitary and terminal on stem and branches. Sepals 5 or more; petals several to none; when present reduced to little more than stalked nectaries; stamens many. Carpels 5 or more, multiovuled, coherent to varying degrees, forming coherent follicles or a capsule.

About 20 species of southern Europe, western Asia, and North Africa.

Nigella damascena **L.**   **Love-in-a-Mist**

Stem slender, glabrous, 20–75 cm tall. Leaves bipinnately divided into linear divisions, without stipules. Basal leaves petioled, with rather wider divisions than have the more or less sessile stem leaves. Involucral leaves present, similar to the stem leaves, arising close under each flower base, and curving up to surround the flower.

Flowers 3–6 cm across. Sepals several, conspicuous, typically blue, but varying to pink or white in horticultural strains; ovate, acuminate, and clawed; spreading to slightly reflexed. Petals several to none; 5–6 mm long, notched apically, sparsely hispid, bearing 2–3 nectar glands and a scale-covered pouch. Stamens many. Carpels 5–10, coherent nearly to their tops except for the outwardly arching styles and stigmas. Fruit a globular capsule. 2n = 12.

Sometimes grown as an ornamental garden plant, this species has occasionally been found as an escape along roadsides on southern Vancouver Island (Victoria, Cowichan Bay).

RANUNCULUS L.

Buttercup, Crowfoot

Herbs with cymosely arranged, or terminal and solitary, flowers; the usually 5 sepals often deciduous early. The 3 or 5 to many petals bearing nectaries near their bases; each nectary commonly covered or surrounded by a scale. Stamens numerous to sometimes few; the anthers

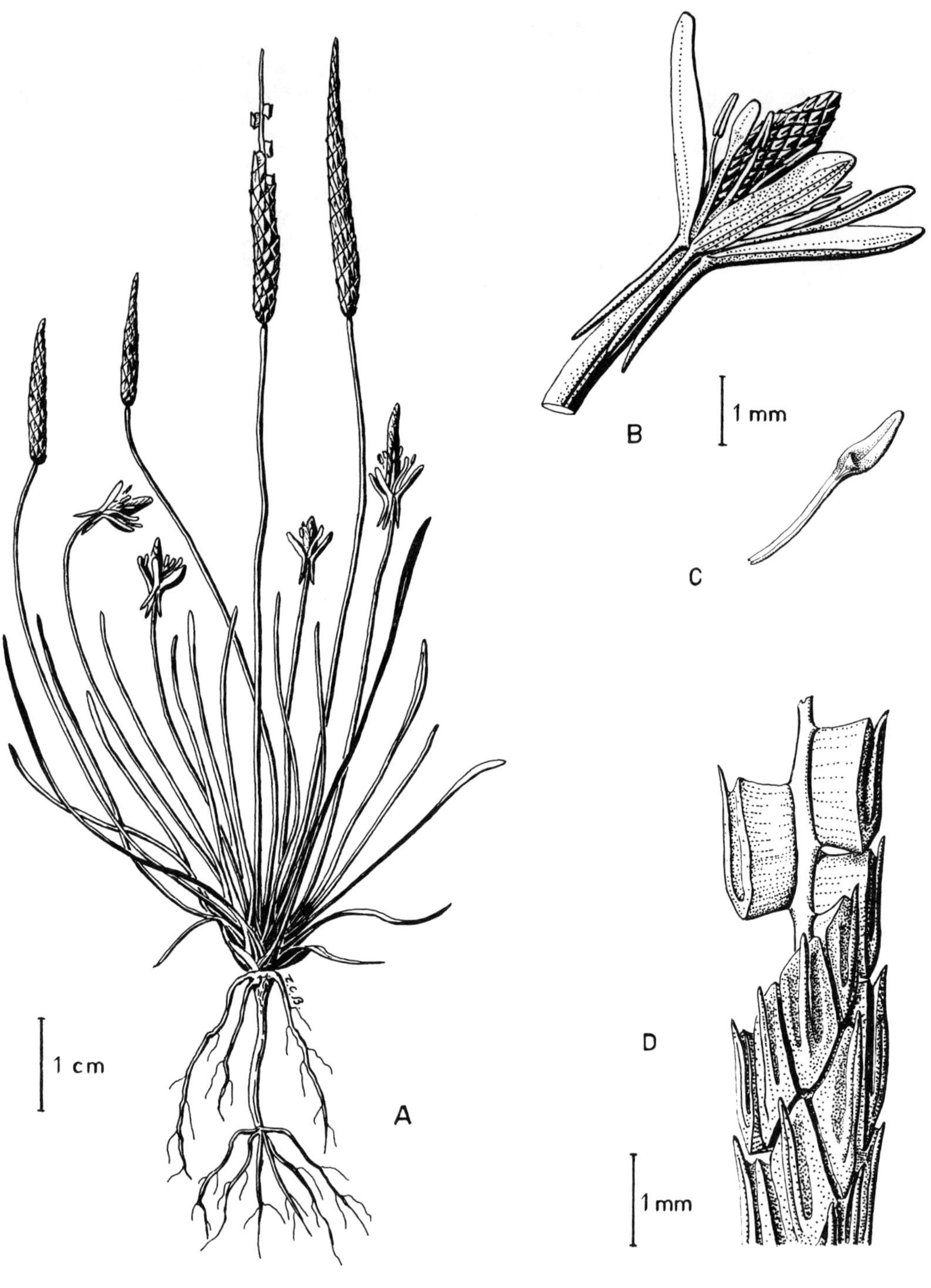

Figure 48. *Myosurus minimus*

A. flowering and fruiting plant.
B. flower.

C. petal.
D. segment of fruiting spike, with achenes.

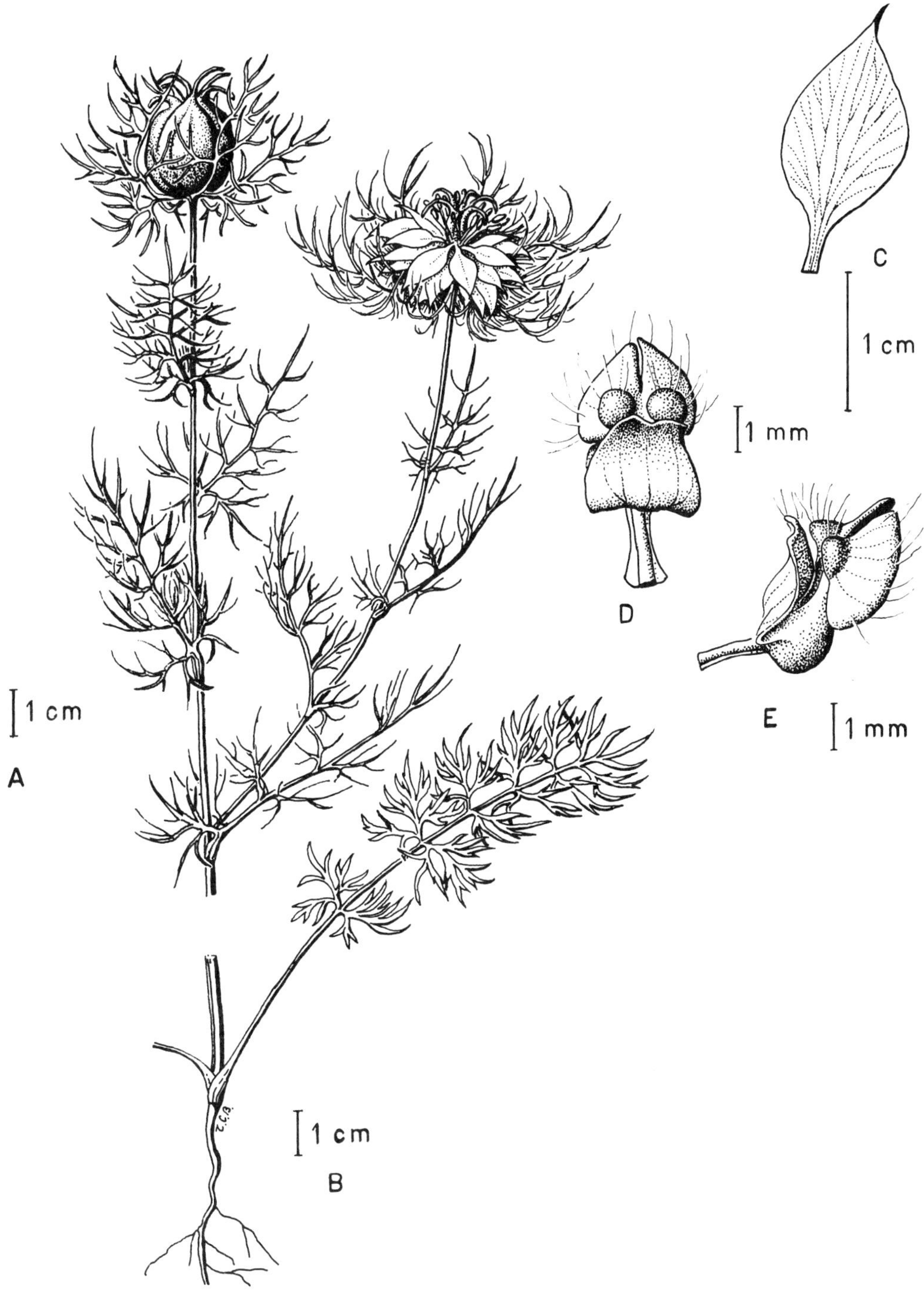

Figure 49. *Nigella damascena*

A. fruiting and flowering shoot.
B. base and basal leaf.
C. sepal.
D. petal: seen from above (ventral view).
E. petal: side view.

extrorse. Carpels 5 to many, separate, basally one-ovuled, on a globose to columnar receptacle. The fruit a head or spike of small achenes, often marginally keeled, and beaked by the usually persistent styles. x = 8 in most sections; but x = 7 in one series of species in Section *Ranunculus*, and in Section *Ceratocephalus*.

Species of *Ranunculus* are highly diverse in form and in preferred habitat; but many species are associated with moist but open sites. *Ranunculus* is a worldwide genus of some 400 species, mainly north temperate to arctic in distribution. Most species are poisonous. The flowers of *Ranunculus* are visited by a wide variety of insects; including bees, flies, butterflies, beetles and ants (Leppik, 1964; and Heimburger, personal communication).

TABLE 4: *RANUNCULUS* SPECIES OF BRITISH COLUMBIA, GROUPED BY SECTIONS

Section *Ranunculus*
  x = 8
  Ranunculus pensylvanicus L.f.
  R. macounii Britton
  R. repens L.
  R. orthorhynchus Hooker
  R. bulbosus L.
  R. sardous Crantz

  x = 7
  R. acris L.
  R. uncinatus D. Don
  R. occidentalis Nuttall
  R. californicus Bentham

Section *Epirotes*
  x = 8
  R. abortivus L.
  R. inamoenus Greene
  R. pedatifidus J.E. Smith
  R. cardiophyllus Hooker
  R. glaberrimus Hooker
  R. rhomboideus Goldie
  R. eschscholtzii Schlechtendal
  R. eximius Greene
  R. gelidus Karelin & Kirilov
  R. verecundus Robinson
  R. pygmaeus Wahlenberg
  R. nivalis L.
  R. sulphureus Solander

Section *Flammula*
  x = 8
  Ranunculus flammula L.
  R. alismifolius Geyer

Section *Ficaria*
  x = 8
  R. ficaria L.

Section *Coptidium*
  x = 8
  R. lapponicus L.

Section *Hecatonia*
  x = 8
  R. gmelinii DC.
  R. flabellaris Rafinesque
  R. sceleratus L.
  R. hyperboreus Rottboell

Section *Batrachium*
  x = 8
  R. aquatilis L.
  R. lobbii (Hiern) Gray
  R. trichophyllus Chaix
  R. subrigidus W.B. Drew

Section *Halodes*
  x = 8
  R. cymbalaria Pursh

Section *Arcteranthus*
  x = 8
  R. cooleyae Vasey and Rose

Section *Ceratocephalus*
  x = 7
  R. testiculatus Crantz

KEY TO SECTIONS AND SPECIES

1. All leaves alike: entire to finely glandular-serrulate, not lobed; or if shallowly lobed or crenate, plant prostrate or stoloniferous **2**

   At least some leaves distinctly lobed, or plant not stoloniferous **5**

2. Leaves lanceolate to linear .... **Section *Flammula* 3**

   Leaves broader, cordate to truncate at base **4**

3. Stem strictly erect.. ***R. alismifolius***
    Stem stoloniferous or tending to recline and root at nodes...................................... ***R. flammula***

4. Leaves truncate to subcordate at base, crenate. Plant stoloniferous. Sepals and petals 5 each. Achenes longitudinally ribbed with raised veins....... **Section *Halodes: R. cymbalaria***
    Leaves nearly reniform and deeply cordate, crenate-margined. Sepals 3, Petals 8 or more. Achenes not longitudinally ribbed.. **Section *Ficaria: R. ficaria***

5. Some or all submersed leaves divided to base into thread-like or hair-like divisions. Petals white with basal yellow spots. Nectary minute, the scale not free at the sides. Achenes usually transversely corrugated. Submersed (or occasionally stranded) aquatics................
    .... **Section *Batrachium* 6**
    Leaf divisions always flat in section, at least near the base, even when submersed and very slender. Petals yellow.. **9**

6. Leaves of two or more distinct kinds: finely divided like tufts of hairs, and broad, flat (lamellar), and palmately rounded-lobed.. **7**
    Leaves all alike: finely divided.. **8**

7. Summer-flowering perennial. Achenes 15–22, 1.3–2.4 mm long................................ ***R. aquatilis***
    Spring-flowering annual. Achenes 2–7, 2–2.5 mm long.. ***R. lobbii***

8. Upper leaves sessile on their stipular sheaths. Achenes 30–80, 1–1.5 mm long................
    .... ***R. subrigidus***
    All leaves normally petioled. Achenes 12–32, 1.5–2 mm long.................. ***R. trichophyllus***

9. Small, grey-tomentose annual of sagebrush steppe. Achene 5 mm or more long, densely floccose-tomentose, with a prominent beak and a pair of dilated lateral chambers near base
    .... **Section *Ceratocephalus: R. testiculatus***
    Achene shorter, or glabrous, without dilated lateral chambers. Plant not grey-tomentose annual or not in Sagebrush steppe.. **10**

10. Achene longitudinally ribbed with raised veins, and with a long hooked beak. Petals 7–15, oblanceolate. Nectary scale divided & V-shaped, its outer margins adnate to petal
    .... **Section *Arcteranthus: R. cooleyae***
    Achene not longitudinally ribbed. Nectary scale not divided.. **11**

11. Plant with pale, thread-like, horizontal buried rhizome bearing scattered, deeply three-lobed leaves. Flower solitary on erect scape. Sepals 3. Petals 5–8..
    .... **Section *Coptidium: R. lapponicus***
    Plant with ascending, erect, or stoloniferous stems.. **12**

12. Achene with pale, swollen, marginal and basal zone. Nectary scale collar-like or cres-centic. Aquatic or amphibious plants.. **Section *Hecatonia:* 13**
    Achene without pale swollen marginal zone. Nectary scale various. Mainly terrestrial **16**

13. Erect annual. Achenes very many in cylindrical heads. Marginal swelling completely encircling achene.. ***R. sceleratus***
    Perennials, commonly with floating or prostrate stems. Achene with incomplete marginal swelling.. **14**

14. Plant stoloniferous, with often rather shallowly 3-lobed leaves. Sepals and petals 3 each. Achene margin variably swollen near base. Nectary scale pouch-like, crescentic...............
    .... ***R. hyperboreus***
    Plant sprawling in shallow water, or sometimes rooting from nodes in contact with mud. Leaf blades deeply parted. Sepals and petals 5 each. Nectary scale collar-like, encircling the gland.. **15**

15. Leaf blades twice to three times ternate; the floating ones 2–3 cm long. Achene swollen around the lower ½–⅔ of margin; distal margin not distinctly keeled. Achene beak 0.5–1 mm long.. ***R. gmelinii***

Leaf blades 3–5 times ternate, up to 8 cm long. Achene swollen around ⅔ to ¾ of margin; the distal margin keeled; achene beak 1–2 mm long.......................... ***R. flabellaris***

16. Leaves rather sinuately lobed or toothed; the lobes often rounded, sometimes angular, their margins not secondarily toothed. Nectary scale pouch-like, adnate along its sides to the petal. Achenes plump. Receptacles often elongating conspicuously in fruit... **Section *Epirotes* 17**

Leaves angularly lobed; the margins of the lobes commonly laciniately toothed. Nectary scale flap-like; its lateral margins free at least along their distal half. Achenes discoid and thin. Receptacles usually not conspicuously elongating in fruit (except *R. pensylvanicus*) .... **Section *Ranunculus* 29**

17. Achene head globose; the receptacle not notably elongating in fruit. Basal leaf blade longer than wide, cuneate to rounded at base. Plants of dry grassland and parkland....... **18**

Achene head ovoid to cylindric; the receptacle elongating in fruit. Basal leaf blade various.. **19**

18. Basal leaf blade rhombic to obovate, crenate-margined. Achene beak very short (0.1–0.2 mm). Prairies east of Rocky Mountains.......................... ***R. rhomboideus***

Basal leaf blade ovate to orbicular or broadly lanceolate, entire or shallowly 3- to 5-lobed toward apex. Achene beak about 0.5 mm long. Semi-arid interior valleys and prairie region.. ***R. glaberrimus***

19. Sepals pubescent with dark brown hairs.. **20**

Sepals pubescent with white or yellowish hairs, or glabrous...................................... **21**

20. Basal leaf blade shallowly crenately lobed, glabrous. Receptacle puberulent with brown hairs... ***R. sulphureus***

Basal leaf blade deeply 3- to 7-lobed, with round-tipped lobes, ciliate. Receptacle glabrous, or with a small apical tuft of white hairs.. ***R. nivalis***

21. Basal leaf blade orbicular, shallowly lobed or crenate. Petals shorter than to slightly longer than the sepals. Receptacle thinly puberulent... **22**

Basal leaf blade deeply lobed (a third to half way or more to base), or petals much longer than sepals, or both... **23**

22. Achenes 20–50, glabrous; the beak 0.1– 0.2 mm long. Petals shorter than sepals.......... .... ***R. abortivus***

Achenes 40–100, canescent with straight white hairs; the beak 0.2–0.9 mm long. Petals equalling or slightly exceeding the sepals.. ***R. inamoenus***

23. Basal leaf blade cut no more than half way to base. Petals, when present, notably larger than sepals... **24**

Basal leaf blade cut more than half way to base. Petals various................................... **25**

24. Basal leaf blade ovate, cordate-based, crenately toothed to serrate. Nectary scale ciliate. Sepals white-pubescent. Achene puberulent................................... ***R. cardiophyllus***

Basal leaf blade orbicular, cuneate-based, acutely lobed. Nectary scale not ciliate. Sepals yellow-pubescent. Achene glabrous.. ***R. eximius***

25. Small to minute subscapose plant. Petals shorter than to equalling the sepals. Achene body 1 mm or less long... ***R. pygmaeus***

Larger plants, usually with stem leaves. Petals exceeding sepals. Achene body more than 1 mm long when mature.. **26**

26. Sepal pubescent with yellow hairs. Achene beak straight to slightly curved, 0.8–1.5 mm
     long... ***R. eschscholtzii***
     Sepal pubescent with white hairs. Achene beak recurved, shorter................................... **27**

27. Basal leaf lobes linear to lanceolate, pedate. Achene beak 0.5–1 mm long.... ***R. pedatifidus***
     Basal leaf lobes cuneate. Achene beak 0.5 mm or less long... **28**

28. Small plant, 4–10 cm tall in flower. Achene body 2–2.5 mm long, the profile typically not
     concave ventrally near the base. Ultimate basal leaf lobes usually tapering to rounded tips
     .... ***R. gelidus***
     Taller plant, up to 20 cm tall. Achene body 1–2 mm long; the profile typically concave
     ventrally near the base. Tips of ultimate lobes of basal leaves usually broadly rounded ...
     .... ***R. verecundus***

29. Stems stolonate or arching and commonly rooting at the nodes. Leaves trifoliolate........ **30**
     Stems erect, or if sprawling, not rooting at the nodes.. **31**

30. Plant strongly stoloniferous. Petals 6–17 mm long, much larger than the sepals... ***R. repens***
     Plant with arching and sprawling stems. Petals 3–8 mm long, a little larger than the sepals
     .... ***R. macounii***

31. Stem base commonly enlarged and bulbous. Leaf ± pinnately dissected. Petals well
     exceeding the sepals, broadly obovate to orbicular. Achenes with thickened rims, the faces
     smooth.. ***R. bulbosus***
     Stem base not bulbous... **32**

32. Achene with thickened rims, papillose on the faces. Annual................................ ***R. sardous***
     Achenes often keeled, but the rims not thickened and the faces not papillose.................. **33**

33. Petals equalling or slightly larger than the sepals. Annuals or biennials............................ **34**
     Petals conspicuously larger than the sepals.. **35**

34. Basal leaf blade simple. Achene head globose; the receptacle scarcely elongating in fruit.
     Achene beak strongly hooked.. ***R. uncinatus***
     Basal leaf blade trifoliolate. Achene head cylindrical; the receptacle notably elongating in
     fruit. Achene beak deltoid, stout, straight, flat... ***R. pensylvanicus***

35. Petals orbicular. Lower leaves simple but deeply palmately and laciniately parted, with
     acuminate lobes and teeth. Achene beak very short, recurved............................... ***R. acris***
     Petals elliptic to oblanceolate, commonly 2 to 3 or more times as long as wide. Leaves and
     achenes various... **36**

36. Achene beak 3–5 mm long: as long as the achene body or longer, straight. Leaves
     pinnately compound... ***R. orthorhynchus***
     Achene beak shorter: usually less than 2 mm long... **37**

37. Petals 5–7, elliptic, 2–3 times as long as wide. Achene beak 0.7–2 mm long, straight or
     recurved. Erect perennial. Widespread.. ***R. occidentalis***
     Petals 8–24, oblanceolate to elliptic... **38**

38. Petals oblanceolate, 3–4 times as long as wide. Achene beak 0.4 –1 mm long, recurved.
     Plant a low, often sprawling perennial or annual. Puget Sound region........... ***R. californicus***
     Petals oblanceolate to elliptic, 2–4 times as long as wide. Achene beak longer (1.5–2 mm)
     and straight. Erect robust perennial of the Queen Charlotte Islands...............................
     .... ***R. occidentalis*** **var.** ***hexasepalus***

Section *Ranunculus*

Basal and lower stem leaves cleft into lobes whose flanks are variously, often laciniately, acutely lobed and toothed. Petals yellow. Each nectary covered by a basally attached, flap-like scale. Achenes flattened, usually unornamented on the faces. x = 8 or 7.

Ranunculus pensylvanicus L.f.   **Pennsylvania Buttercup** / **Bristly Crowfoot**

Erect, hispid annual with fibrous roots and a single hollow stem up to a metre tall.

Basal and lower stem leaves petioled, compound, the 3 leaflets with unequally lobed and laciniately toothed margins.

Flowers on initially short pedicels that elongate as the fruit matures. Sepals 5, reflexed, sparsely hirsute, 3.5–7 mm long. Petals 5, obovate to suborbicular, 4–8 mm long, generally slightly longer than the sepals, yellow, with the nectary covered by a truncate scale attached only at its base or sometimes also by the lower third of the lateral margins. Stamens 15–25; carpels 50–100 on an ellipsoidal hairy receptacle that elongates up to 14 mm as the fruit matures.

Achenes in a cylindrical head up to 15 mm long by 8 mm thick, each achene glabrous, with a flat, discoid body 2–3 mm long, inconspicuously bordered and keeled, and bearing a flat, deltoid, straight stylar beak 0.7–0.9 mm long. 2n = 16.

Widespread in the forested parts of North America, usually in wet ground along streams or ditches.

Ranunculus macounii Britton   **Macoun's Buttercup**

Annual or perennial with erect to sprawling, hispid to nearly glabrous hollow stems and a cluster of somewhat thickened roots. Stems, when in contact with ground, sometimes rooting at the nodes, 2 up to 9 dm long.

Basal and lower stem leaves trifoliolate, but upper bract-leaves often reduced and simple: the leaflets variously lobed and toothed.

Flowers on pedicels that may elongate after anthesis. Sepals 5, sparsely hispid, spreading to reflexed, 3.5–7 mm long. Petals 5, yellow, 3–8 mm long, usually slightly larger than the sepals, the nectary covered by a truncate to notched scale about ½ mm long, free except at the base and lower lateral edges. Stamens 20–40. Carpels 30–60.

Achenes in an ovoid cluster on an ovoid hairy receptacle; the achene body flattish, obovoid, 2.5–3.5 mm long, with a broad-based beak ⅓–½ as long, and straight to slightly hooked at the tip. 2n = 32, 48.

KEY TO VARIETIES

1. Plant hispid.. **var.** *macounii*
   Plant glabrous or almost so... **var.** *oreganus*

Ranunculus macounii is a plant of moist sites, as along lake shores and ditches, from Alaska to the Atlantic coast of Canada, and southward in the west to New Mexico and California. Variety *macounii* occurs all over British Columbia, but is seldom abundant in any place.

Variety *oreganus* Gray is found from southeastern British Columbia to southern and southeastern Washington State. It is known in British Columbia at Crawford Bay, on Kootenay Lake, and at Paterson, south of Rossland.

Ranunculus pacificus **(Hulten) Benson,** distinguished by petals 8–13 mm long and achenes 3–4 mm long with beaks half as long, occurs in southeastern Alaska. It has not yet been found in British Columbia, but may be looked for in the north coastal areas. It is questionably a distinct species, since it differs only quantitatively from *R. macounii*. When better known, it may be reduced to a subspecies or variety of *R. macounii*.

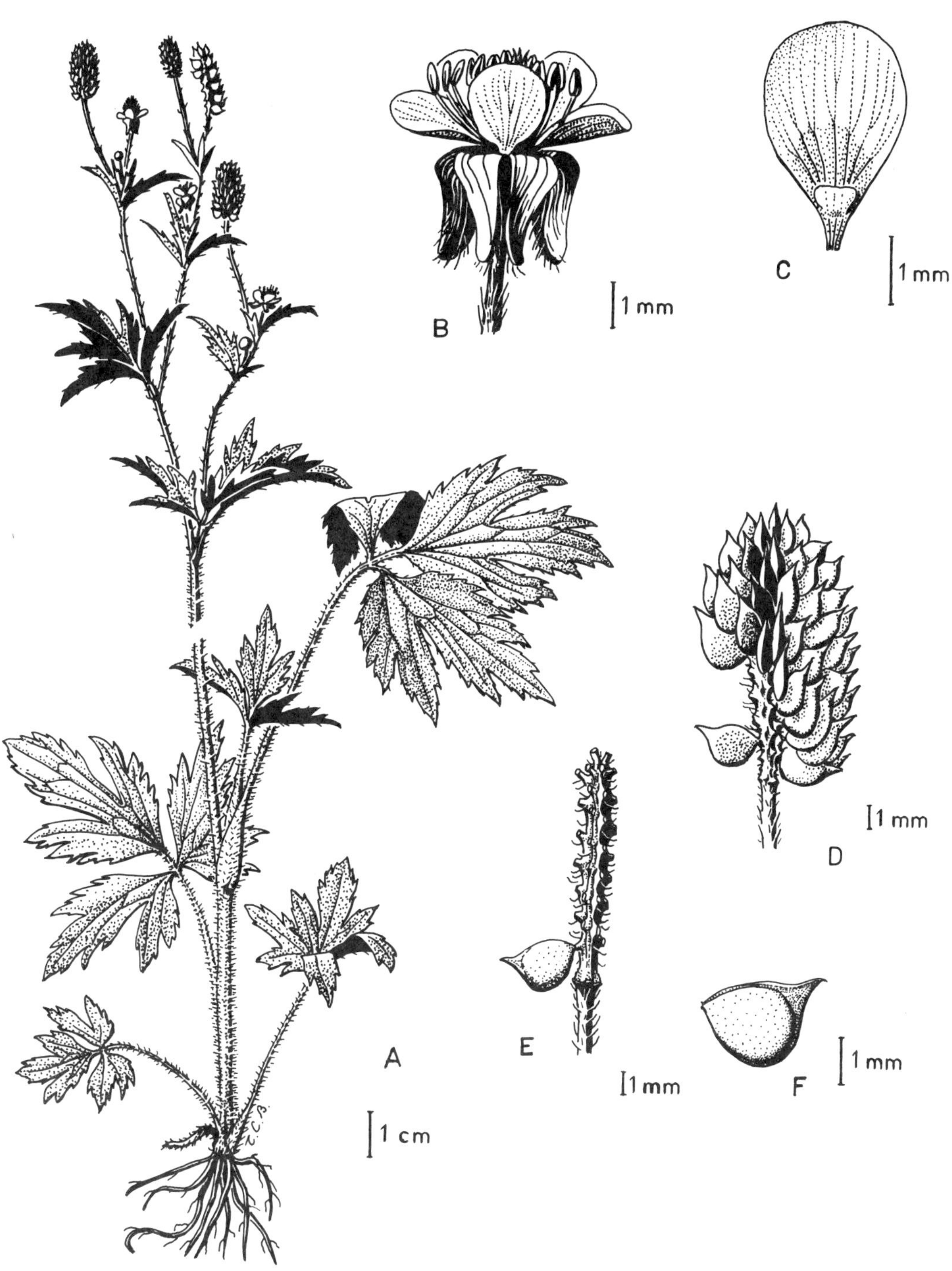

Figure 50. *Ranunculus pensylvanicus*

A. flowering and fruiting plant.
B. flower.
C. petal with nectary scale.
D. head of achenes.
E. receptacle.
F. achene.

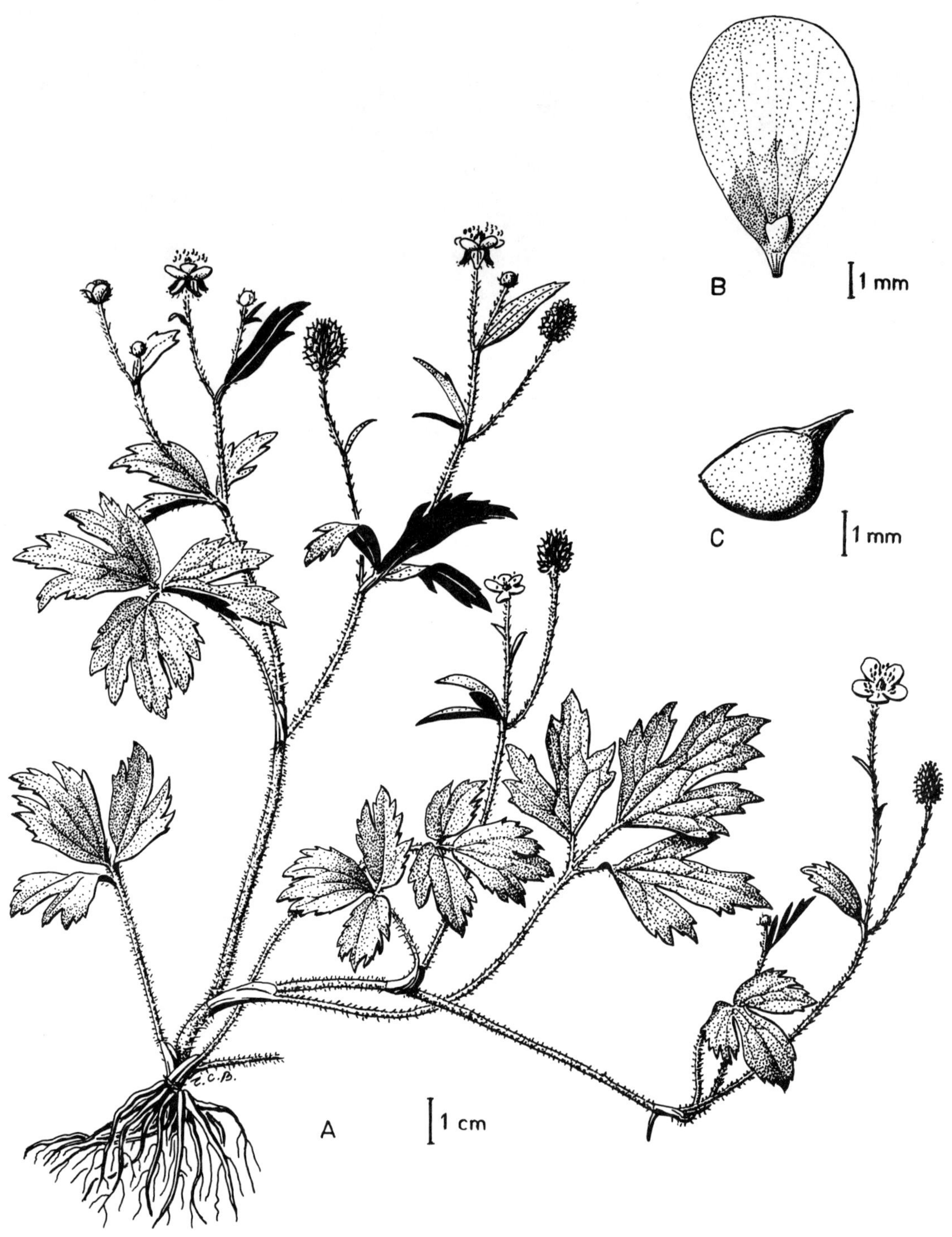

Figure 51. *Ranunculus macounii*

A. flowering and fruiting plant. B. petal. C. achene.

Ranunculus repens **L.**   **Creeping Buttercup**
                                                                     Swamp Buttercup

Stoloniferous perennial, rooting at the nodes, with fibrous roots and typically hirsute stems and leaves. Prostrate stem growth proceeds from an axillary bud at rooting nodes that give rise to erect, terminal flowering shoots.

Leaves pinnately trifoliolate, the terminal leaflet longer-stalked than the lateral ones. The blades variously lobed and toothed, often with a pale spot at the base of each sinus. Leaves from rooting nodes and the bases of aerial shoots are long-petioled, but those further up the aerial stems are sessile or nearly so.

Flowers terminal on erect stem segments, or on their branches. Sepals spreading, hirsute. Petals 5 to many, broadly obovate to nearly orbicular, 6–17 mm long, often shallowly emarginate, of a deeper, richer yellow than in most other species of this genus. Nectariferous scale obovate, 1–1.5 mm long, free for ¾ of its length.

Carpels 20–50. Achenes in a globose to ovoid head. Achene body lens-shaped, broadly obovate, smooth, marginally keeled, and beaked by the style and stigma; the beak varying from strongly recurved to nearly straight. 2n = 32.

Native of Europe, introduced and now widely naturalized in North America, *Ranunculus repens* shows a preference for ditches and other moist habitats. It has shown itself also to be a very persistent garden weed.

Several varieties have been described, mainly on the basis of European material. For those found in North America the following key has been provided by Fernald (1919).

A. Middle leaflet of the basal leaves cuneate to subtruncate at base: petals 5–9: stamens numerous B.

    B. Lobes and teeth of the leaves deltoid or ovate to oblong, obtuse or bluntish.
        Trailing or repent branches or stolons present.
            Stems and petioles distinctly pubescent.
                Pubescence appressed .... ***R. repens* L.** (typical).
                Pubescence wide-spreading .... **var. *villosus* Lamotte**
            Stems and petioles glabrous or nearly so .... **var. *glabratus* DC.**
        Trailing or repent branches wanting .... **var. *erectus* DC.**
    B. Lobes and teeth of the much-cleft leaves lanceolate to linear, acuminate ....
                                                     var. *linearilobus* DC.

A. Middle leaflet of the basal leaves rounded or subcordate at base: petals very numerous, forming a "double" flower .... **var. *pleniflorus* Fernald**

Of these varieties, to date, var. *repens* (i.e. typical *R. repens*), var. *villosus*, and var. *pleniflorus* have been found in British Columbia. Variety *villosus* seems barely worthy of recognition. Many of our plants bear spreading hairs on their lower stems and appressed hairs above. Variety *pleniflorus* is a cultivated sterile ornamental variety that sometimes escapes and colonizes ditches by its stoloniferous growth. A double-flowered variety with cuneate leaflets in the basal leaves is known in horticulture as var. *florepleno* (Fernald 1917a).

Ranunculus orthorhynchus **Hooker**   **Straight-Beaked Buttercup**

Erect perennial herb from a cluster of thickish roots, with stems branched or unbranched, 2–6 dm tall, variously pubescent to glabrous.

Leaves pinnately compound, or trifoliolate with the terminal leaflet long-stalked, or sometimes the basal leaves simple and trilobate. Basal and lower stem leaves long-petioled, with cuneate to elliptic leaflets; the middle to upper leaves short-petioled, the bracts sessile, their leaflets narrower and dissected into linear lobes. Leaves pubescent beneath, sometimes very sparsely so.

Flowers terminal, long-pedicelled, the usually pubescent sepals reflexed. Petals 5, oblong to elliptic or obovate, more or less twice as long as wide, twice or more as long as the sepals. Nectariferous scale broad and truncate, ½ to 1½ mm long, the edges free for ⅔ or more of their length. Stamens 50–70.

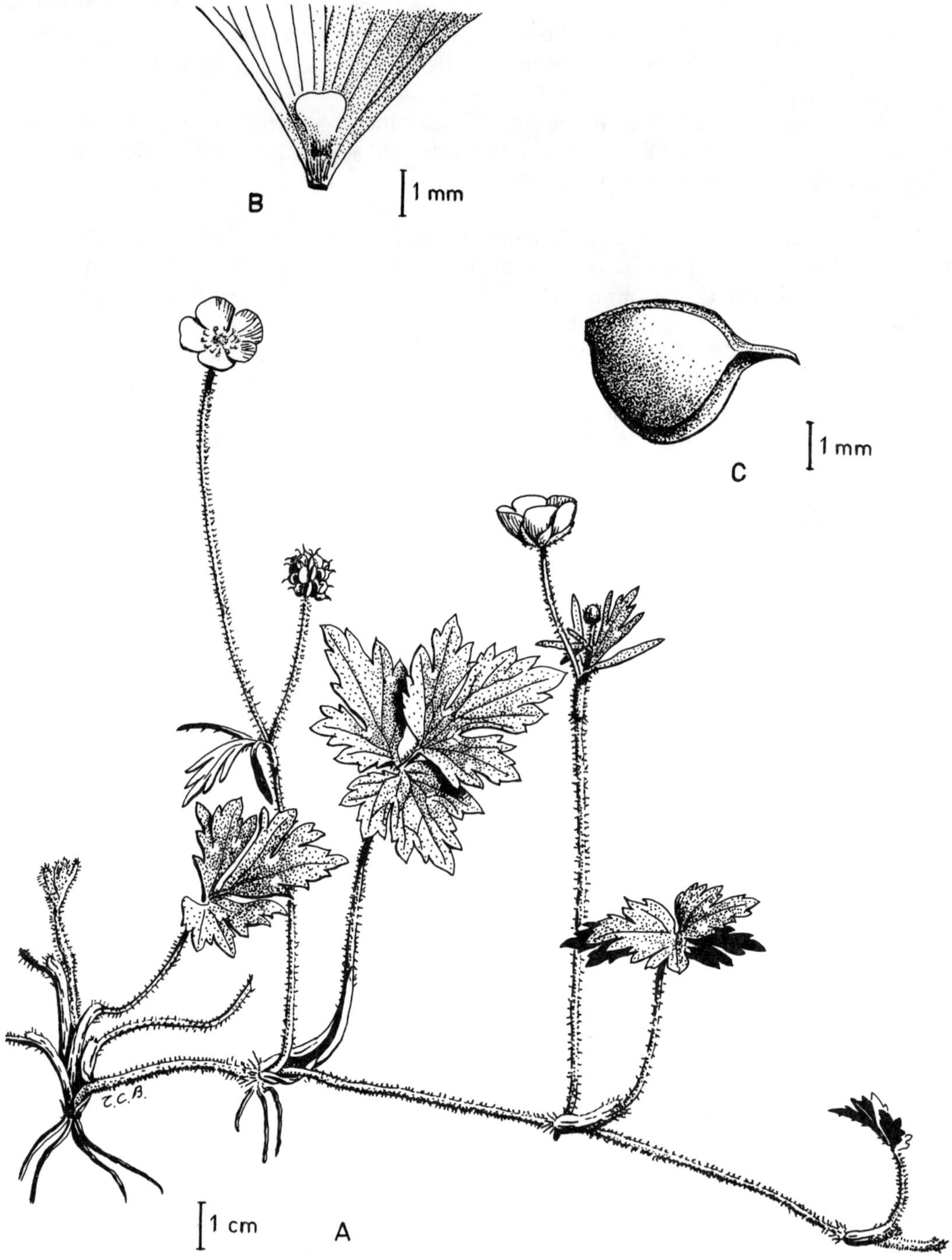

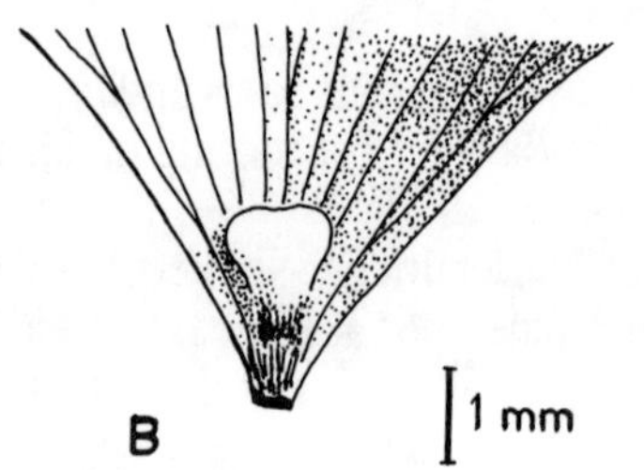

Figure 52. *Ranunculus repens*

A. part of plant.   B. petal base and nectary scale.   C. achene.

Carpels 5–50 on an ovoid hairy receptacle, with long styles tipped by short stigmas.

Fruit an ovoid head of achenes on a receptacle 1½–11 mm long, commonly becoming columnar as the fruit matures; the achene body 3–4 mm long, and the straight stylar beak 3- 5 mm long. 2n = 32.

KEY TO VARIETIES

1. Sepals glabrous .... **var. *alaschensis***
    Sepals pubescent **2**

2. Petals usually less than twice as long as wide (commonly about ⅔ as wide as long). Achenes 25 or more. Fruiting receptacle 4–11 mm long. Stylar beak less than 3.5 mm long. Leaflets obovate, broadly cuneate-based. East of Cascade and Coast Ranges except northward ......
    .... **var. *platyphyllus***
    Petals twice or more as long as wide. Achenes 5–30. Fruiting receptacle 1.5–5 mm long. Stylar beak 3.5–4 mm long. Leaflets linear to oblanceolate, narrowly cuneate-based. Coastal .... **var. *orthorhynchus***

Ranunculus orthorhynchus is distributed from southeastern Alaska to California, and eastward to Montana and Wyoming. It typically inhabits moist meadows, including those along the sea shore, and stream banks.

Variety *orthorhynchus* is the common variety on the coast, but has been collected more commonly on Vancouver and Queen Charlotte Islands than on the mainland.

Variety *platyphyllus* Gray is found mainly in inland areas of the western United States; and occurs at Nelson and Trail in southeastern British Columbia, with an apparently isolated area including Bella Coola and Cascade Inlet on the coast.

Variety *alaschensis* L. Benson, originally discovered in southeastern Alaska, is not sharply distinguished from var. *orthorhynchus*. It represents the culmination of a trend toward reduction of the hairiness of the sepals and other parts of the plant that is widespread in the northwestern parts of the range of this species. Some plants with a few basal hairs on the sepals in bud have shed these hairs by the time the flower is open. A few of our plants have sepals that are glabrous from the start. Plants with sepals on which no hairs can be found, and which can be assigned to var. *alaschensis*, have been found in widely scattered locations on the coast: at Long Arm, Graham Island, Queen Charlotte Islands, and at Alberni, Port Renfrew, and Comox on Vancouver Island.

Some botanists feel that var. *alaschensis* is not sufficiently strongly distinguished to warrant formal recognition. However, the information needed for this decision is presented here, so as to allow the reader to draw his own conclusion.

Ranunculus bulbosus L.   **Bulbous Buttercup**
R. tuberosus **Hornemann**

Hirsute perennial herb with fibrous roots, and erect stems usually with thickened, bulbous bases, 1.5–7 dm tall.

Basal and lower stem leaves long-petioled, the blades 2–4 cm long, 3-lobed to pinnately ternate, the divisions deeply lobed; the upper stem leaves transitional to the sessile, 3- to 5-divided bracts.

Flowers terminating stem and branches. Sepals 5, hirsute without, reflexed. Petals 5, glossy yellow, showy, broadly obovate, 8–15 mm long; the nectary scale broad and rounded to rather truncate, the lateral edges free for about ⅔ of their lengths. Receptacle ovoid, glabrous, up to 3 mm long in fruit. Carpels 15–40.

Achenes disc-like, obovate, usually glabrous; the body 2–2.5 mm long, with a thickened and keeled margin; the beak stout, recurved at the tip, and ½–1 mm long. 2n = 16.

Native of Europe, this species has been found at widely scattered locations across North America, commonly in open, rather dry, grassy habitats. In British Columbia it has been recorded only once; having been found by John Macoun in damp places at Revelstoke in 1890. A flowering

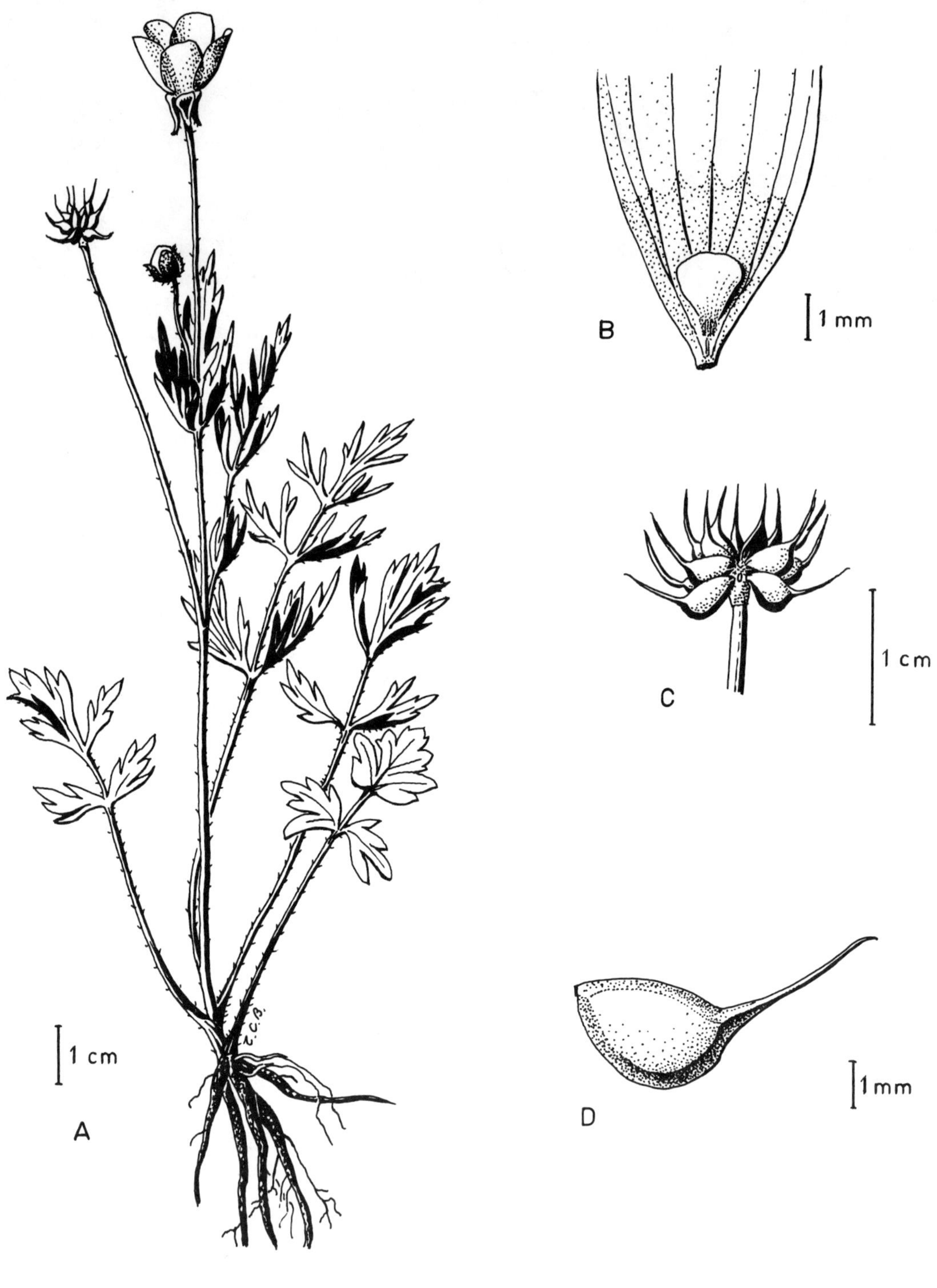

Figure 53. *Ranunculus orthorhynchus*

A. plant.
B. petal base and nectary scale.
C. head of achenes.
D. achene.

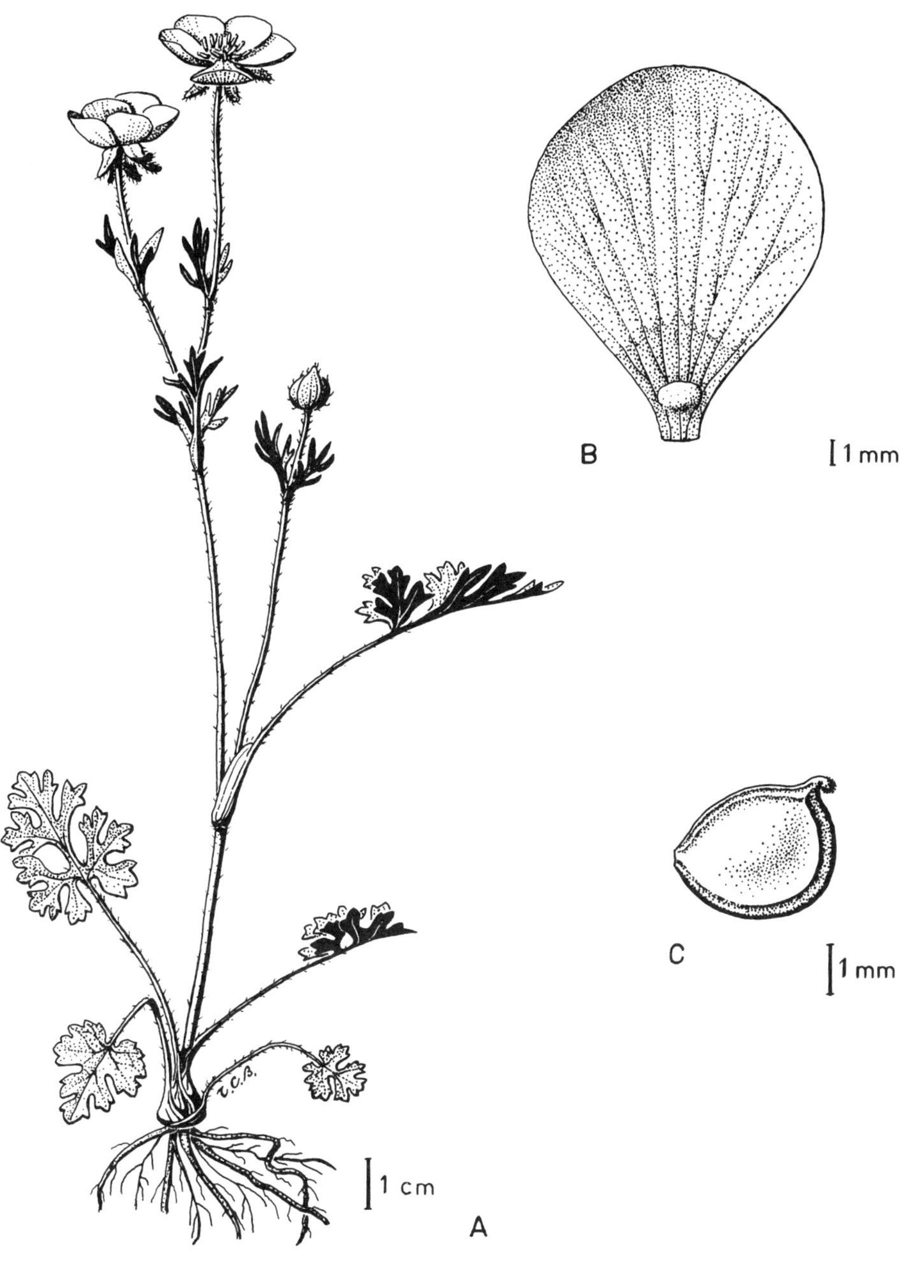

Figure 54. *Ranunculus bulbosus*

A. flowering plant. B. petal. C. achene.

specimen in the Royal British Columbia Museum herbarium from Port Alberni lacks the bulbous base, and also lacks achenes; and its identity is uncertain. Fruiting material from that area should be sought, to confirm or otherwise its tentative identification. Since this is one of the commonest species of *Ranunculus* in northwestern Europe, its re-introduction here may be expected.

Ranunculus sardous Crantz
Incl. *R. parvulus* L.

Slender hirsute annual herb with slender fibrous roots and a sparingly branched stem 5–40 cm tall.

Earliest basal leaves petioled, simple and 3-lobed; these withering early. Later basal and lowest stem leaves somewhat larger than the first leaves, petioled, the blades 1–2.5 cm long, trifoliolate, with lobed and toothed leaflets, the terminal leaflet stalked. Upper stem leaves and bracts sessile or subsessile, narrowly trifid or simple and lanceolate.

Flowers 1–2 cm wide. Sepals 5, sparsely hirsute, attenuate, green becoming pale yellow, reflexed, soon deciduous. Petals 5, obovate, shiny yellow, multi-veined, short-clawed. Nectary covered by a broadly truncate scale 1–1.5 mm long and wide; the lateral margins free for $\frac{2}{3}$–$\frac{3}{4}$ of their lengths. Stamens 25–50. Receptacle ovoid to cylindrical, hairy. Carpels 10–30, glabrous, with short, triangular stylar beaks.

Achenes in a globose to ovoid head; the achene body flattened, nearly circular in outline, 2–3 mm long, at maturity with pale brownish lateral faces with a few to many small submarginal papillae, green, prominently thickened and keeled margins, and with a very short (about 0.5 mm long or less), triangular, uptilted beak and stigma. $2n = 16, 32$.

Native of Europe; introduced into North America, and established locally as a weed of open fields and meadows. In British Columbia, it has been found on Galiano and Saltspring Islands.

Ranunculus acris L. **Tall Buttercup**
Meadow Buttercup
Tall Crowfoot

Herbaceous perennial with fibrous roots, sometimes rhizomatous, the hollow erect stem 2–11 dm tall, branching above, hairy.

Foliage mostly concentrated at the base and on the lower stem. Basal leaves long-petioled, the blades pentagonal in outline, deeply palmately cleft or divided into 3 segments that are further cleft and toothed. Lower stem leaves similar, on shorter petioles. Upper, bract leaves in the diffuse inflorescence reduced to 3 linear, entire or sparingly toothed segments sessile on the sheath.

Inflorescence diffuse, cymose, flowers 15–25 mm wide. Sepals 5, not reflexed, hairy. Petals 5(–9) glossy yellow, orbicular to broadly obovate, with an obovate truncate nectary scale free at the sides for $\frac{2}{3}$ its length. Stamens 30–70.

Achenes 15–40 on an ellipsoid glabrous receptacle, rather plump, glabrous, 2–3 mm long, with very short hooked beaks. $2n = 14$.

Native to Eurasia this species is now established widely across North America, growing in ditches and in drier soils in meadows and pastures. In British Columbia it is particularly abundant now in the farming districts of the Kootenay region and the lower Fraser Valley.

Ranunculus acris occasionally crosses with *R. uncinatus*, generating a hybrid offspring, which, however is sterile. The plant has the palmately dissected leaves of *R. acris*, petals intermediate in shape and size between the two parent species, and carpels with the prominent hooked styles of *R. uncinatus*. The carpels of this hybrid, however, do not appear to develop into ripe achenes; and the pollen grains are empty, and many of them are collapsed. A specimen of this hybrid form in the Royal British Columbia Museum was collected in a ditch at Paterson, near Rossland, where it was growing in company with both its parent species.

102

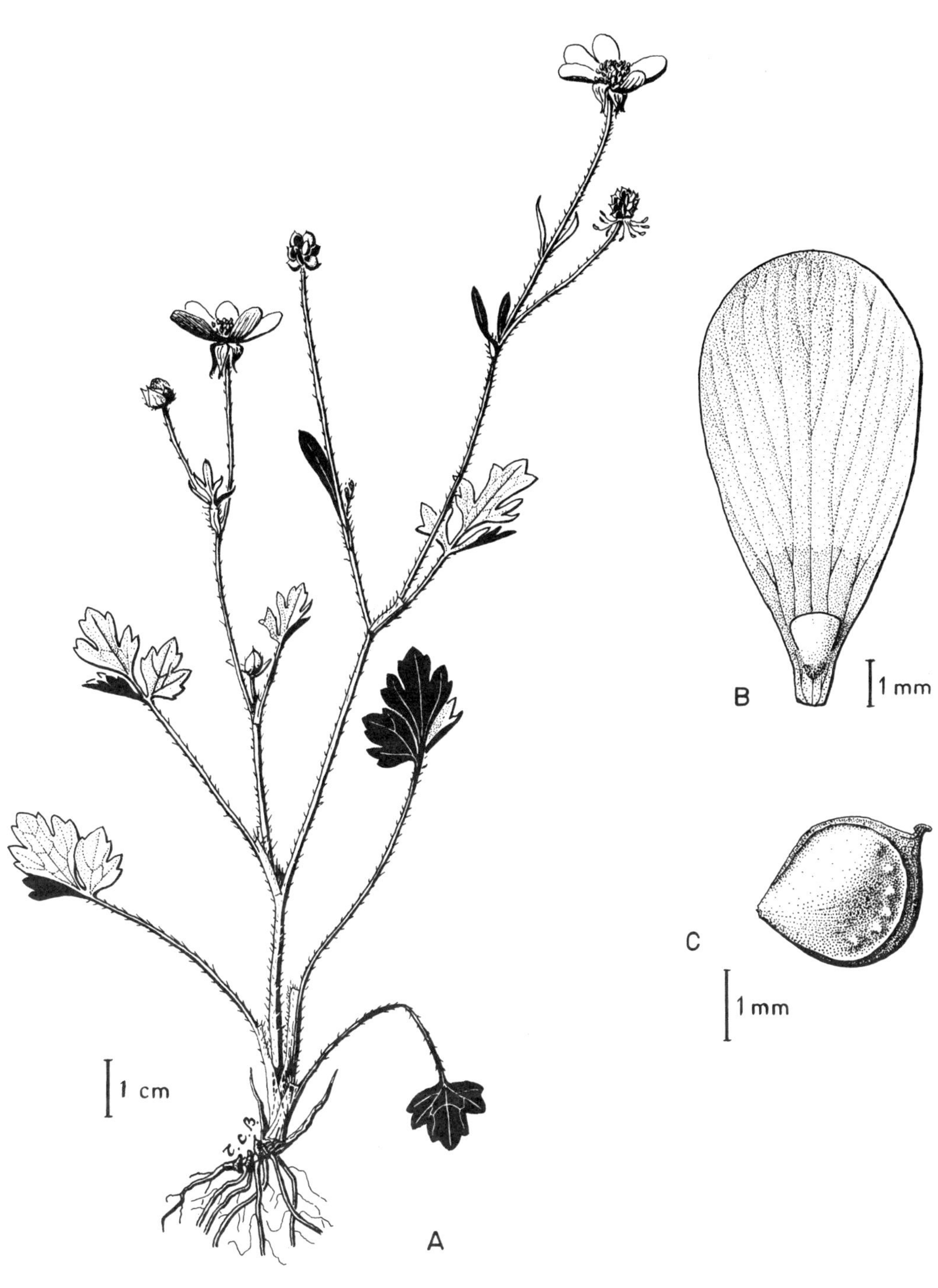

Figure 55. *Ranunculus sardous*

A. plant.   B. petal.   C. achene.

103

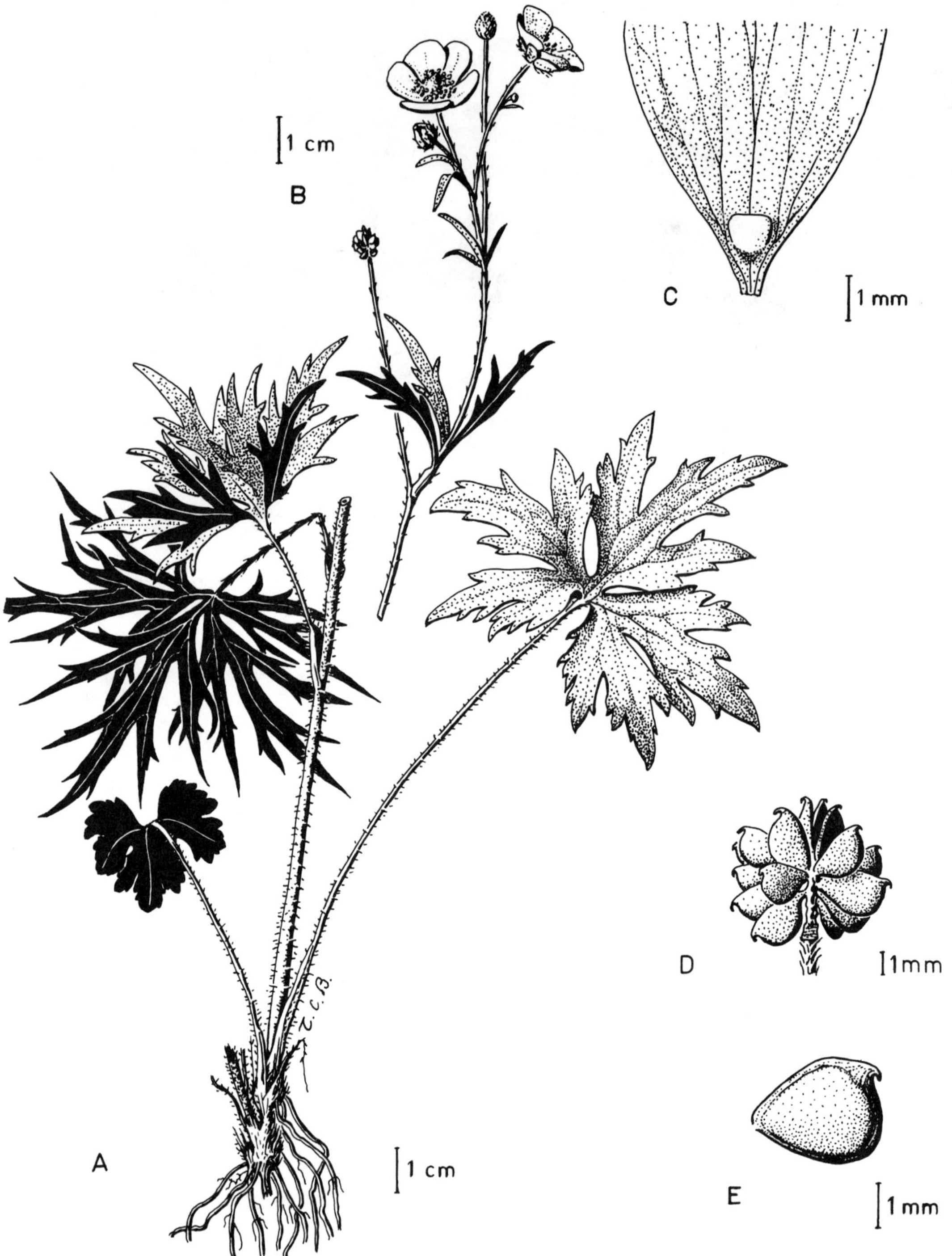

Figure 56. *Ranunculus acris*

A. base of plant.
B. part of inflorescence.
C. base of petal, with nectary scale.
D. head of achenes.
E. achene.

***Ranunculus uncinatus* D. Don *in* G. Don**   **Small-Flowered Buttercup**
** *R. bongardii* Greene**

Erect annual, biennial, or short-lived perennial, arising from a tuft of coarsely fibrous roots; with a hollow, branching stem, 2–9 dm tall, hispid, with spreading hairs on the lower stem and oblique or appressed hairs above.

Basal leaves on long hispid petioles, the blade simple but shallowly to deeply 3-cleft. Stem leaves short-petioled, deeply cleft or divided. The upper, bract-leaves reduced to 1 to 3 lanceolate blades sessile on the stipular sheaths.

Flowers on short pedicels that elongate as the fruits mature. Sepals 5, spreading or becoming reflexed near the midpoint, hispid, 1.5–3 mm long. Petals 5, oblanceolate to obovate, yellow, 2 to 5 mm long, usually slightly longer than the sepals. Stamens 10–20, carpels 10–30, on a glabrous receptacle, themselves glabrous to sparsely hispid, with strongly recurved styles and stigmas.

Fruit a head of obovate, flattish achenes; the achene body 2–2.5 mm long, the beak strongly hooked, 1–2.5 mm long. The achene from glabrous to hispid in varying degrees, sometimes hispid at anthesis but becoming glabrous as it matures. This character is used as a criterion for separating our two varieties. 2n = 28.

KEY TO VARIETIES

1. Achenes glabrous, plants sometimes glabrous... **var. *uncinatus***
    Achenes hispid, especially on the keel, the plant distinctly hispid...................... **var. *parviflorus***

Variety *parviflorus* (Torrey) Benson, (*R. bongardii* Greene) is the commoner one in British Columbia. There is no clear geographic separation of these varieties, both of which occur scattered widely across the province.

Ranunculus uncinatus occurs from Alaska to California and New Mexico, more or less throughout the Cordilleran region. It grows in moist soil, and commonly in shady wooded sites.

Of the specimens with roots in the Royal British Columbia Museum Herbarium none shows distinct evidence of having produced flowering stems in previous seasons. It appears that this species flowers only once here, and is either an annual or a biennial.

***Ranunculus occidentalis* Nuttall *in* Torrey & Gray**   **Western Buttercup**

Fibrous-rooted perennial of rather diverse aspect, with erect or decumbent, hollow stems.

Basal leaves withering early, long-petioled, the blade diverse in shape, varying from simple and shallowly lobed and toothed to trifoliolate, but normally deeply dissected into three cuneate divisions, which are further divided and coarsely toothed. Cauline leaves alternate, short-petioled below, where often like the basal ones, sometimes trifoliolate, and transitional to the divided lower bracts and the simple, sometimes opposite, uppermost bracts.

Flowers terminal on pedicels up to 10 cm long. Sepals normally 5, 4.5–8 mm long, hirsute, reflexed or sometimes spreading, usually shed early. Petals typically 5, but sometimes more, especially in the first flower to open; and up to 14 in var. *hexasepalus*; oblong-obovate, 5–12 mm long, 1.5–2.5 times as long as wide, yellow. Nectary scale broadly obovate, free for about ¾ of its length, about 1 mm long and wide on the average, but with considerable variation about this proportion. Stamens 30–60. Carpels 5–20, usually glabrous. Receptacle ovoid, glabrous, up to 2 mm long in fruit.

Achenes in a globose head, flattened, smooth or rarely thinly strigose, obovate to nearly orbicular; the body 2.5–3.5 mm long, with a stylar beak 0.5–2.4 mm long and up to 0.5 mm wide at base, and varying from recurved to, commonly, straight with a terminal minute hook. 2n = 28.

KEY TO VARIETIES

1. Stem slender: 1–3 mm thick by 1.5–4 dm tall. Basal leaves 2–4.5 cm wide. Nectary scales
    longer than wide. Sepals 5. Petals 5... **2**

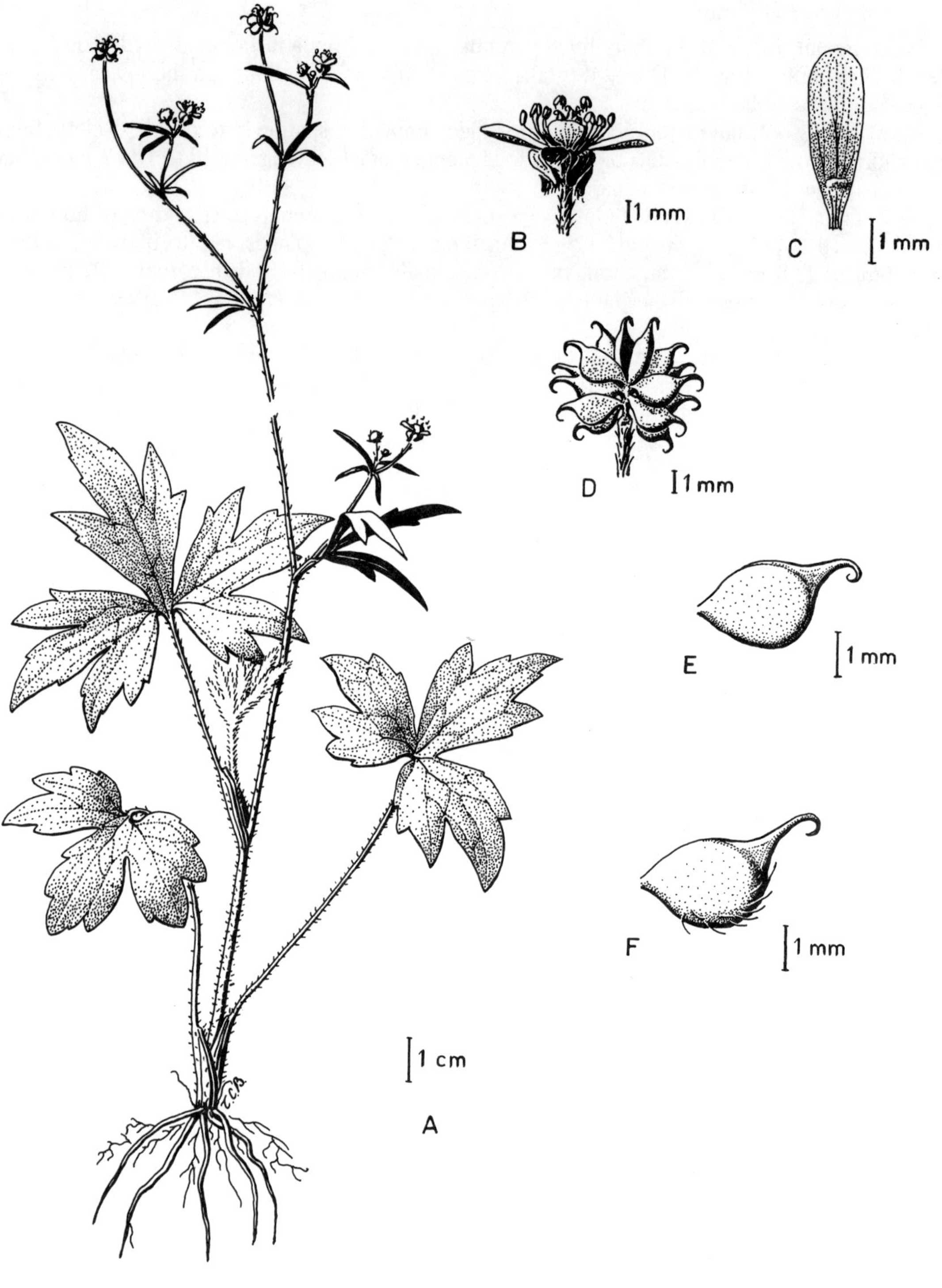

Figure 57. *Ranunculus uncinatus*

A. flowering and fruiting plant.
B. flower.
C. petal.

D. head of achenes.
E. achene of var. *uncinatus*.
F. achene of var. *parviflorus*.

Stem stout and tall: 3–5 mm thick by 4–6 dm tall; and strongly hirsute. Basal and lower stem leaves alike, 5–7 cm wide. Nectary scale as wide as or wider than long. Sepals 5 or 6. Petals 5 to many... **3**

2.   Stem hirsute. Achene beak straight with a terminal hook, 1.2–1.8 mm long.........................
.... **var. *occidentalis***
Stem softly pubescent to glabrous. Achene beak recurved, 0.5 -1.3 mm long.....................
.... **var. *brevistylis***

3.   Petals 5 or 6, 7–10 mm long, 1–2 times as long as wide. Nectary scale very wide: 3–5 times as wide as long. Achene beak distinctly recurved, prolonging the ventral achene margin, and around 2 mm long. Plant with spreading brownish hairs..................... **var. *nelsonii***
Petals 8–14, 2–3 times as long as wide, rounded or emarginate at apex. Nectary scale about as wide as, or slightly wider than, long. Achene beak nearly straight, upturned, 1.3–1.7 mm long. Pedicels rather short and stout, 1–7 cm long... **var. *hexasepalus***

This is a very variable species, inhabiting open or lightly wooded, dry or seasonally moist sites, from coastal bluffs to alpine areas. It is distributed from Alaska to California, and inland to the Rocky Mountains in west central Alberta; but is absent from much of the southern and southeastern interior of British Columbia, and from inland Washington State.

On southern Vancouver Island this species is summer-dormant: growth starting with the appearance of new basal leaves in September, and ceasing with seed-set and dormancy in late May or June. Flowering occurs in April and early May.

Variety *occidentalis* extends from Vancouver Island southward to Oregon west of the Cascade Mountains, on the Queen Charlotte Islands, where it intergrades with var. *hexasepalus*, and sporadically on the mainland, where intergrading with var. *brevistylis*.

Variety *brevistylis* Greene occurs along the southeastern coast of Alaska, in the Yukon, and central and northern British Columbia and the northern Rocky Mountain region of Alberta, where ascending to alpine levels (Benson 1948). This is the predominant variety on the British Columbia mainland. Individuals are also found on the Queen Charlotte Islands and Vancouver Island.

Variety *nelsonii* (DC.) L. Benson occurs on the Aleutian Islands and the islands along the southeastern coast of Alaska (Benson 1948). It has not been found in British Columbia in its typical form, but specimens approaching it in character have been found on the northwestern Queen Charlotte Islands. Further collecting in that area may reveal the presence of typical var. *nelsonii*.

Variety *hexasepalus* L. Benson (1941) [*R. hexasepalus* L. Benson (1948)], in spite of its name, has five sepals more often than six. It is found on coastal bluffs and islets of the Queen Charlotte Islands, in at least some cases on limestone, and in at least some cases on rocks inhabited by colonies of seabirds. There is a suspicion that the tall robust growth it commonly shows may be in part a consequence of nitrification of its soil by the droppings of the seabirds inhabiting these sites.

The intergradation between var. *hexasepalus* and the other varieties suggests that the multi-petalled characteristic is subject to genetic determination, but is not associated with a barrier to breeding: hence the preference here, for varietal, rather than specific rank for *hexasepalus*.

From a regional standpoint, we see in this province three populations in as many geographic regions. These are: (a) a generally uniform population on the central and northern mainland. In contrast to this, two insular areas are characterized by conspicuous variability. These are: (b) southern Vancouver Island and the adjacent Gulf Islands and islets; and (c) the Queen Charlotte Islands.

On the mainland of British Columbia there is general uniformity in habit and petal number; five petals being consistently displayed; though a general trend is seen toward shorter average achene beak length as one progresses toward the northwestern extremity of the province. This population as a whole can be assigned to var. *brevistylis*, though beak length, while averaging 1 mm or less, varies continuously from short to long, with some individuals typical of var. *occidentalis*.

In the insular areas, diversity of character expression involves: habit, from short and almost prostrate to tall and robust; petal number, flowers with from 5 to 14 petals being found; petal shape,

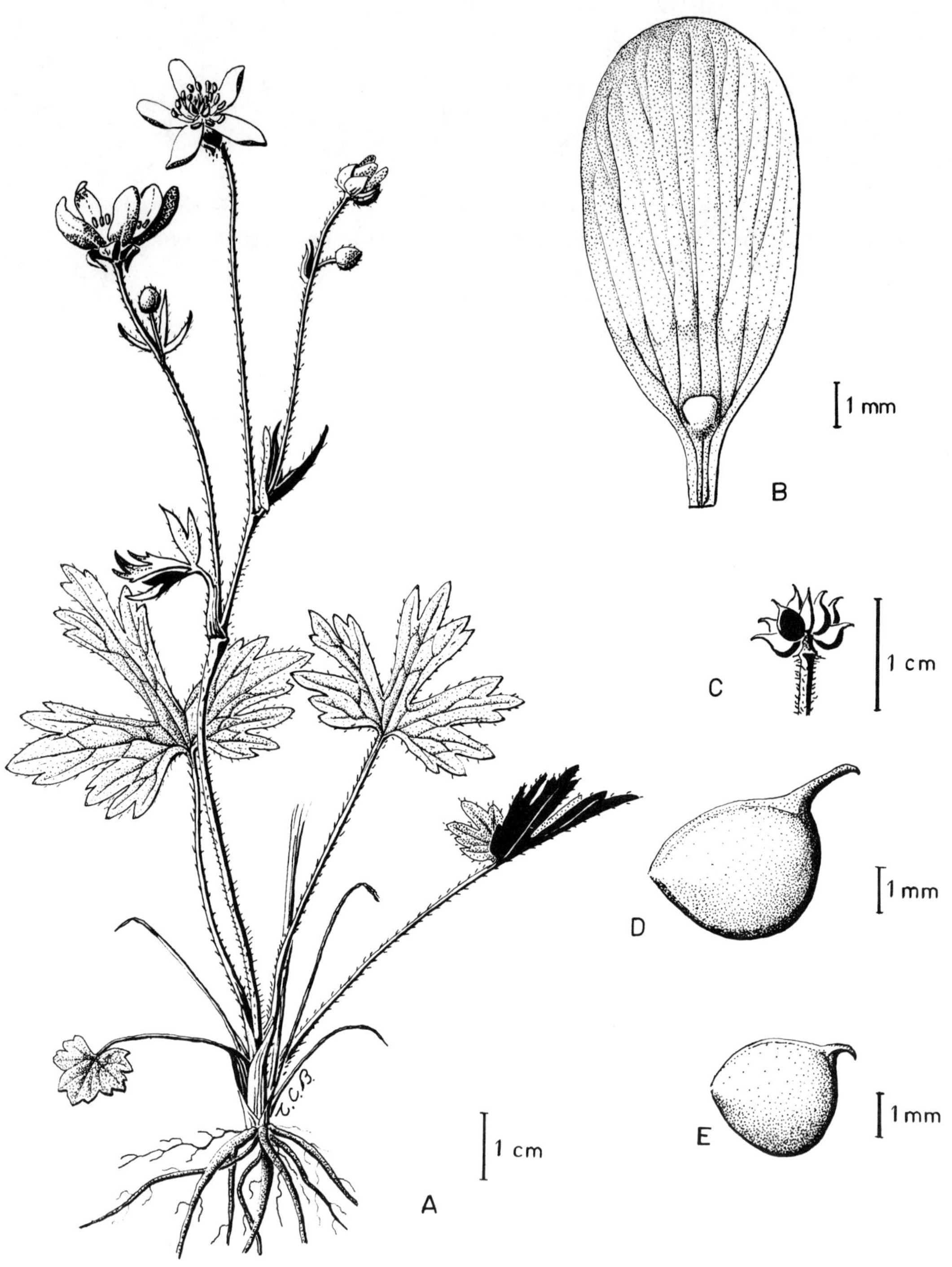

Figure 58. *Ranunculus occidentalis*

A. flowering plant.
B. petal.
C. head of achenes.

D. achene of var. *occidentalis*.
E. achene of var. *brevistylis*.

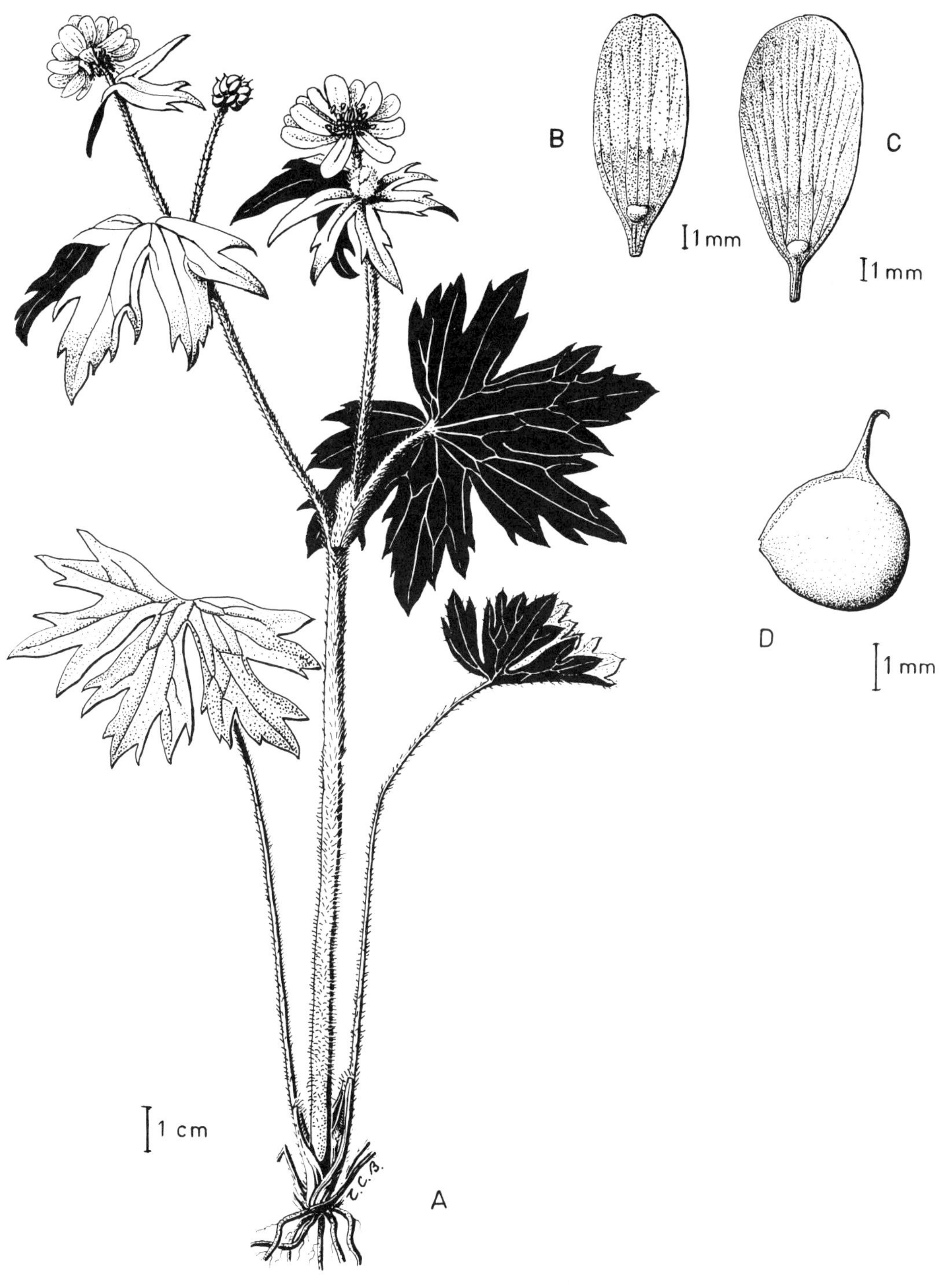

Figure 59. *Ranunculus occidentalis* var. *hexasepalus*

A. flowering plant.
B. petal of paratype specimen (CAN 56609).
C. petal of another specimen.
D. achene of paratype.

from almost linear to nearly circular; and length, curvature, and attitude of the achene beak.

On southern Vancouver Island and adjacent islands it can be postulated that this diversity marks a hybrid swarm produced by interbreeding of the local population of *R. occidentalis* with nearby populations of *R. californicus*, which have contributed genes for high petal number, narrow petals, short, recurved, stylar beaks, and some tendency toward a depressed growth habit.

Three observations support this explanation: (a) Plotting petal numbers on a map reveals a progressive increase in number as one approaches Oak Bay, in the southeastern corner of Vancouver Island, where colonies of *R. californicus* are found. (b) The late Dr. Margaret Heimburger experimentally cross-bred these two species, and found that they are interfertile; and that the hybrid offspring resemble the extra-petalled forms of *R. occidentalis* that are found growing wild in that area, in having 5 to 14 petals per flower, achenes that combine features of both parent species, and often a rather depressed, sprawling growth habit like that of many plants of *R. californicus*. It is interesting to note that the wide variability in basal leaf form that is exhibited by *R. californicus* is inherited by these hybrid offspring. In the case of the above experiment, the pollen was obtained from several plants of *R. californicus*; and the pistillate parent was a plant of *R. occidentalis*. (c) Breeding experiments by the same person, involving the local, wild, extra-petalled form of *R. occidentalis*, have shown that they have very reduced fertility, as indicated by the proportion of abortive pollen grains: a condition that can be expected in a population of hybrid origin.

The author is pleased to name this hybrid after Dr. Margaret L. Heimburger, whose experiments produced it artificially, and thus demonstrated the probable hybrid origin of the suspect natural population.

Ranunculus X *heimburgerae* T. C. Brayshaw *nom. nov.* (=*R. occidentalis* pistillate X *R. californicus* staminate, with respect to the first artificial cross). Hybridae hortenses et naturales, e *Ranunculo occidentali* Nuttall et *R. californico* Bentham genita, characteribus parentales variabiliter combinatis. Caulis erectus vel decumbens. Folia basalia simplicia, trifoliolata, pinnatisecta, bipinnatisectave, ut in *R. californico* visus. Petala 5–14, elliptici, 6–14 mm longa, 2-vel 3-plo longiora quam latiora. Achenia 9–40, frequentissime 19–26, orbicularia, stylis subrectis vel valde curvatis.

Garden and natural hybrids, arising from *Ranunculus occidentalis* Nuttall and *R. californicus* Bentham, with the parental characters variously combined. Stem erect to decumbent. Basal leaves simple, trifoliolate, pinnatisect or bipinnatisect, as seen in *R. californicus*. Petals 5–14, elliptic, 6–14 mm long, twice to thrice as long as wide. Achenes 9–40, most frequently 19–26, orbicular, with nearly straight to strongly curved styles (see Figure 61: F to J).

Representative specimens of the progeny of this experiment are filed in the herbarium at the Royal British Columbia Museum (nos. V 127472–127523 & 127525–127547). The type specimen representing the artificial cross is selected as *M. Heimburger #11 (V 127522)*: Albert Head pistillate *R. occidentalis* X Alpha Islet staminate *R. californicus*: sample collected June 6, 1984. The type representing the wild hybrid swarm is *Connell & Hardy, V 14618-C*, from Foul Bay, Victoria, May 1, 1942. Samples of the artificial hybrids are now in cultivation at the Royal British Columbia Museum. Unusually for *Ranunculus* hybrids, these are partially fertile; the reduced fertility being indicated by the proportion (30–70%) of their pollen grains that are empty.

The diversity in *R. occidentalis* that is found in the Queen Charlotte Islands is harder to explain. There, the high petal number of var. *hexasepalus* is commonly associated with long, straight, erect stylar beaks and a robustly upright habit of growth. It is a rather long hop (about 700 km) from the northernmost limit of *R. californicus* in the Puget Sound region to the nearest location of *R. occidentalis* var. *hexasepalus* in the Queen Charlotte Islands.

Ranunculus californicus **Bentham**   **California Buttercup**

Perennial, or sometimes annual, herbs with hollow, hirsute to glabrous, prostrate to erect stems with ascending pedicels.

Leaves extremely diverse in form and texture: much more conspicuously so than those of *R. occidentalis*. Basal and lower stem leaves with blades variably simple and 3-lobed to pinnately compound with 3 or 5 or more leaflets, or the lower stem leaves even biternate or bipinnate, the divisions or leaflets again lobed, the overall blade outline ovate to orbicular and cordate. Cauline leaves alternate; the lowermost ones petioled, and sometimes compound even when the basal leaves are simple. Bracts divided, sessile, alternate or sometimes subopposite.

Flowers on ascending pedicels 2–11 cm long. Sepals 5, ovate, pilose dorsally, reflexed, often deciduous early, about half as long as the petals. Petals 9–26, obovate to oblanceolate, (2.5–) 3–4 times as long as wide. Nectary scales truncate, free almost their full length, about 1 mm long. Stamens 30–60. Carpels (5–) 20–40: "more numerous than in *R. occidentalis*." (Benson, 1948). Receptacle cylindrical, 1–2 mm long in flower, up to 3 mm long in fruit, glabrous or sometimes hispid.

Achenes broadly obovate, flattened, noticeably margined; the beak recurved, stout, 0.4–1 mm long; the body surface glabrous or rarely hispid, and otherwise smooth or rarely minutely reticulate. 2n = 28.

Ranunculus californicus ranges from an isolated area in the Puget Sound region of British Columbia and Washington State (Denton, 1978) along the coast of Oregon and California to Baja California. It inhabits open coastal bluffs and prairies.

Like *R. occidentalis* on Vancouver Island, this species is summer-dormant. Its growing season extends from September to May; with flowering occurring in April and early May, concurrently with neighbouring populations of *R. occidentalis*.

Northern populations of *R. californicus* have been assigned to var. *cuneatus* Greene, which is distinguished as having prostrate stems, '3-lobed' to 3-parted but simple basal leaf blades with cuneate divisions, and 20–30 achenes, and to behave sometimes as an annual.

Examination of the type specimen of var. *cuneatus* (No. 238185 at Pomona College herbarium, a neotype designated by L. Benson in 1954), reveals a plant with rather thick, cordate-orbicular, moderately incised, simple blades on the basal and lower cauline leaves. While agreeing with Greene's description, it can hardly be considered typical of the population here at the northern limit of the species' range.

The plants of *R. californicus* found here vary in number and shape of petals, length of achene beaks, and attitude of stem. Commonly decumbent, the attitude appears to be plastic, and modifiable by the depth and relative luxuriance of the surrounding vegetation. This plant appears occasionally to be an annual, as noted for var. *cuneatus* by Greene (1892, as quoted by Benson, 1948). Lower cauline leaves are commonly pinnately trifoliolate. The basal leaves may have withered and almost disappeared at flowering time; so that the lower cauline leaves above relatively short internodes may be taken for basal leaves, to which they are not necessarily identical in form.

Basal leaf blades here vary from simple, shallowly to deeply, yet bluntly lobed, nearly circular, often rather thick and leathery in texture, as in var. *cuneatus* Greene (var. *crassifolius* Greene) to pinnately trifoliolate to 5-foliolate, or even biternate or bipinnate leaves with narrow, lobed leaflets, as in var. *californicus* (see Figure 61: A to E).

At present it seems questionable to try to force our population, as a whole, into any one variety; but individuals within it could be assigned to one or other of the above varieties according to the following key, derived, in part, from that of Munz & Keck (1963).

KEY TO VARIETIES

1. Basal leaf blades pinnately 3- to 5-foliolate, ovate in outline. Stem typically erect ....
    .... **var. *californicus***

    Basal leaf blades simple. Stems prostrate to decumbent .... **var. *cuneatus***

In British Columbia, *Ranunculus californicus* occurs on Trial, Alpha and Griffin Islands, near Oak Bay, just east of Victoria, and on Saturna Island. On Trial Island it occurs adjacent to plants of *R. occidentalis*. Both species flower together, and are interfertile (see discussion under *R. occidentalis*). Many plants with mixed combinations of characteristics are found there, and also on nearby points of Vancouver Island, from which Trial Island is separated by a channel not more than 400 metres wide. Even the plants that appear most like *R. californicus* embody some introgressant,

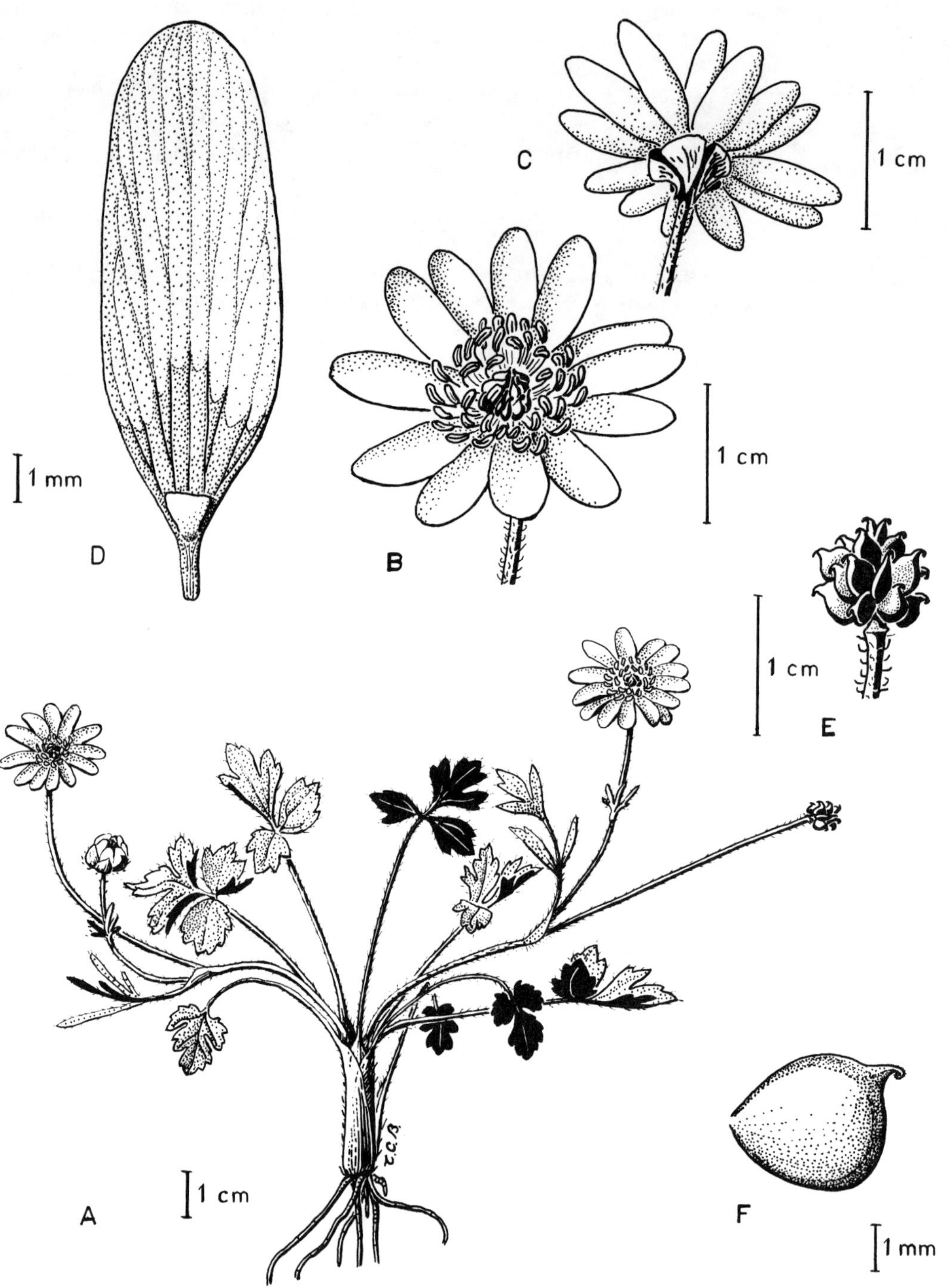

Figure 60. *Ranunculus californicus*

A. flowering plant.
B. flower, from above.
C. flower, from below.
D. petal.
E. head of achenes.
F. achene.

112

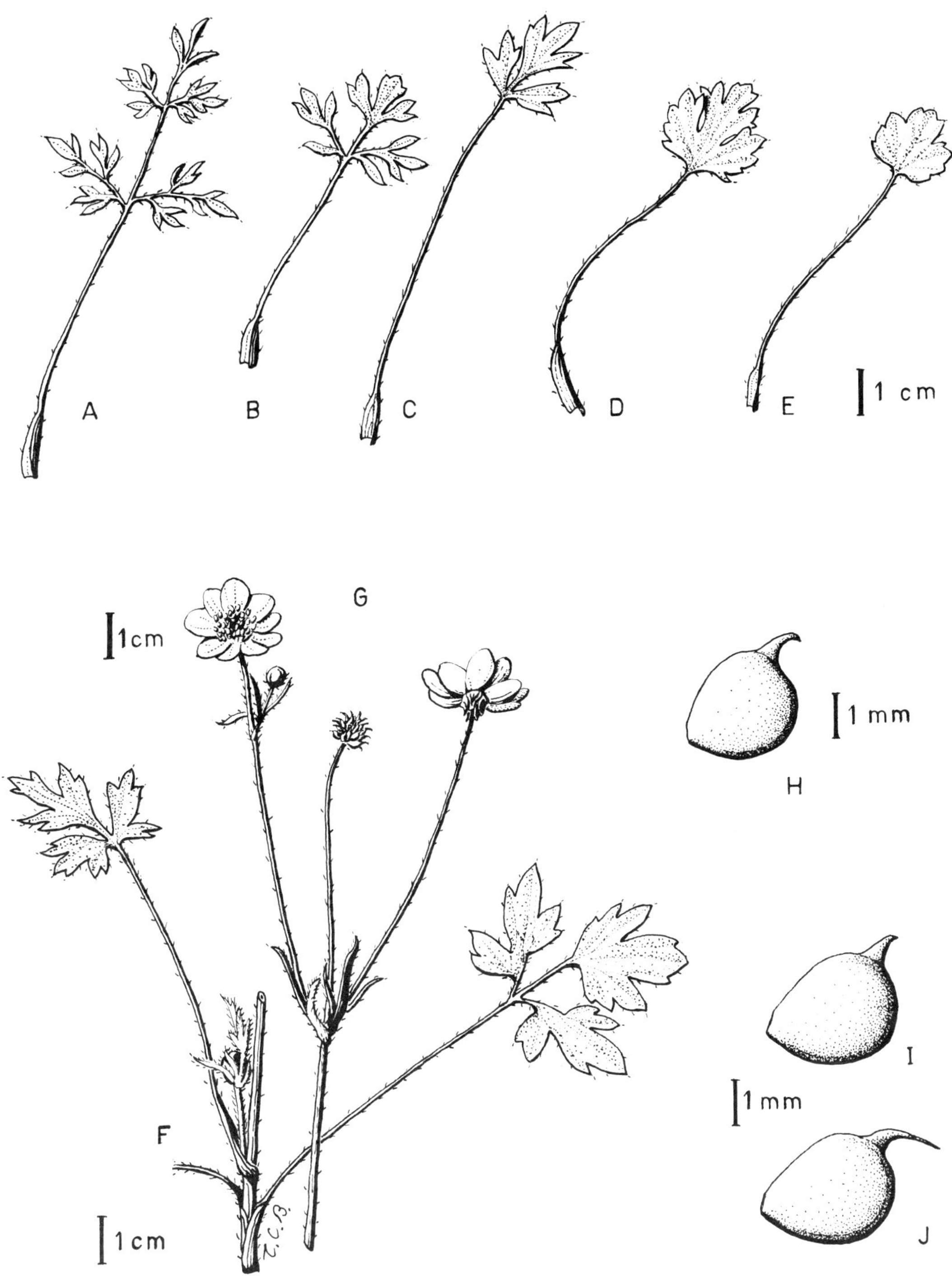

Figure 61. A–E: *Ranunculus californicus*: Range of form of basal and near-basal leaves in the British Columbian population.
A., B. and C. var. *californicus*; D and E. var. *cuneatus*.
F–J: *Ranunculus* X *heimburgerae* (*R. occidentalis* X *californicus*): F, basal and near-basal leaves; G. flowers; H. achene; I and J. other achene forms seen in the hybrid swarm. F, G, and H based on the type specimen (V 127522).

occidentalis-like features. On Alpha Islet, which is a small, low, grassy and treeless islet, *R. californicus* is the only *Ranunculus* species present; and it is more typical of the species and more uniform than on Trial Island. More plants have compound basal leaves, and the fruits are typical. Both annual and perennial plants have been observed.

On the nearby Griffin Island, *R. californicus* populates an area of grassland almost exclusively, while *R. occidentalis*, which on the mainland of Vancouver Island often occurs in open grassland, is here confined to a small deciduous wood and its immediate border, where a few intermediate specimens are also found. The segregation of these species by habitat here is very pronounced; and suggests that the strong exposure to sun, wind, and perhaps salt spray, may be the factors that have enabled *R. californicus* to maintain its identity in a geographic region where it is vastly outnumbered by a species with which it is interfertile.

It can be postulated that *R. californicus* reached this region prehistorically — perhaps during Hypsithermal time; subsequently being eliminated generally at this latitude through competition with the hardier, and better woodland-adapted *R. occidentalis*, except in a few exposed, near-littoral habitats where it still has some slight competitive advantage.

Section *Epirotes* (*Auricomus*)
Basal leaves pedately lobed to nearly entire or toothed, but never both lobed and toothed. Flowers yellow. Nectary pocket-like, with a crescentic scale. Achenes plump, glabrous or hairy. x = 8.

Ranunculus abortivus L.   **Kidneyleaf Buttercup**
Biennial or short-lived perennial, with one or more erect hollow stems from a cluster of fibrous roots; glabrous or finely puberulent, 1–5 dm tall.

Basal leaves with long slender petioles and orbicular to reniform, commonly cordate-based blades 1–4 cm long, with crenate margins; sometimes with deeply 3-lobed leaves transitional to the upper stem leaves and bracts; the latter divided into 3–5 linear segments.

Flowers small on pedicels up to 10 cm long. Sepals 5, glabrous or puberulent, deciduous early, 2.5–4 mm long. Petals 2–3.5 mm long, narrowly rhombic, shorter than the sepals, yellow fading white, with a pocket-like nectary, the scale crescentic. Stamens 15–30. Receptacle ovoid, elongating to 5 mm in fruit, finely hairy.

Achenes 20–50, glabrous, ovoid and moderately compressed, the body 1–1.5 mm long, with a short beak 0.1–0.2 mm long. 2n = 16.

KEY TO VARIETIES

1. Plant glabrous .... **var. *abortivus***
   At least the upper, younger parts of the plant finely pilose or puberulent .... **var. *acrolasius***

Widespread across North America, including British Columbia; in moist woodlands, stream banks, and subalpine clearings. Most of our material belongs to var. *acrolasius* Fernald. The distribution of this variety shows no clear regional segregation.

Ranunculus inamoenus Greene
    R. alpeophilus A. Nelson
    R. inamoenus Greene var. *alpeophilus* (A. Nelson) L. Benson

Erect perennial herb with fibrous roots, 10–15 cm tall; the stem pubescent above.

Basal leaves on petioles sometimes much longer than the blades; the blades orbicular to broadly obovate, crenate to shallowly round-lobed, sparsely ciliate. Stem leaves, if any, short-petioled, the blade deeply cut into 3–5 oblanceolate lobes. Bracts sessile, divided into 3–5 oblanceolate divisions.

Flowers one to several, small, the pedicels puberulent, commonly no longer than their subtending bracts at flowering time, elongating in fruit. Sepals 5, 2½–6 mm long, spreading to deflexed, pubescent, persistent through the flowering stage. Petals 5, oblanceolate to narrowly

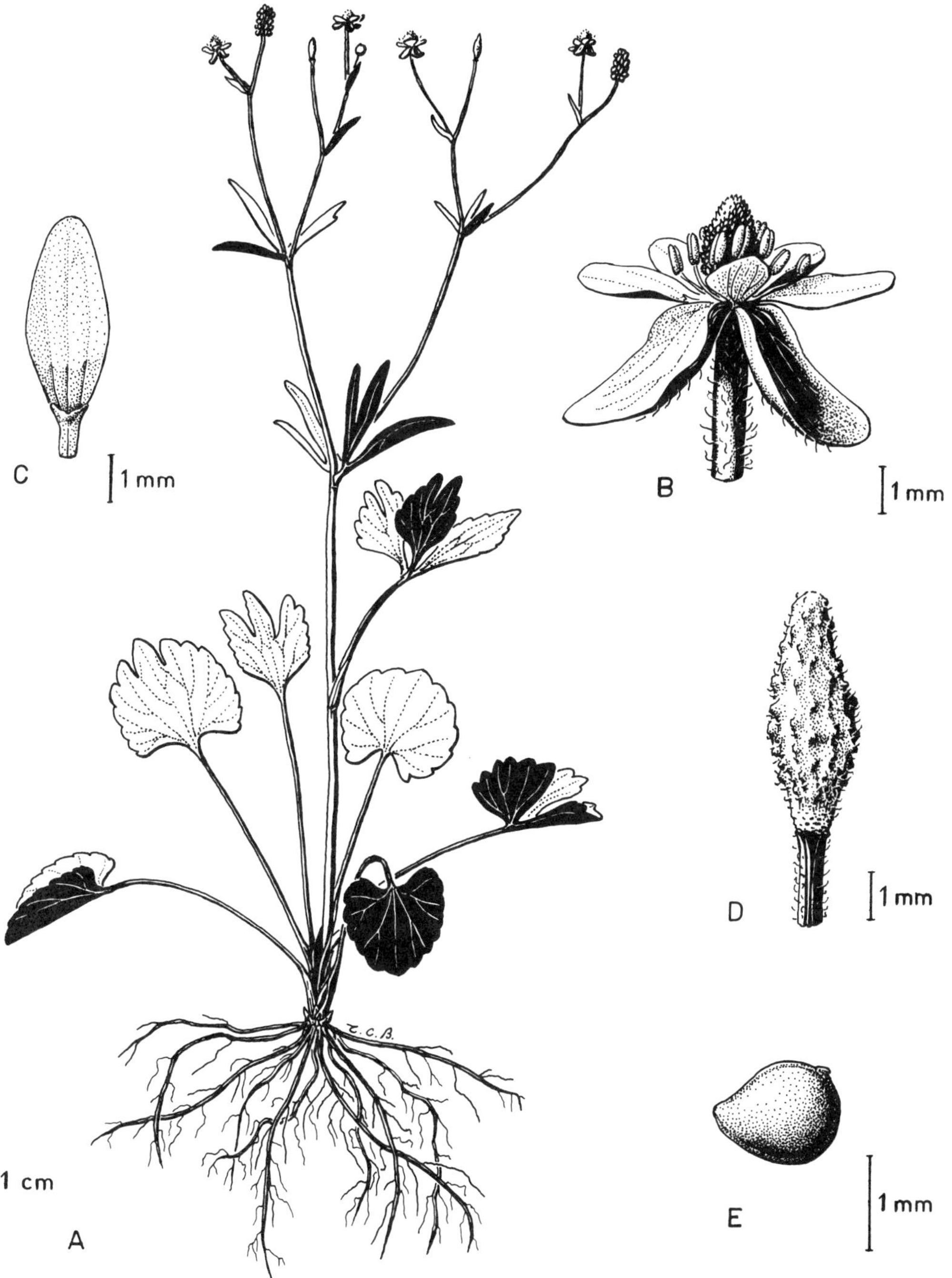

Figure 62. *Ranunculus abortivus* var. *acrolasius*

A. plant.　　B. flower.　　C. petal.　　D. receptacle.　　E. achene.

obovate, 2–8 mm long, mostly a little longer than the sepals, yellow; the nectary pouch-like, the free upper edge of the scale concave. Stamens 25–50. Carpels 40–100, usually greyish-puberulent. Receptacle columnar, normally thinly pubescent.

Achenes broadly obovoid, the body 1½–2 mm long, and canescent with short, straight, white hairs; the beak short (0.2–0.9 mm long), and recurved. 2n = 48.

Specimens with glabrous achenes may be distinguished as variety *alpeophilus* (A. Nelson) L. Benson. Benson (1948) reported this variety from the Yoho Valley, in eastern British Columbia; but no specimen has been seen.

Ranunculus inamoenus is a species mainly of the Rocky Mountains and Great Basin regions; ranging from Alberta and British Columbia southward through Idaho to Utah, Colorado, and Arizona. Records from British Columbia are scarce and widely scattered in the interior. It may be looked for in the Rocky Mountains in eastern British Columbia, since a record from Alberta near Mount Assiniboine is very close to the Provincial boundary (and is shown on the range map).

The few records from British Columbia give no information on the habitats in which the plants were found; but elsewhere the species is reported from open mountain meadows.

Ranunculus pedatifidus J.E. Smith *ex* Rees   **Birdfoot Buttercup**

R. affinis R. Brown
R. pedatifidus ssp. *affinis* (R. Brown) Hulten
R. pedatifidus var. *affinis* (R. Brown) Benson = var. *leiocarpus* (Trautvetter) Fernald
R. affinis var. *leiocarpa* Trautvetter
R. pedatifidus var. *leiocarpus* (Trautvetter) Fernald

Variably pilose perennial from a cluster of fibrous roots, with one or more erect, slender, sparsely branched stems 1–3 dm tall.

Basal leaves long-petioled, the blade rather reniform in outline, cordate-based, and pedately divided into 5–9 narrow, lanceolate or oblanceolate lobes, which are further dissected into many linear lobes in var. *pedatifidus*. Stem leaves short-petioled or sessile, similarly divided, but with narrower lobes; the uppermost ones reduced and three-lobed.

Flowers 1–4, terminal on pedicels up to 15 cm long. Sepals 5, spreading, 5–6 mm long, tomentose to villous, often deciduous early. Petals usually 5, obovate to nearly orbicular, yellow, 5–8 mm long in our material (8–12 mm in var. *pedatifidus*); the nectary shallowly pouch-like, small, close to the base of the petal. Stamens 25–50, their filaments often short. Receptacle ovoid to cylindric, elongating up to 8 mm in fruit, hairy. Carpels 25–80.

Achenes obovate, the body up to 2 mm long, glabrous or short-hairy; the strongly recurved beak up to 1 mm long but commonly not over 0.5 mm long, sometimes deciduous. 2n = 48.

Circumpolar, and southward in the mountains of Asia; and in North America, in the Rocky Mountains south to Colorado and New Mexico. Found on dry grassy sites and alpine and arctic tundra.

KEY TO VARIETIES

1. Achene glabrous .... **var. *leiocarpus***
   Achene pilose .... **var. *pedatifidus*** in broad sense

The commoner variety in British Columbia resembles var. *pedatifidus* in the character of its fruit. However, the typical var. *pedatifidus* in Asia has more dissected leaves and larger flowers than our material [as seen in photos of type material in the Linnaean herbarium, one of which is reproduced by Fernald (1934, plate 279)]. This suggests that the British Columbian population must be something else yet.

In British Columbia the 'var. *pedatifidus*' is progressively replaced northward by var. *leiocarpus* (Trautvetter) Fernald (*R. affinis* R. Brown), which is the dominant variety in the Arctic. The southernmost specimen of var. *leiocarpus* known in British Columbia (V:92027) is from Mt. Kostuik (54°54′N., 121°01W.).

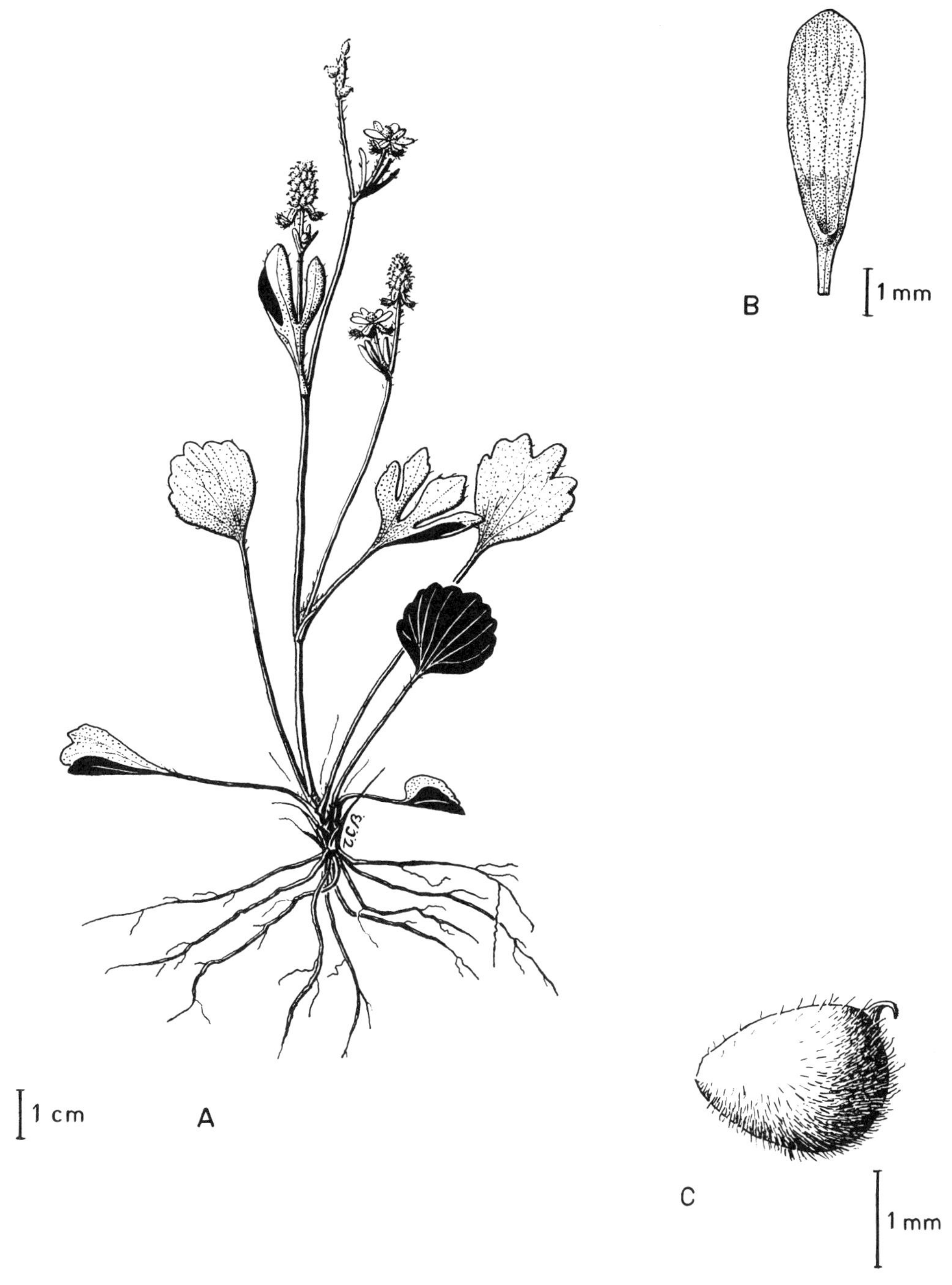

Figure 63. *Ranunculus inamoenus*

A. flowering and fruiting plant.   B. petal.   C. achene.

117

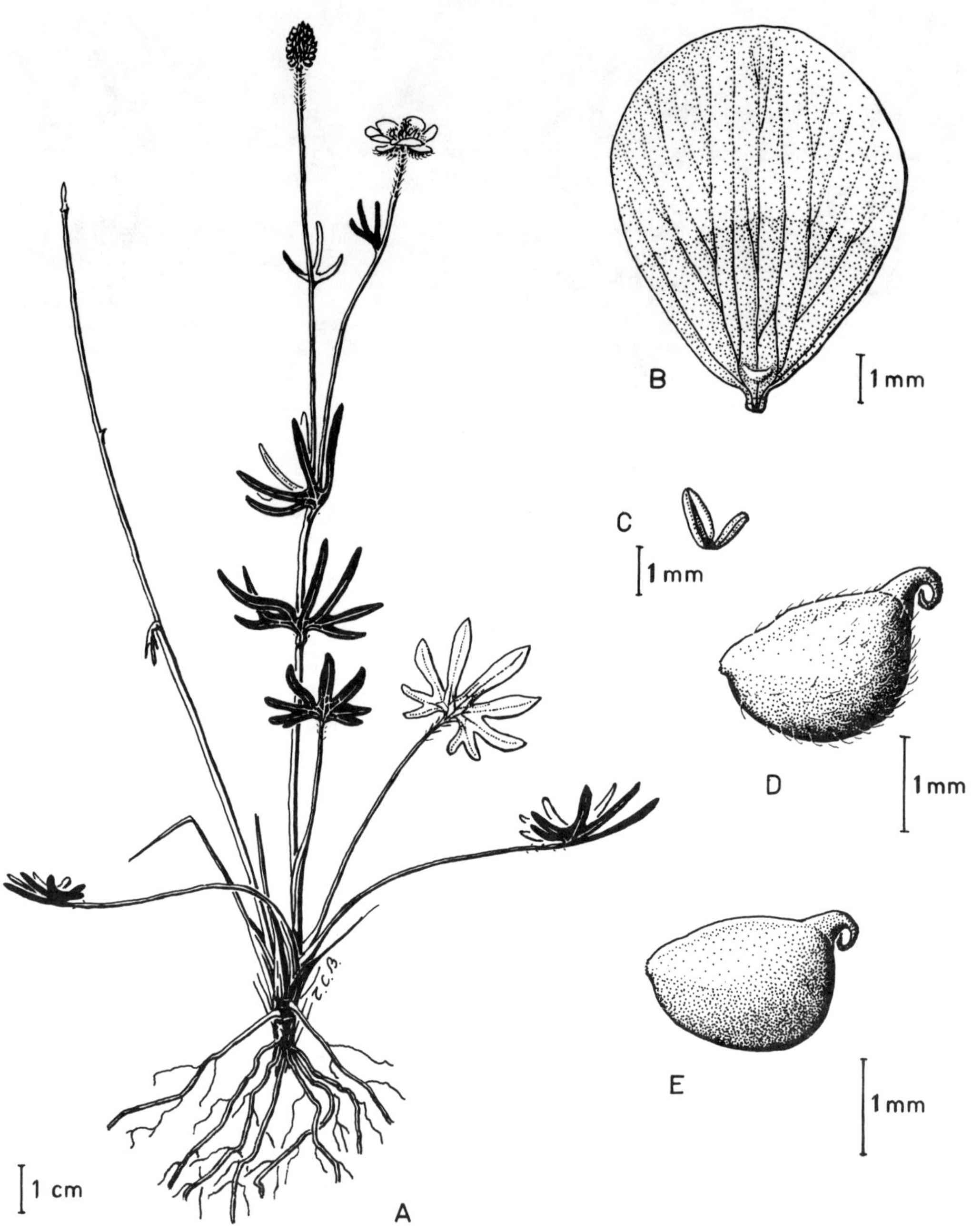

Figure 64. *Ranunculus pedatifidus*

A. plant.
B. petal.
C. stamen.
D. achene of var. *pedatifidus*.
E. achene of var. *leiocarpus*

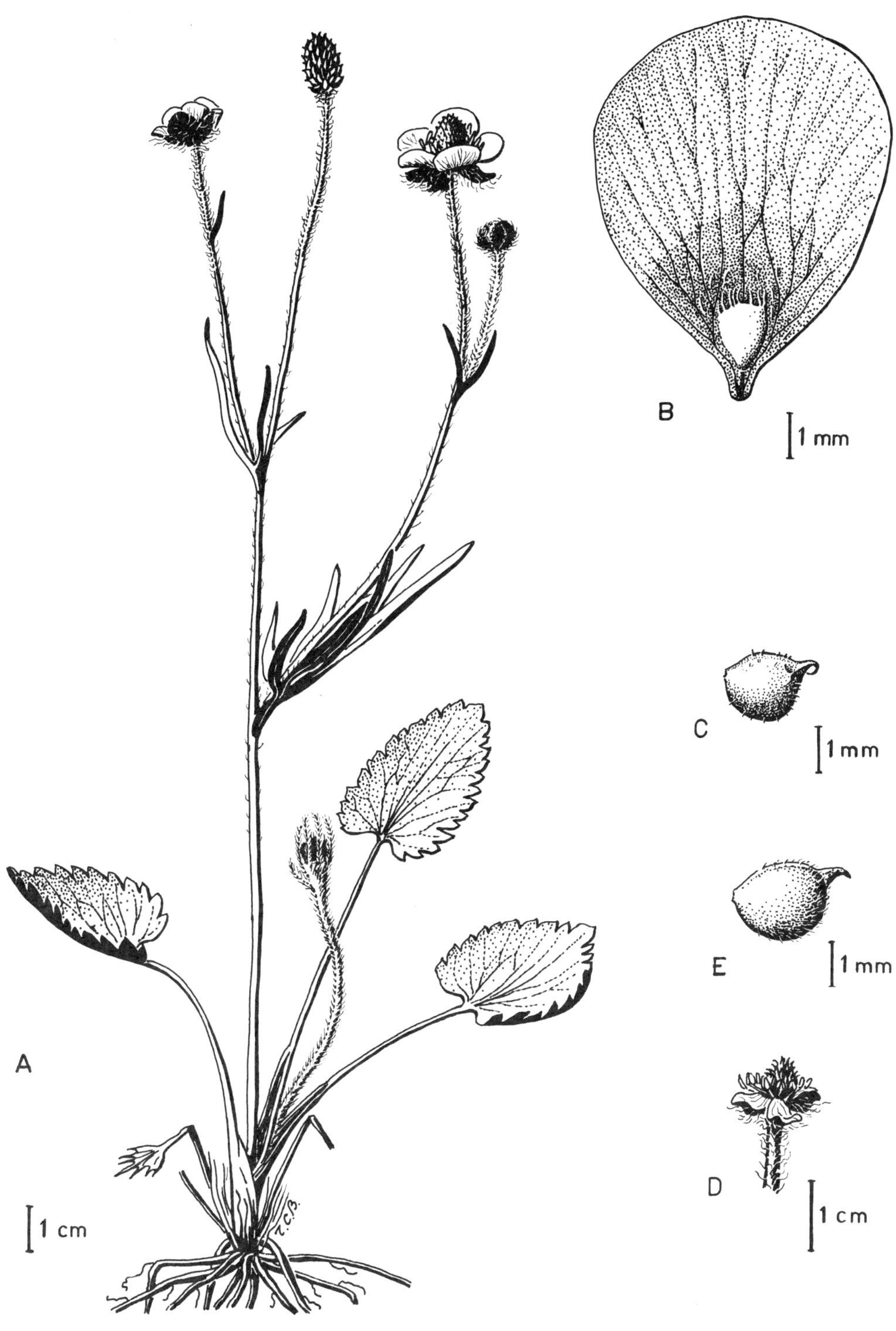

Figure 65. *Ranunculus cardiophyllus*

A. plant.
B. petal, with ciliate nectary scale.
C. achene of forma *cardiophyllus*.

D. flower of forma *apetalus*.
E. achene of forma *apetalus*.

Ranunculus cardiophyllus **Hooker**   **Heartleaf Buttercup**
   R. affinis R. Brown var. *cardiophyllus* (Hooker) Gray
   R. affinis var. *lasiocarpus* Torrey
   R. apetalus Farr
   R. pedatifidus var. *cardiophyllus* forma *apetalus* (Farr) Boivin
   R. cardiophyllus forma *apetalus* (Farr) Boivin

Perennial from coarsely fibrous roots, the hollow, erect stems 15–40 cm tall and usually strongly pilose to villous, rarely glabrous.

Basal leaves with stout petioles up to 12 cm long; the blades broadly ovate, cordate-based, and crenately toothed to occasionally lobed, rather thick and leathery in texture. Stem leaves deeply lobed to 5–7 parted, short-petioled to sessile.

Flowers few, terminal on pedicels up to 14 cm long. Sepals 5, spreading, 6–10 mm long, pilose, early deciduous. Petals 5 or more [absent in forma *apetalus* (Farr) Boivin], yellow, obovate, 8–15 mm long; the nectary deeply pocket-like, the truncate upper margin of scale (adnate at the sides) long-ciliate. Stamens 35–75. Receptacle ovoid, up to 12 mm long in fruit, hairy. Carpels 20–100 in a conical head.

Achenes in a cylindric to ovoid head, obovate, compressed, the body 1.5–2 mm long, finely puberulent and often reticulate; the beak straight or curved, 0.6–1.0 mm long.

This species ranges from the Northwest Territories to New Mexico, and from eastern British Columbia to Saskatchewan; growing in grassland and in open glades in forest. The forma *apetalus* (Farr) Boivin occurs rarely in the Rocky Mountains of southwestern Alberta and southeastern British Columbia, and differs from the typical *R. cardiophyllus* in lacking petals and (in the few specimens at V from the Crowsnest Pass area), in having more deeply lobed basal leaves.

Ranunculus glaberrimus **Hooker**   **Sagebrush Buttercup**

Low, generally glabrescent perennial herb, 5–20 cm tall, from a branched, rather fleshy root. Stem soft, glabrous to thinly villous; few-flowered.

Leaves mainly basal, petioled, with slender petioles and lanceolate to ovate or orbicular blades blunt at apex, shallowly 3-lobed or sometimes crenate, to entire, glabrous. Stem leaves sessile or subsessile, 3-parted, or the upper (bracts) reduced and entire. Foliage often darkening on drying.

Flowers 1½ to 2½ cm across. Sepals 5, 5–8 mm long, green, thinly villous, sometimes the only parts of the plant with hairs, reflexed, and shed before the petals. Petals 5, or sometimes up to 8, 8–15 mm long, obovate, bright shiny yellow, multi-veined, sometimes absent; the nectary scale covering the gland, joined to the petal along most of its lateral margins, forming a pocket 1.0–1.7 mm long, the upper, free edge often shallowly lobed, and occasionally ciliate. Stamens many, the anthers extrorse. Carpels very numerous (75 to 150), glabrous or finely puberulent, on a globose, usually glabrous receptacle.

Achenes obovoid, rather plump, slightly to strongly keeled ventrally, glabrous or thinly puberulent, 1.5–2.3 mm long; the beak slender, and straight or recurved, almost ventrally placed. 2n = 128.

Occasional plants are staminate; producing a naked receptacle in the centre of the flower.

KEY TO VARIETIES

1. Basal leaves broadly ovate to orbicular, 3-lobed to notched. Stem leaves often entire ....
    .... **var. *glaberrimus***
    Basal leaves narrowly elliptic to lanceolate, entire or almost so. Stem leaves often 2- or 3-
    cleft **2**

2. Carpels glabrous .... **var. *ellipticus***
    Carpels finely puberulent .... **var. *buddii***

Ranunculus glaberrimus occurs from British Columbia southward to California, eastward across the Prairie Provinces to the Dakotas, and southward to New Mexico. It is found abundantly in semi-arid grassland and sagebrush steppe, where it is one of the first plants to flower in Spring, and extends upward into the Ponderosa Pine and Douglas-fir parklands. The varieties are intermingled over the same general range, geographically. According to Hitchcock and Cronquist (1964) var. *glaberrimus* tends to occupy steppe habitats at low elevations, while var. *ellipticus* is more montane in distribution. British Columbian records are not sufficiently detailed to provide confirmation of this statement.

Variety *ellipticus* Greene was described by its author (Greene, 1890; originally as *R. ellipticus*) as having glabrous carpels. This statement is accepted here, as it was by Boivin (1968–1969), rather than that by Benson (1948) that they are canescent. Boivin's (1968–1969) establishment of var. *buddii* Boivin as having, by inference, hairy carpels, is also acceptable. Unfortunately, in British Columbian material, the carpel vesture does not correlate well with the leaf characters used by other authors to distinguish varieties. Almost all plants in this province, regardless of variety as defined by leaf characters, have glabrous carpels. The three specimens seen with canescent carpels are from Oliver, Vaseaux Lake, and Nimpo Lake. Of these, the Nimpo Lake specimen has crenate basal leaves, the others have more or less entire ones. All have at least some stem leaves trifid. These are here accepted as var. *buddii* Boivin.

Ranunculus rhomboideus Goldie

Low perennial, from a tuft of somewhat thickened fibrous roots. Stems 1 or more, 1–2.5 dm tall, commonly villous when young, often becoming glabrescent in age; rather stout for their height.

Basal leaves with short to long petioles and generally rhombic blades varying to suborbicular or obovate, cuneate and entire at base, crenate above. Stem leaves flabelliform and deeply divided into 3–5 divisions below, or simple, lanceolate, reduced and bract-like among the flowers.

Sepals ± reflexed, long-villous, 3.5–8 mm long. Petals deep yellow, 5–9 mm long, obovate. Nectary deeply pocket-like, the covering scale adherent full length along its sides to the petal. Stamens about 50. Receptacle obovoid, about 3 mm long in fruit, hairy. Carpels 40–50.

Achenes in a globose head; obovoid, plump, smooth and glabrous, 1.5–2 mm long, with a very short slender straight stylar beak. 2n = 16.

A species of the Great Plains and central Canada, from the eastern foot of the Rocky Mountains to Ontario, and southward to Nebraska. In British Columbia it has been found growing in grassy prairies in the Peace River basin. The most recent collection known to the author was made in 1938; suggesting that this species may now be extinct in this province.

Ranunculus eschscholtzii Schlechtendal   **Mountain or Snowpatch Buttercup**
(as *Eschscholzii* originally)
R. nivalis L. var. *eschscholtzii* (Schlechtendal) S. Watson

Herbaceous perennial from a compact erect caudex 1–3 cm high, covered with persistent withered leaf bases, and with a tuft of slender fibrous roots. Stem 5–25 cm tall, usually glabrous, scapose or with one or two stem leaves.

Basal leaves long-petioled; the blade usually 3-parted nearly or quite to the base; the central division 3-lobed to entire, the lateral divisions 3-lobed or 4-lobed; the margin ciliate but the surface normally glabrous. Stem leaves, when present, short-petioled. Bracts 3-parted, with linear or lanceolate divisions.

Flower commonly solitary. Sepals 5, thinly villous with yellow, wavy, shiny hairs, rarely glabrous, 3–8 mm long, deciduous early or persistent. Petals 5, or sometimes absent, yellow, 1½–2 times as long as the sepals, broadly obovate, the nectary scale pouch-like, its upper margin concave in outline. Stamens 20–100 or more. Carpels 20–80 in an ovoid head, generally glabrous, with erect styles 0.8–1.5 mm long. Receptacle more or less cylindrical, glabrous or with a terminal tuft of hairs, rarely long-pilose.

Achenes forming a cylindrical head, usually glabrous, rarely thinly long-pilose; the body obovoid, sometimes slightly keeled, 1–2 mm long, with a terminal or subterminal, slender erect or curved stylar beak 0.8–1.5 mm long. 2n = 16, 32, 48.

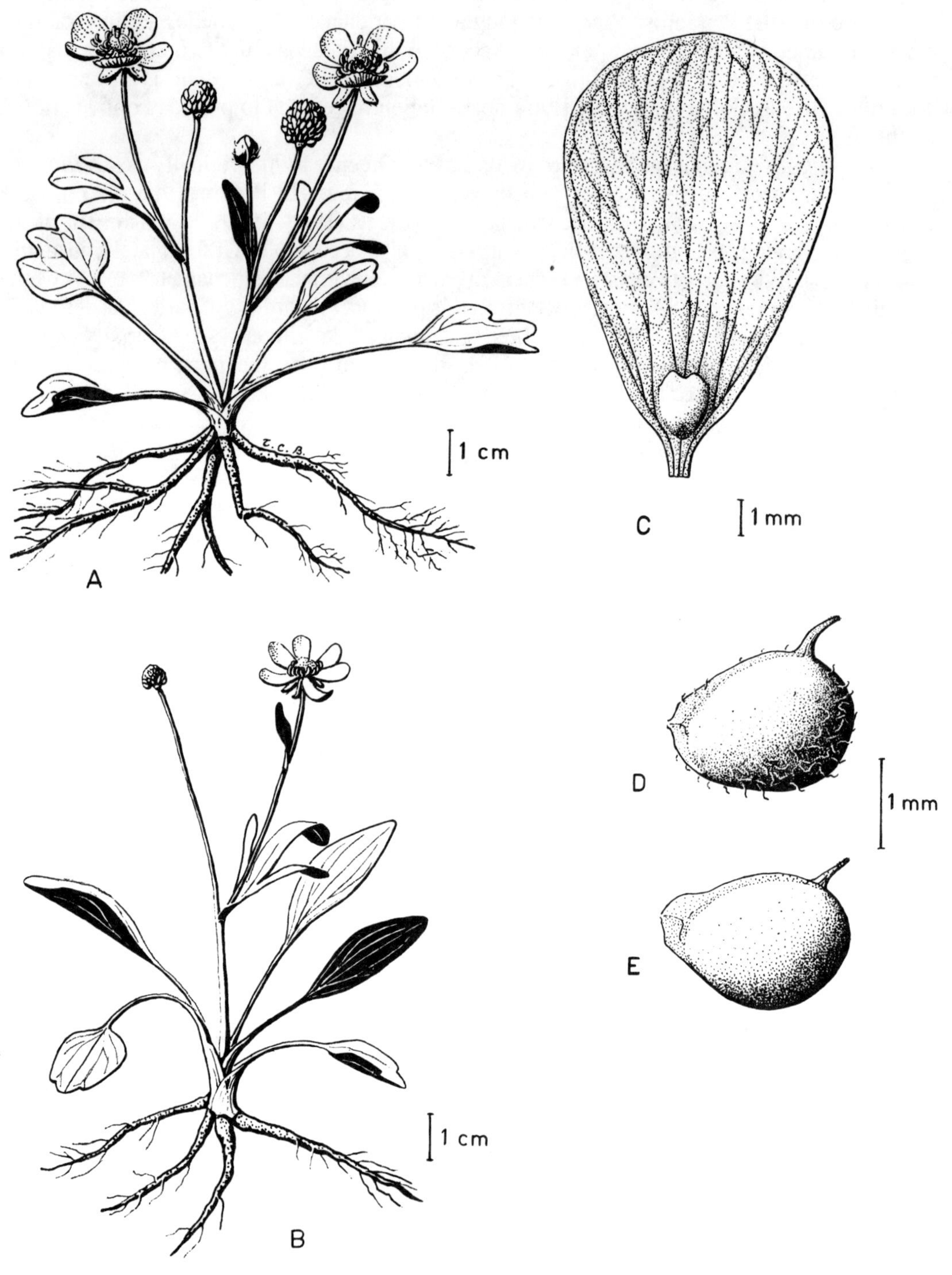

Figure 66. *Ranunculus glaberrimus*

A. plant of var. *glaberrimus*.
B. plant of var. *ellipticus*.
C. petal.

D. achene of var. *buddii*.
E. achene of var. *glaberrimus*.

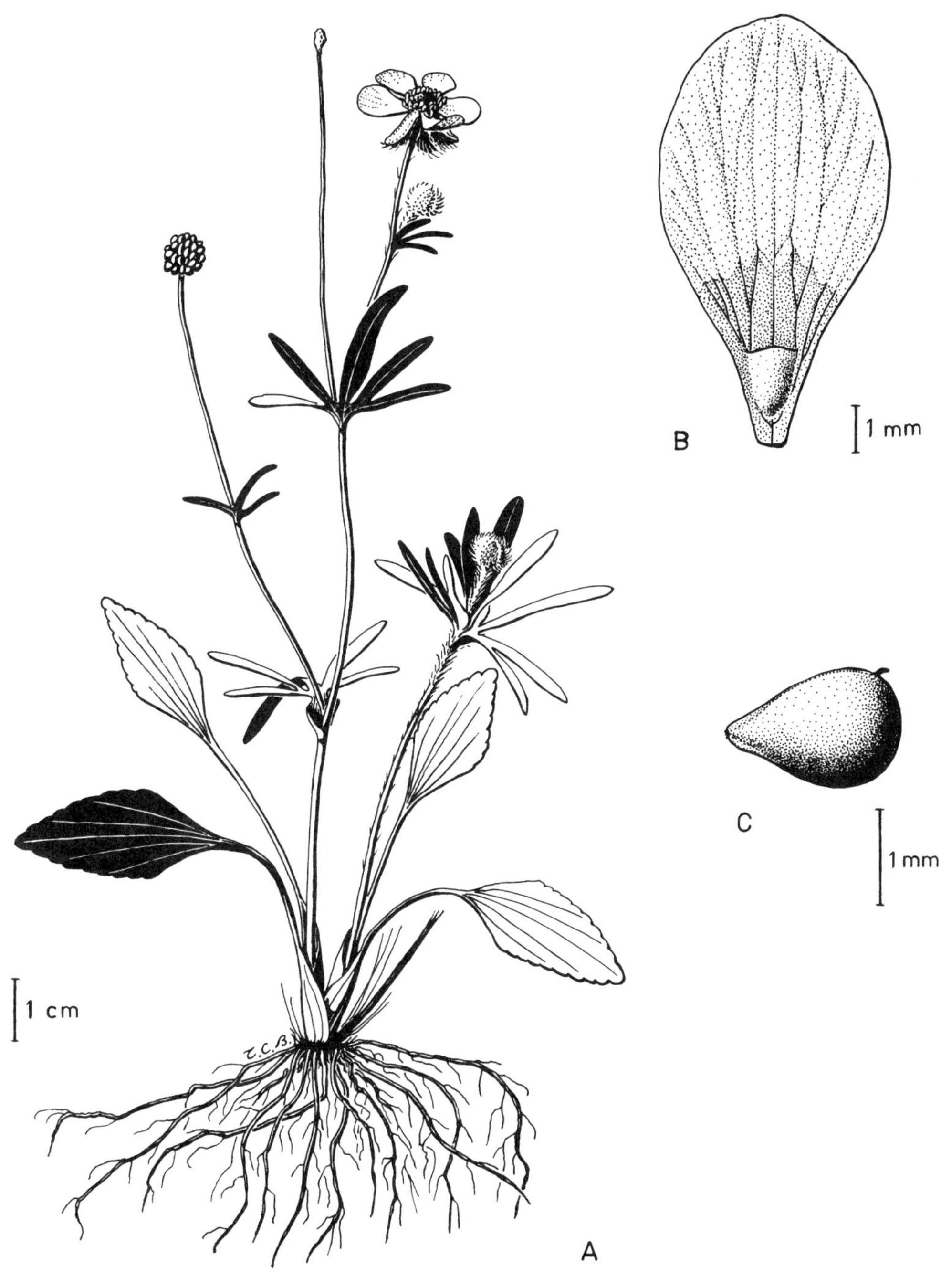

Figure 67. *Ranunculus rhomboideus*

A. plant. B. petal. C. achene.

This species differs from the related *R. nivalis* L. in its 3-parted basal leaf with a ciliate margin, and paler, yellow, wavy calyx hairs.

Ranunculus eschscholtzii is a plant mainly of the western Cordilleran region of North America. It is found in subalpine to alpine environments, commonly on the sites of late-melting snow patches. In the southern part of its range it is often found in shady situations, but northward it occurs in fully exposed but commonly moist sites in the tundra.

Ranunculus eschscholtzii is a heterogeneous complex of several closely related entities that differ in a number of characters, including chromosome numbers. The several component races, while genetically isolated from each other by their chromosomal differences, are often so weakly distinguished morphologically that only a microscope can reveal clear distinctions. Different authors treat these races variously as distinct species or as varieties of one inclusive species, according to taste; and with some justification for either interpretation. The following treatment is a compromise between extremes.

KEY TO SUBSPECIES

1. Lobes and sinuses of basal leaf blade rounded to obtuse. Middle division of basal leaf entire to shallowly lobed. Pollen grains more than 3 micrometres in diameter. 2n = 48 on mainland, 32 on Queen Charlotte Islands.. **ssp. *eschscholtzii***

    Lobes and sinuses of basal leaf distinctly acute. Middle division mostly deeply 3-lobed. Pollen grains smaller: 2.5–3 micrometres in diameter. 2n = 16 .... **ssp. *suksdorfii***

Subspecies *eschscholtzii* occurs over most of Alaska, including the Aleutian Islands, and throughout most of the Cordilleran region of western North America — most of the range of the whole complex, except in the extreme south, where it is replaced in part by other subspecies.

With respect to the basic chromosome number of this complex (x = 8), ssp. *eschscholtzii* is mostly hexaploid (2n = 48) (Fisher, 1973). Material from the Queen Charlotte Islands (Taylor and Mulligan, 1968) and one specimen from the Northwest Territories have been found to be tetraploid (2n = 32), but show no external morphological distinctions.

Ranunculus eschscholtzii* Schlechtendal subspecies *suksdorfii* (Gray) Brayshaw, *stat. nov.
>   **R. *suksdorfii* Gray, Proc. Am. Acad. 21:371. 1886.**
>   **R. *eschscholtzii* Schlechtendal var. *suksdorfii* (Gray) L. Benson, Am. Jour. Bot. 23:170. 1936.**

This subspecies occurs in a belt across northern Washington State, from the Olympic Mountains, into Idaho and Montana, and southward in the Cascade and Rocky Mountains [Fisher *et al.*, 1973; (as *R. suksdorfii*)], and has been found in the Flathead Valley, in extreme southeastern British Columbia.

Where ssp. *suksdorfii* is sympatric with the southern populations of ssp. *eschscholtzii*, it shows a preference for sunny habitats, while ssp. *eschscholtzii* tends to occupy more shady habitats. Where the two plants occur together, hybrids may be found. However, such hybrids have been found to be sterile (Fisher, 1973). There are thus good grounds for regarding *suksdorfii* as a species distinct from *R. eschscholtzii*. It is retained here at the highest subspecific level within *R. eschscholtzii* because of the rather vaguely defined (or vaguely described) morphological differences between them; making unambiguous diagnoses of plants, as found, difficult.

***Ranunculus eximius* Greene**
>   **R. *eschscholtzii* Schlechtendal var. *eximius* (Greene) L. Benson**
>   **R. *saxicola* Rydberg**

Perennial, similar in habit to *R. eschscholtzii*, with stems, in our material, averaging rather taller (12–25 cm), and more slender.

Basal leaf blades orbicular to broadly obovate, cuneate at base; apically cut about half way to base into acute lobes and sinuses. Stem leaves, if present, similar but shorter-petioled. Bracts sessile, divided into 3–5 linear to lanceolate, acuminate divisions.

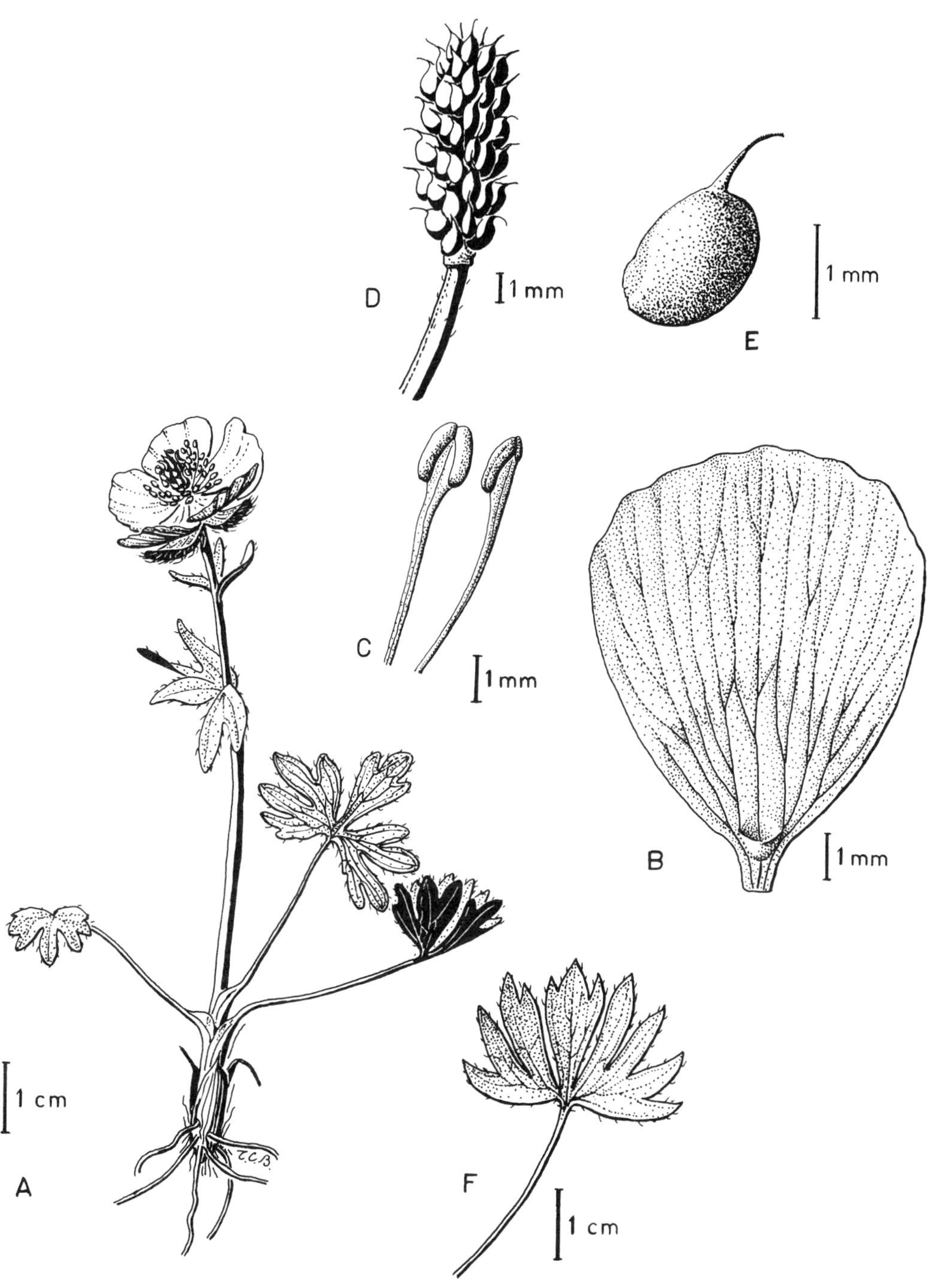

Figure 68. *Ranunculus eschscholtzii*

A. flowering plant.
B. petal.
C. stamens.
D. head of achenes.
E. achene.
F. basal leaf blade of ssp. *suksdorfii*.

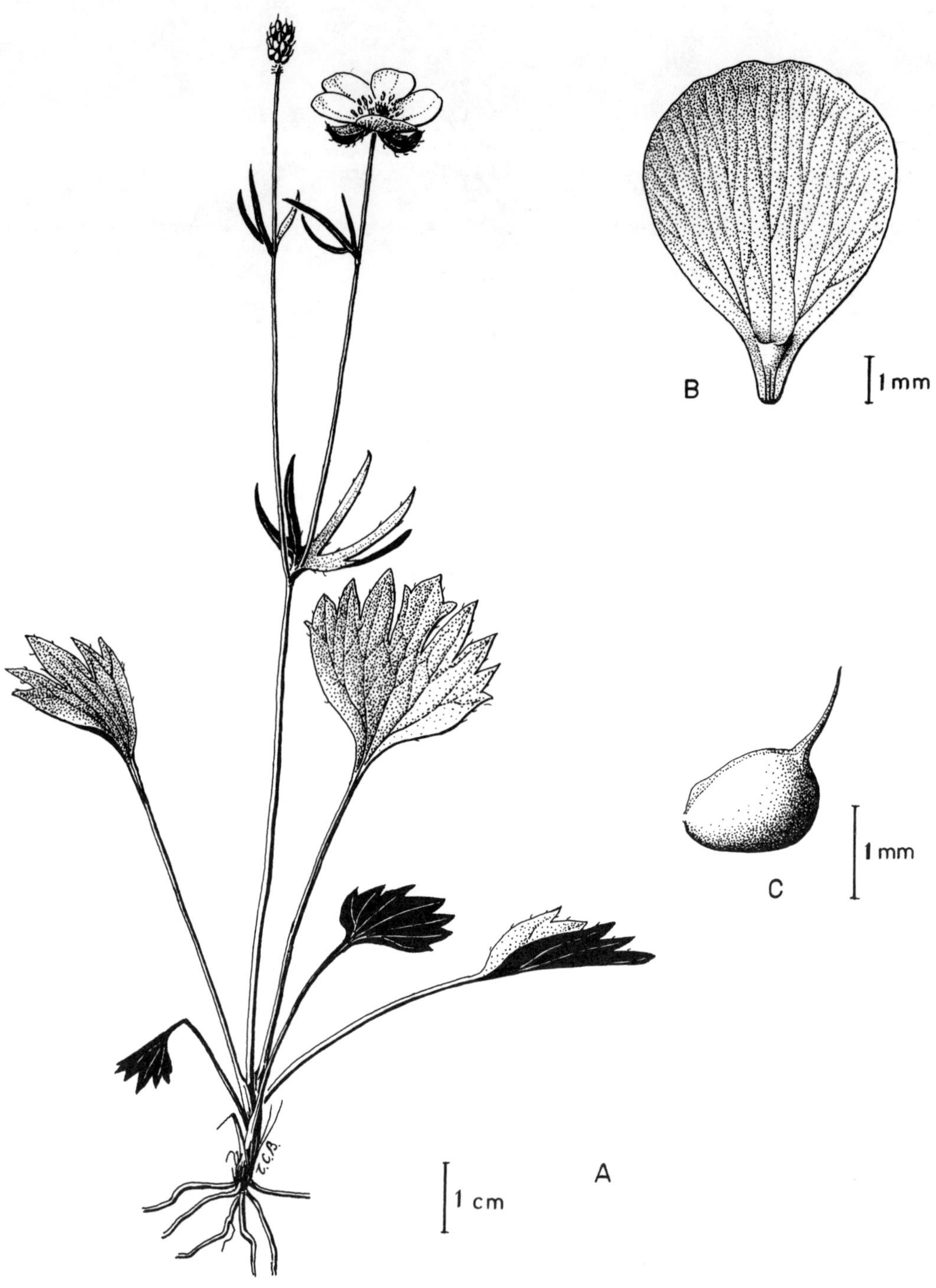

Figure 69. *Ranunculus eximius*

A. plant. B. petal. C. achene.

Flowers as in *R. eschscholtzii* but averaging rather larger; the hairs on the sepals rather sparse, and a paler yellow than in *R. eschscholtzii*. Receptacle glabrous to thinly puberulent. Carpels glabrous or puberulent; in the latter case the hairs sometimes shed as the achenes mature.

Achenes plump, glabrous or occasionally thinly puberulent; the body 1–2 mm long; with a very slender, more or less straight beak of about the same length. $2n = 32$.

This species, closely related to *R. eschscholtzii*, ranges from southwestern Alberta, through the Rocky Mountains of Montana, to Wyoming. It enters British Columbia locally in the Flathead Valley, in the extreme southeastern corner of the province. It is found on alpine rock slides and ledges, and in meadows, normally where drainage is efficient.

R. gelidus Karelin & Kirilov
non R. gelidus Hoffmansegg *ex* Reichenbach *nomen nudum*.
R. grayi Britton

Small perennial herb, 4–10 cm tall in flower, up to 15 cm tall in fruit; from a small, deeply buried caudex with fibrous roots. Stem glabrous below, puberulent above, often flexed at ground level.

Leaves mainly basal and long-petioled; the blades 1–2 cm long, and up to twice as wide, reniform to orbicular in outline, deeply 3-parted into divisions that are in turn 3- to 5-lobed, glabrous or sparingly ciliate, rarely with a few hairs beneath.

Pedicels pubescent with pale hairs. Flowers 1–3, sepals 5, broadly elliptic to obovate, 4–5 mm long, purplish tinged above, pubescent with whitish hairs, and deciduous early or persistent to the end of the flowering period. Petals obovate, rather larger than the sepals, 4½–5½ mm long, yellow; the nectary small and pouch-like, up to 0.5 mm deep. Stamens 20–50. Carpels 20–80, generally glabrous; the style short, stout, and strongly recurved. Receptacle ovoid to columnar, glabrous or sometimes sparsely white-hairy.

Achenes in an ovoid head; the achene body 2–2.6 mm long, plump, its upper (ventral) margin convex, or sometimes concave near the base; with a short (about 0.5 mm long) recurved, stout beak. $2n = 16$.

This species is found in widely scattered areas in Siberia, and in northwestern North America, where it ranges from Alaska southward along the Rocky Mountains and other mountain ranges to Colorado. In British Columbia it has been found on the Spatsizi Plateau, and is reported from the Marble Mountains by Benson (1948). The plant inhabits alpine and arctic tundra, and is often associated with late-melting snow-patches and unstable soil.

Ranunculus verecundus Robinson *ex* Piper

Perennial with one or more slender, erect or spreading stems from a cluster of fibrous roots; not rooting at the nodes; up to 20 cm tall, glabrous or finely puberulent above.

Leaves mostly basal; the basal leaves on long petioles, reniform to nearly circular, but wider than long, shallowly to deeply lobed or divided palmately into 3–5 divisions, the lobes usually rounded at the tips but occasionally acute. Lower stem leaves with similar blades on short petioles. Upper stem leaves (bracts) sessile and reduced to 1–3 narrowly lanceolate segments.

Flowers on pedicels 2–4 cm long, that elongate after anthesis to 6–10 cm long at maturity of the fruit. Sepals 5, ovate, puberulent, purplish tinged, deciduous with the petals, 3–5 mm long. Petals 5, yellow, obovate, 4–6 mm long; the nectary shallow and pocket-like, with a crescentic scale. Stamens 20–30. Receptacle glabrous, becoming columnar and up to a centimetre long in fruit.

Achenes 30–70 or more in a spike; obovoid, the body 1–2 mm long, its upper edge concave near the base; with a short (up to ½ mm long) straight or curved beak.

Ranging from southern British Columbia and Alberta to Oregon and Idaho, this species is found in rocky alpine slopes and tundra.

Ranunculus pygmaeus Wahlenberg   Dwarf Buttercup

Small to minute perennial herb, 1.5–10 cm tall in flower, up to 20 cm tall in fruit. Stems glabrescent below, finely puberulent above.

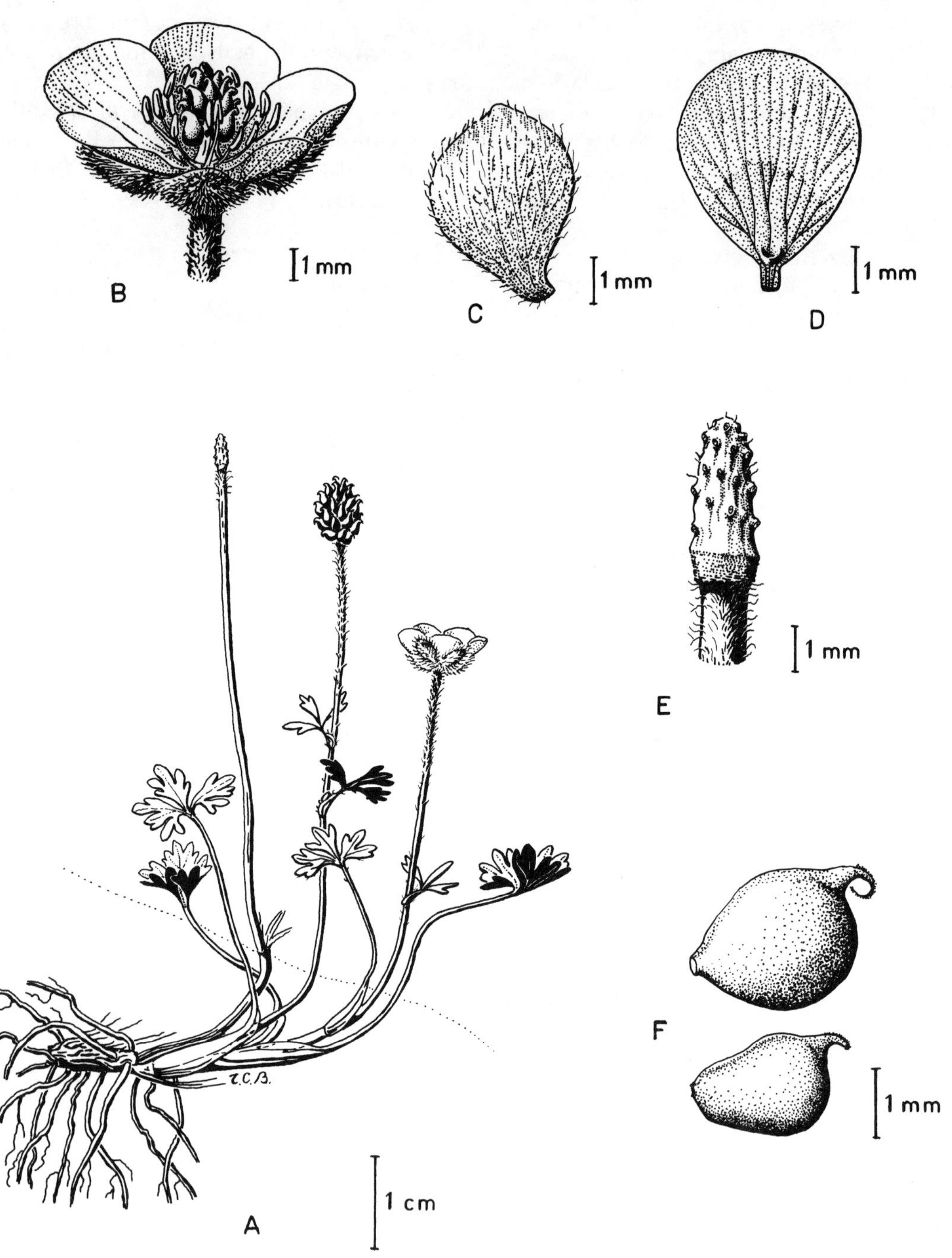

Figure 70. *Ranunculus gelidus*

A. plant.
B. flower.
C. sepal.
D. petal.
E. receptacle.
F. achenes.

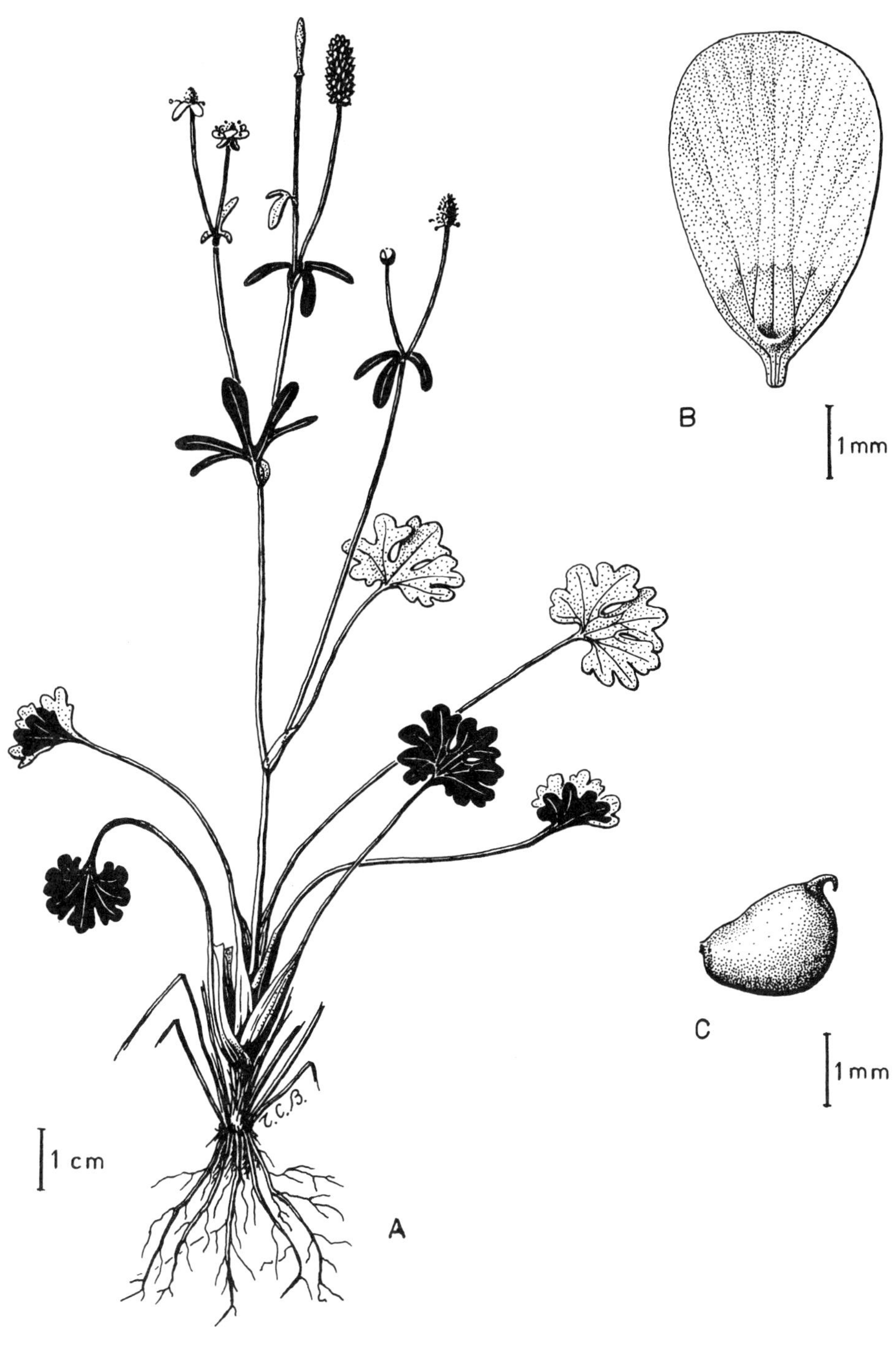

Figure 71. *Ranunculus verecundus*

A. plant.   B. petal.   C. achene.

129

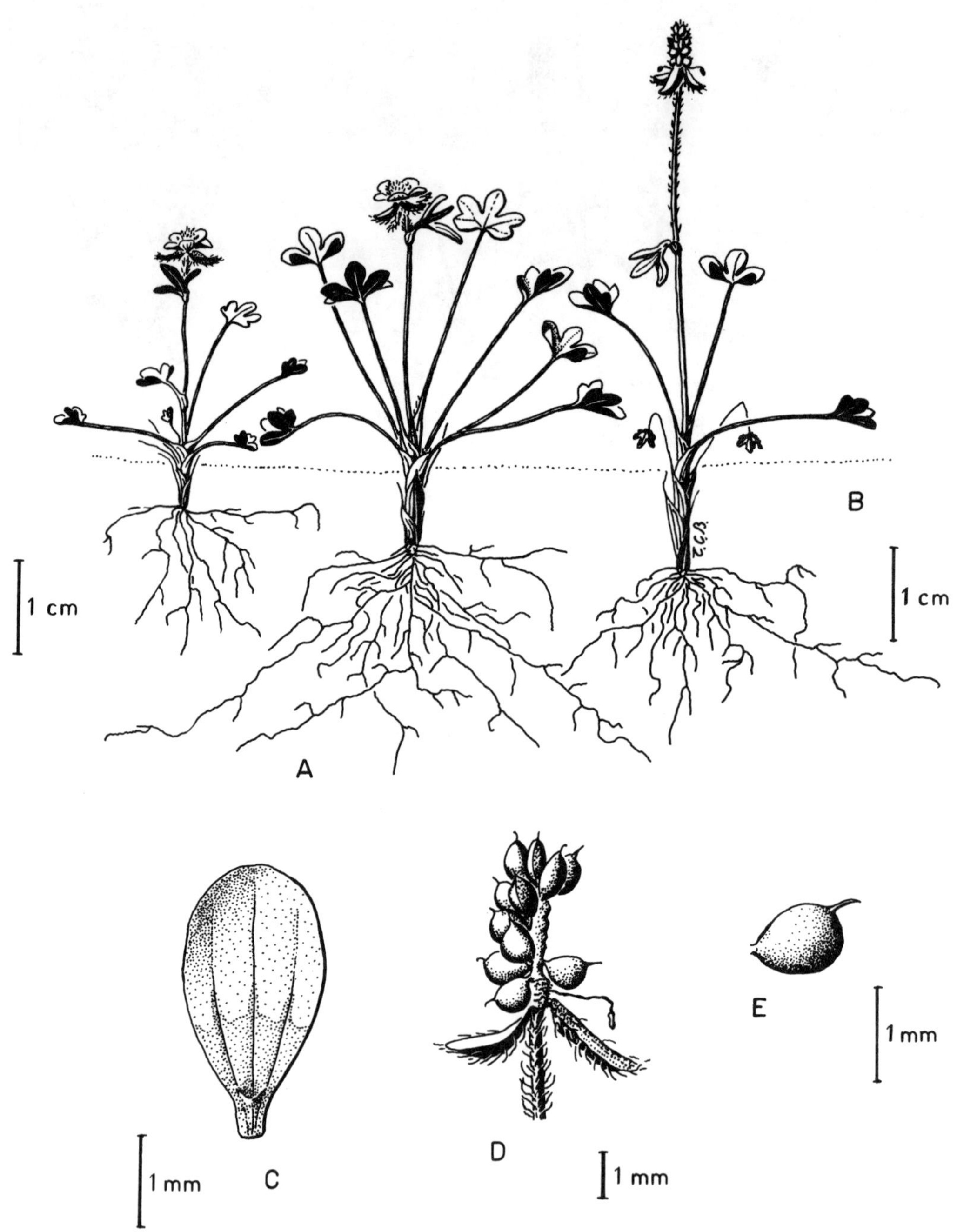

Figure 72. *Ranunculus pygmaeus*

A. 2 flowering plants.   D. head of achenes.
B. fruiting plant.   E. achene.
C. petal.

130

Leaves mainly basal, or with 1 or 2 subsessile stem leaves. Petiole of basal leaf 1–4 cm long; blade 3–10 mm long (or longer in var. *langeanus*) and wider than long, cut at least half way to its base into 3–5 rounded main lobes, commonly ciliate. Bract normally one, divided into 3 lanceolate divisions.

Pedicels usually yellowish-hairy, but not uncommonly glabrous. Flower solitary, small. Sepals 5, ovate, often purplish-tinged, generally yellowish-hairy, 2–4 mm long, deciduous early or persistent through the flowering stage. Petals 5, obovate, yellow, commonly shorter than to about equalling the sepals, rarely slightly longer; the nectary scale pouch-like, 0.3 mm or less deep. Stamens 10–20. Carpels 25–50 or more, in an ovoid head, glabrous, the style short, erect at first, becoming straight or hooked, 0.3–0.5 mm long.

Achenes in an ovoid or cylindrical head on a more or less cylindrical, glabrous receptacle; the achene body obovoid, plump, 0.8–1 mm long; the beak slender, 0.4–0.8 mm long, its attitude variable. 2n = 16.

KEY TO VARIETIES

1. Leaf lobes entire or the lateral lobes shallowly bilobed. Head of achenes 2.5–4 mm long. Achene beak curved or hooked, 0.4–0.5 mm long.. **var. *pygmaeus***
    Leaf blade deeply parted or divided to base into 3 main, sometimes petiolulate divisions, of which the central one is 3-lobed and the laterals 2- to 4-lobed. Achene head 5–7.5 mm long by 3–4.5 mm thick. Achene beak 0.4–0.8 mm long, straight or nearly so **var. *langeanus***

Ranunculus pygmaeus has a circumpolar distribution; and in North America extends southward along the Rocky Mountains to Colorado. It is found in arctic and alpine tundra, commonly around late-melting snow-patches. It is rarely subalpine.

Variety *langeanus* Nathorst is a generally larger, coarser, more leafy plant than the typical variety. It has been found in widely scattered areas; from Greenland and Labrador to the Gaspe Peninsula of Quebec, and in Glacier National Park, Montana. In British Columbia, the collections by J.A. Calder and R.L. Taylor (#23701) in DAO from Mosquito Mtn., Moresby Island, Queen Charlotte Islands, contain some specimens that resemble this variety, in addition to some more typical representatives of *R. pygmaeus*.

Ranunculus nivalis L.   Snow Buttercup

Modest-sized erect perennial herb, 10–25 cm tall, with stout but rather soft and fleshy stems arising from short rhizomes bearing clusters of fibrous roots; the stem bases wrapped in brown old leaf sheaths. Upper stem glabrous or sparingly pubescent with brownish hairs.

Leaves both basal and cauline, ciliate, but otherwise glabrous, or sometimes thinly puberulent beneath. Basal leaves long-petioled; the blades deeply 3- to 7-lobed; the lobes rounded at tips. Stem leaves 1–3, short-petioled to sessile, deeply divided into 3–5 lanceolate but usually round-tipped lobes.

Flowers 1–few, terminal. Sepals 5, 5–9 mm long, spreading, conspicuously woolly with dark brown hairs; the sepals deciduous early or persistent to the end of the flowering period. Petals 5, 6–12 mm long, distinctly longer than the sepals, yellow; the nectary small, pouch-like. Stamens 70 or more. Carpels many in an ovoid head, glabrous, with rather stout curved styles. Receptacle ovoid to columnar, glabrous.

Achenes in a more or less cylindrical head; obovoid; the body obscurely keeled, plump, the profile slightly indented ventrally, glabrous, 1.5–2 mm long; with a rather stout, subterminal to terminal beak 1–1.5 mm long. 2n = 48, 96.

Ranunculus nivalis is a species of circumpolar and high arctic distribution. It is found in arctic and alpine tundra; typically where cold and moist, as along streams or around patches of late-melting snow. In British Columbia it is found in the Rocky Mountains and other ranges from around 55° North latitude northward.

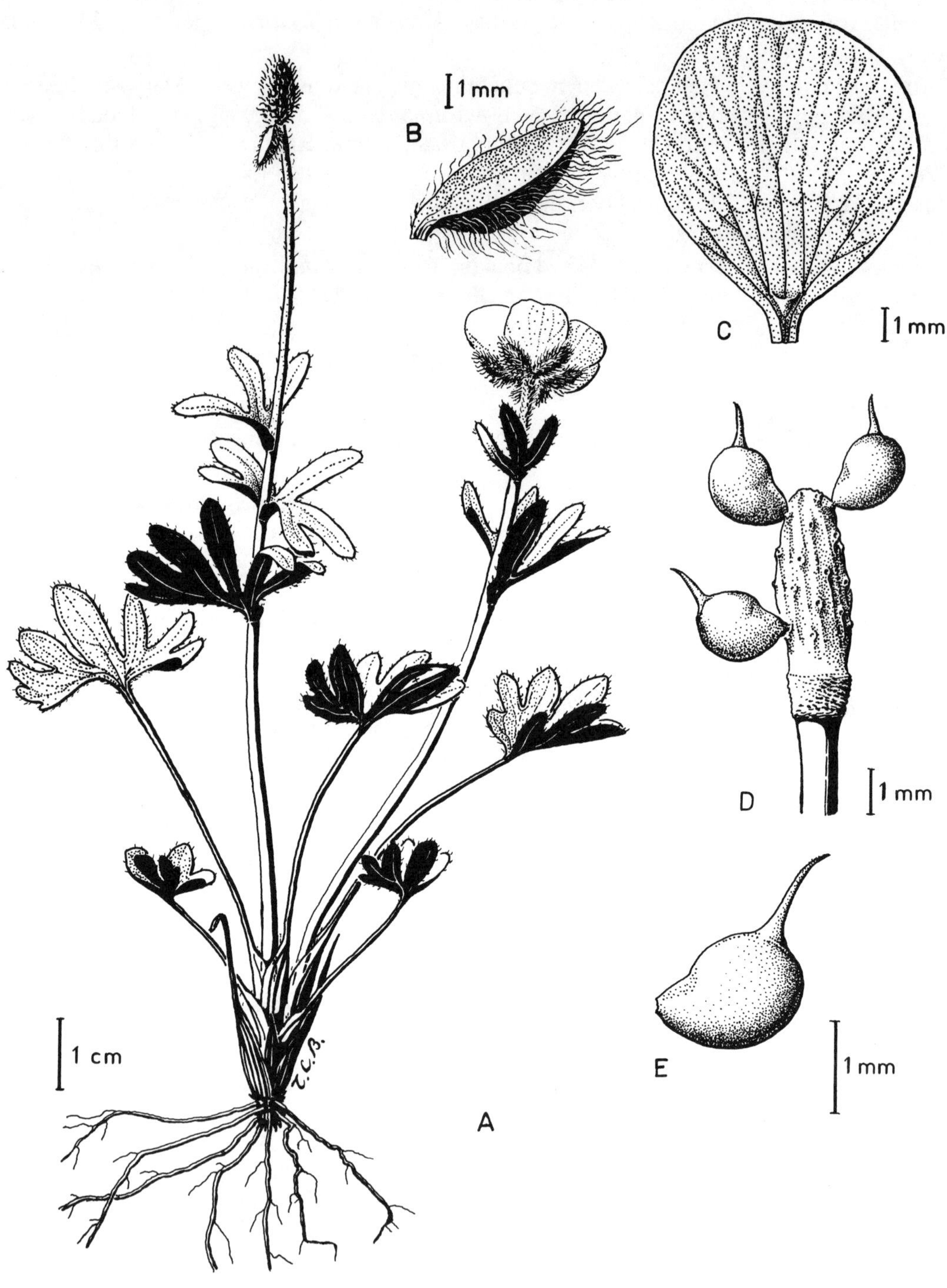

Figure 73. *Ranunculus nivalis*

A. plant.
B. sepal.
C. petal.
D. receptacle with achenes.
E. achene.

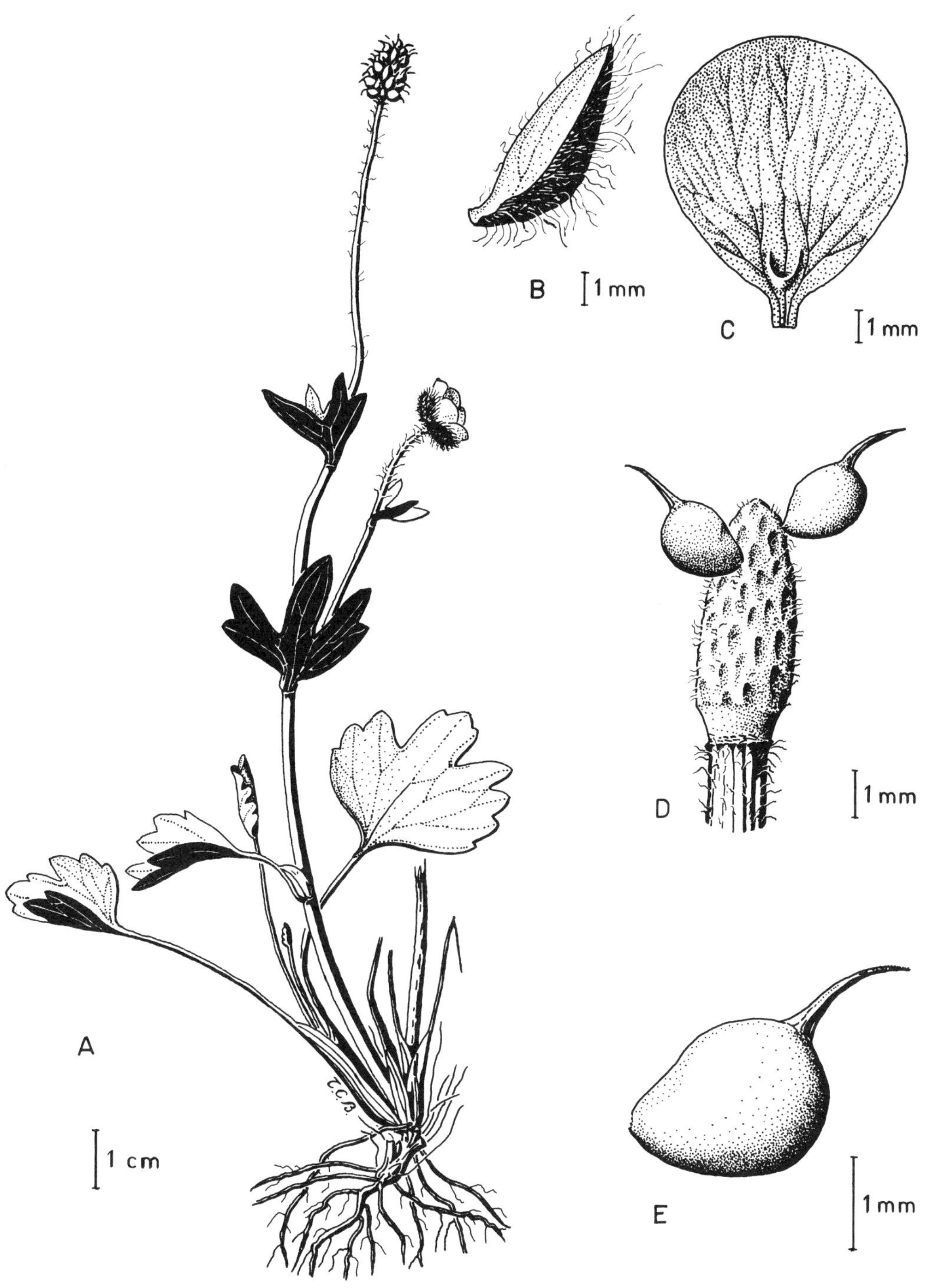

Figure 74. *Ranunculus sulphureus*

A. plant.
B. sepal.
C. petal.
D. receptacle and achenes.
E. achene.

Ranunculus sulphureus **Solander**   **Sulphur Buttercup**

Modest-sized erect, or sometimes sprawling, perennial herb 15–30 cm tall, similar in habit to *R. nivalis*. Stems stout, glabrous or sparsely pubescent with brownish hairs.

Basal leaves long-petioled; the blades more or less orbicular, shallowly lobed with rounded lobes, or crenate; glabrous, even on the margins. Stem leaves 2 or 3, sessile or subsessile, deeply cleft into 3–5 lanceolate or narrowly ovate, round-tipped lobes.

Flower usually single on a stem; the pedicel commonly pubescent with brown hairs. Sepals 5, 6–9 mm long, spreading, moderately to densely pubescent with dark brown hairs; deciduous early or persistent to the end of the flowering period. Petals 5, 7–10 mm long; a little longer than the sepals, yellow; the nectaries small and pouch-like. Stamens 30–50. Carpels many, in a globose to ovoid head, glabrous, with short straight styles. Receptacle puberulent with clusters of short brown hairs.

Achenes in an ovoid to subglobose head; the body obovoid, 1½–2½ mm long, not, or only obscurely, keeled; the beak rather stout, straight, diverging or ascending, 1–2 mm long. Fruiting receptacle ovoid to cylindrical with tufts of short brown hairs among the achene bases. 2n = 96.

Ranunculus sulphureus is circumpolar and high arctic in distribution. It is found in wet, seeping mossy sites in arctic or alpine tundra. In British Columbia it is very rare and local; and has been found only on the Liard Plateau and in the Spatsizi Wilderness area, so far.

Section *Flammula*

Leaves linear to lanceolate, entire or obscurely glandular-serrulate. Flowers yellow. Achenes plump, short-beaked. x = 8. Amphibious.

Ranunculus flammula **L.**   **Spearwort**

Amphibious and highly variable, glabrous to appressed-hirsute perennial, with slender fibrous roots; prostrate to decumbent or erect stems rooting at the nodes that make contact with the ground; often perennating from rooted nodes.

Leaves filiform to lanceolate, elliptic or ovate, with margins usually entire, sometimes shallowly toothed. The basal leaves and those in axillary tufts and on the bases of axillary branches long-petioled. Stem leaves shorter-petioled to commonly subsessile; the branches thus repeating the leaf form pattern of the main stem.

Flowers solitary and terminal, or few in cymose inflorescences, on terminal or axillary peduncles and pedicels that are erect regardless of the attitude of the vegetative part of the stem. Sepals 5, glabrous to strigose, 2–5 mm long. Petals 5 or sometimes more, yellow, obovate, 3–11 mm long, the nectary covered by a short, pocket-forming scale that is broader than high. Stamens 10–50, maturing before the stigmas. Receptacle more or less ovoid, glabrous. Carpels 5–50.

Achenes obovoid, glabrous, smooth to reticulate-surfaced, 1.5–3 mm long, bearing stylar beaks of varying size up to 0.7 mm long. 2n = 32.

KEY TO SUBSPECIES AND VARIETIES

1. Achene beak minute: 0.1–0.2 mm long. Fruiting receptacle ca. 3 mm long. Plant typically decumbent to erect, with lanceolate leaves.. **ssp.** *flammula*
    Achene beak more prominent: 0.2–0.7 mm long. Fruiting receptacle up to 2.5 mm long. Plants commonly prostrate to decumbent.. **ssp.** *reptans* **2**

2. Stems stoloniform, prostrate, filiform. Aquatic leaves filiform, bladeless. Aerial leaves with lanceolate blades up to 2 mm wide. Fruiting receptacle 1 mm long............... **var.** *reptans*
    Stems rather coarser, sometimes ascending. Aerial, and often aquatic, leaves with lanceolate to ovate or elliptic blades 2–15 mm wide. Fruiting receptacle 1–2 mm long **var.** *ovalis*

Subspecies *flammula*

Stems decumbent to erect, or sometimes arching and rooting at the nodes. Leaves with lanceolate to oval blades sometimes obscurely toothed. Flowers often several in a cymose inflorescence. Petals 4–8 mm long and wide. Stamens 25–50. Fruiting receptacle ca. 3 mm long. Achene with minute stylar beak 0.1–0.2 mm long and at least as wide, the stigma commonly strongly recurved.

The common subspecies in Europe, and known from the Atlantic Provinces of Canada; possibly introduced westward. Reported from Queen Charlotte Islands (Calder & Taylor, 1968) but the identity of the material is questionable.

Subspecies *reptans* (L.) Syme *in* Sowerby   **Creeping Spearwort**
    R. reptans L.
    R. flammula var. filiformis (Michx.) Hook. *sensu* Hitchcock and Cronquist (1964) *non*
        Michaux (1803)

Stems commonly prostrate and rooting at nodes, sometimes ascending to erect. Leaves variable in shape, almost always entire. Flowers solitary or cymose. Petals 2–7 mm long. Stamens 10–30. Fruiting receptacle 2 mm long or less. Achenes 5–25, 2–3 mm long with relatively conspicuous beaks 0.2–0.7 mm long, tipped by a relatively inconspicuous stigma.

Circumboreal: the common subspecies in North America, occurring throughout British Columbia. In shallow water of lake margins or ditches, or on moist ground in marshes or meadows, or becoming terrestrial through late season recession of the water.

Var. *reptans*
    R. filiformis Michaux
    R. flammula var. filiformis (Michaux) Hooker *sensu* Michaux *non* Hitchcock and
        Cronquist

Stems filiform, stoloniform, arching and rooting at all nodes. Leaves filiform and bladeless or with linear to narrowly lanceolate blades 1–2 mm wide. Flowers commonly solitary on erect pedicels terminating the stolons. Petals 2–7 mm long. Fruiting receptacle less than 1 mm long.

Var. *ovalis* (Bigelow) L. Benson
    R. filiformis Michaux var. ovalis Bigelow
    R. flammula L. var. intermedius Hooker (as *intermedia*)
    R. unalaschensis Besser
    R. flammula L. var. unalaschensis (Besser) Ledebour

Rather coarser plants than var. *reptans*. Stems sometimes ascending, or very occasionally erect. Leaves with lanceolate to ovate or elliptic blades up to 3 cm long by 2–15 mm wide. Flowers often more than one, and cymose on an erect bracted peduncle. Petals 2–6.5 mm long. Fruiting receptacle 1–2 mm long.

Like many amphibious plants, this complex is made up of very heterogeneous populations. Not only are these populations genetically diverse, but they include both elements that are plastic in their ability to adapt to varying environmental factors, such as the season and duration of flooding, and the overall instability of the habitat, and more rigid elements that are adapted to more stable habitats with narrower ranges of variation. The environment selects the best adapted genotypes from the overall heterogeneous population in the area: the more unstable, early successional situations favouring the more plastic components of the population (Cook and Johnson, 1968). Plastic individuals respond to varying water levels by producing filiform stems and leaves in water and sturdier stems and broader, lanceolate leaves when exposed to air. These two manifestations have been named forma *submersus* Gluck and forma *terrestris* Gluck, respectively. While the genetic code that is embodied in the chromosomes determines the range of leaf form available to the plant, the existing state of the habitat determines the actual leaf form assumed by the plant within that range.

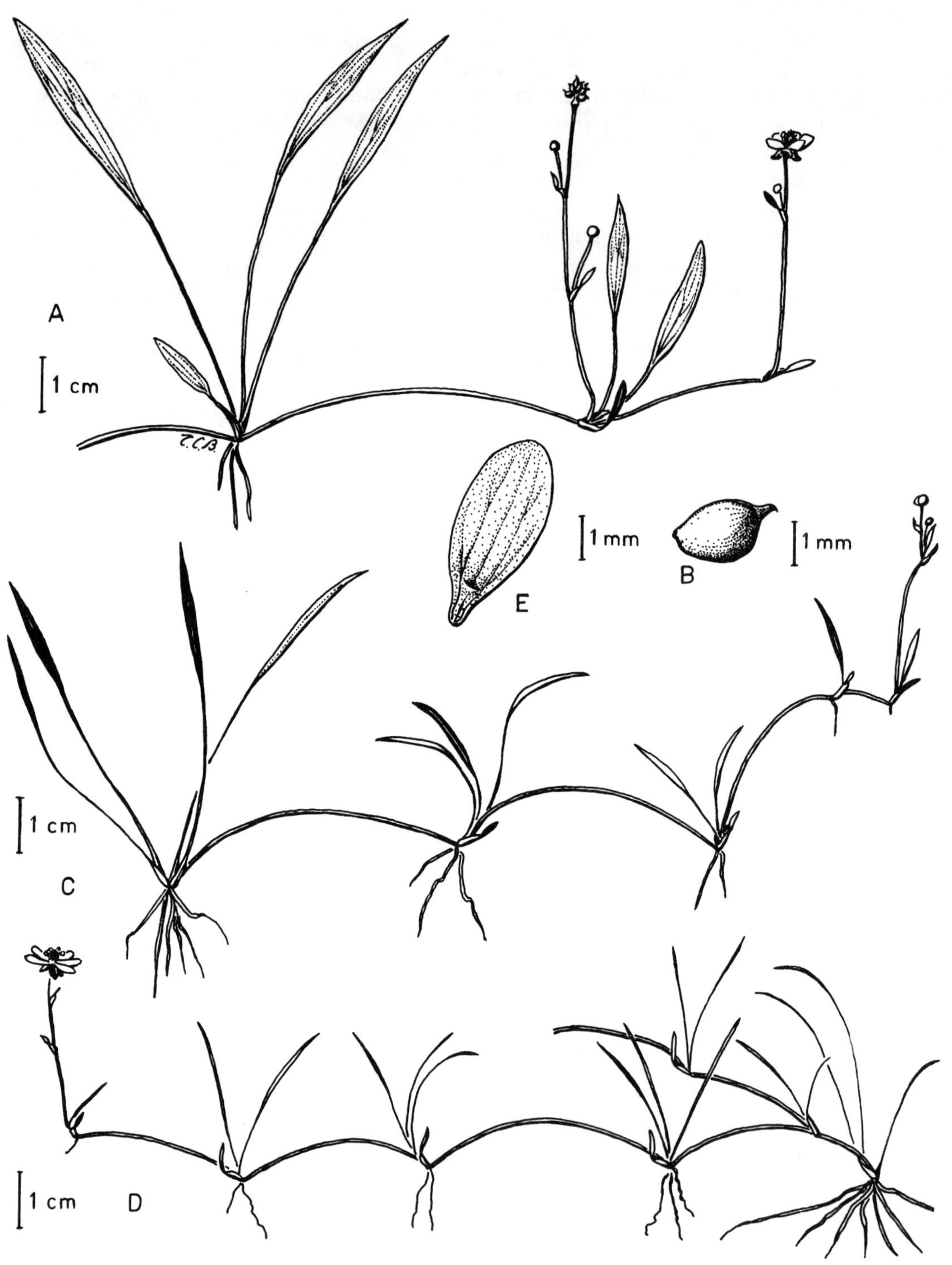

Figure 75. *Ranunculus flammula* ssp. *reptans*

A. var. *ovalis*: part of plant.
B. var. *ovalis*: achene.
C. var. *reptans* forma *terrestris*: plant.

D. var. *reptans* forma *submersus*: plant.
E. var. *reptans*: petal.

Many names have been applied to the resulting diversity of forms, at both subspecific and specific levels. The transfer of varieties from one species to another within the complex has further aggravated the confusion. The above, conservative, treatment is adapted from that of Benson (1948), which appears at present to provide the best means of keeping together, taxonomically and nomenclaturally, the members of the complex that cannot in practice be satisfactorily separated. This treatment will be most helpful in distinguishing among the more narrowly adapted, invariant elements generally associated with the more stable environments. The more plastic individuals of unstable habitats may, because of their variability, be found more difficult to place unambiguously in one variety; and transitional forms can be expected.

Ranunculus alismifolius Geyer *ex* Bentham   **Water-Plantain Buttercup**
(as *alismaefolius* originally)

Erect, glabrous or sometimes strigose perennial with thick but hollow stems branching above, and fibrous or basally swollen roots.

Leaves lanceolate to narrowly ovate, tapering, sometimes abruptly, to the stout petioles, the blades entire or shallowly serrulate, up to 15 cm long by 1–3 cm wide. Basal leaves long-petioled, the stem leaves shorter-petioled to sessile on their sheaths; reduced to bracts above, sometimes opposite in the inflorescence.

Flowers few to several, on stout pedicels elongating in fruit. Sepals 5, glabrous or hirsute, 3–5 mm long. Petals 5, sometimes more, obovate, 5–12 mm long, the nectary covered by a scale about as high as wide, adherent to the petal laterally for about half its height, and sometimes notched above. Stamens 25–90, extrorse. Receptacle globose to obovoid, glabrous.

Achenes 10–60 in a head, glabrous, plump; the body 1.5–2.5 mm long, with a beak up to 1 mm long. $2n = 16$.

A plant of open moist sites such as stream banks and meadows, sometimes of upland sites, from sea level to alpine meadows. Widespread in western North America, but barely entering Canada.

KEY TO VARIETIES

1.   Leaves on thick (2–3 mm thick) petioles, commonly serrulate. Petals 8–10 mm long ....
    .... **var. *alismifolius***

    Leaves on slender (1 mm thick) petioles, entire. Petals 6–8 mm long .... **var. *hartwegii***

Variety *alismifolius* occurs from the coast to Idaho and Montana, and from Vancouver Island southward to California, normally in moist lowland sites. In British Columbia it has been recorded only from the Victoria area, where it is now very rare.

Variety *hartwegii* (Greene) Jepson is found on the east side of the Cascade Range, from northern Washington and Idaho to California, at mid to high elevations, sometimes in well drained sites. It is not definitely known in British Columbia, but may be looked for in the southwestern interior.

Section *Ficaria*

Leaves broadly heart-shaped to reniform, crenate but unlobed. Roots of two kinds: fibrous and tuberous. Sepals 3. Petals 7–12, yellow. Achenes hairy, beakless. $x = 8$.

Ranunculus ficaria L.   **Lesser Celandine**

Glabrous perennial with club-shaped puberulent tubers among the fibrous roots. Stem soft, decumbent and rooting at the nodes, or ascending, sparingly branched.

Leaves basal and cauline with similar, reniform blades but the basal leaves long-petioled, the stem leaves short-petioled. Petioles soft, with dilated, sheathing bases. Blades broadly cordate-based, the margin entire or crenate; the upper stem leaves often subopposite.

Flowers solitary, terminal on each major branch, 2–3 cm wide. Sepals 3, spreading, green, 5–8 mm long, with minute white basal solid spurs. Petals 8, or sometimes more, up to 12, narrowly

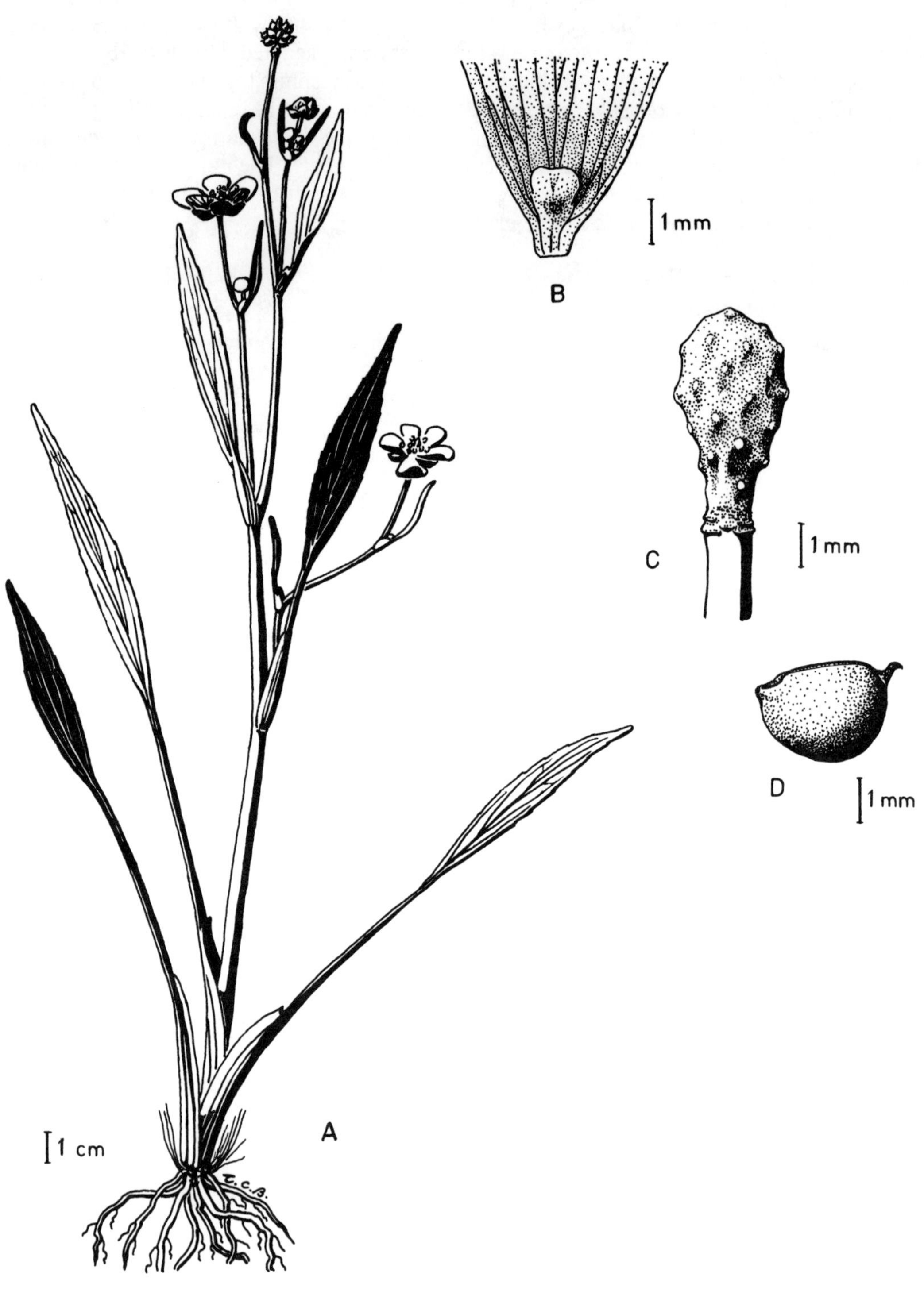

Figure 76. *Ranunculus alismifolius*

A. plant.
B. petal base and nectary scale.
C. receptacle.
D. achene.

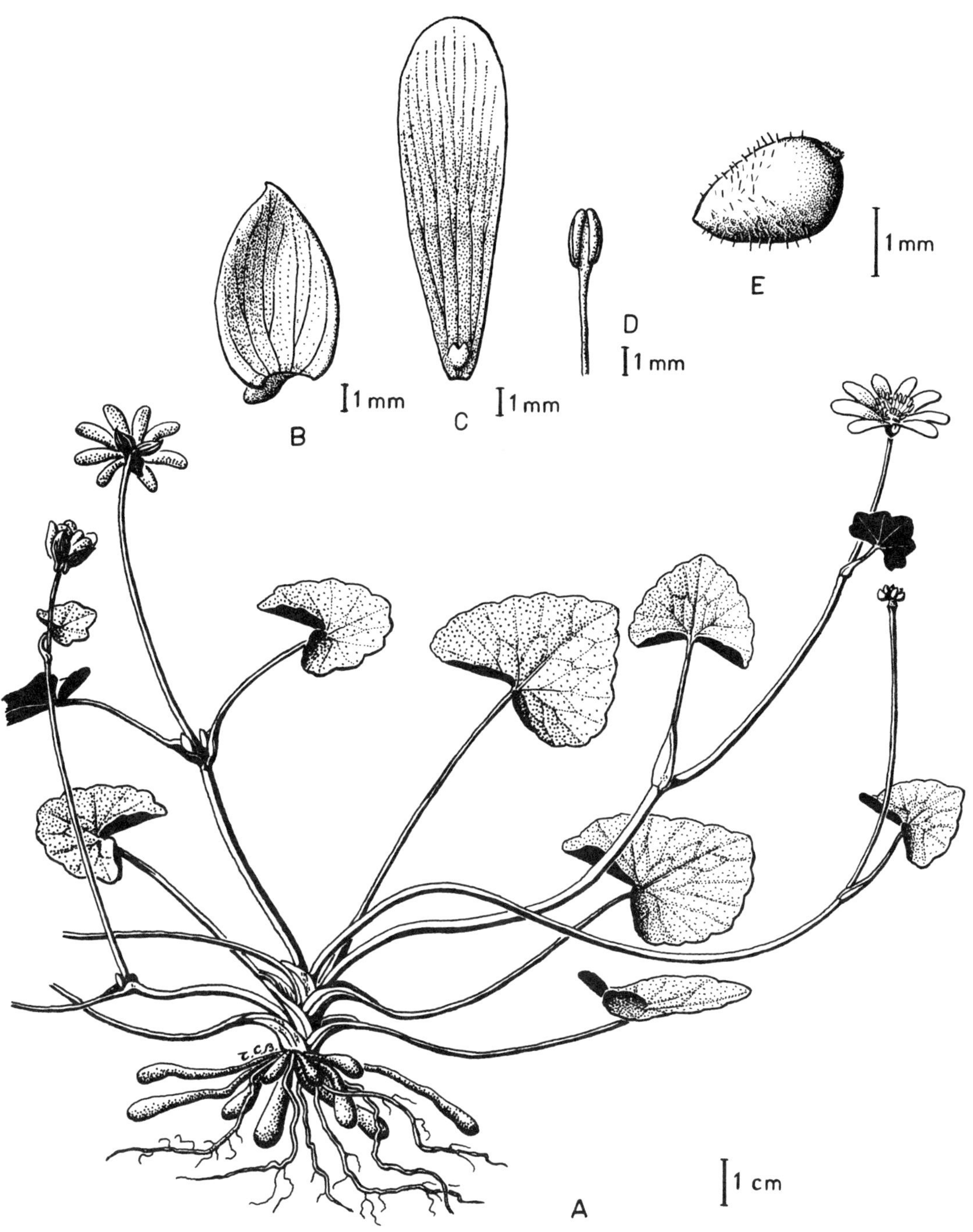

Figure 77. *Ranunculus ficaria* ssp. *bulbifer*

A. plant.   B. sepal.   C. petal.   D. stamen.   E. achene.

obovate, cuneate toward the bases, about twice as long as the sepals, recurving in age, bright yellow, fading whitish, deeper coloured on the basal part above, sometimes greenish yellow beneath, especially toward the apex, due to blue pigment in the epidermis overlying the yellow internal tissues. Nectary scale about as long as wide, 2-toothed, adnate along its sides to the petal, or sometimes absent. Stamens 15–30 in ours. Receptacle hairy. Carpels 15–30 in ours, very sparsely puberulent with erect white hairs.

Achene not keeled or beaked, the stigma sessile and flattened. Embryo with one cotyledon due to suppression of one of the pair. Flowering in April, the aerial parts of the plant dying down in May.

There are two subspecies: ssp. *ficaria* produces no axillary bulblets, but has fertile flowers that can produce many ripe achenes. The plant is a normal diploid, with 2n = 16. Subspecies *bulbifer* (Marsden-Jones) Lawalree *in* Robyns, a largely sterile tetraploid with 2n = 32, sets very few ripe achenes, but produces many axillary bulblets that propagate the plant vegetatively. The flower has slightly narrower, more nearly lanceolate petals, than has ssp. *ficaria.*

Ranunculus ficaria is native of Europe; and has been introduced, and become locally naturalized, in North America. It is found typically in ditches and other moist sites. It has been found in a few shady roadside sites in the Victoria area. One colony, in Oak Bay, is definitely ssp. *bulbifer.* Another colony, near Ten Mile Point, Saanich, needs further examination to decide its subspecific identity.

Section *Coptidium*

Rhizome filiform, white, creeping. Leaves scattered. Flower solitary. Sepals 3, Petals 5–8, yellow. Achene with spongy tissue above the basal seed; its beak long, hooked. x = 8.

Ranunculus lapponicus L.   **Lapland Buttercup**

Low, slender, glabrous perennial, 2–16 cm tall, somewhat resembling *Anemone richardsonii*; with buried horizontal, elongate, thread-like, pale rhizomes that give rise to nodal roots and terminal erect scapose flowering shoots. Prolongation of the rhizome is from axillary buds.

Leaves scattered, mostly basal, arising from nodes of the rhizome, long-petioled; the blade deeply 3-parted, the divisions crenate to shallowly lobed with rounded lobes. Occasional stem leaves arise near the bases of the flowering shoots, at or below ground level, and resemble the basal leaves except for shorter petioles.

Flower solitary on each scape. Sepals 3 (rarely 4), reflexed, narrowly ovate or obovate to lanceolate, 3½–5½ mm long, deciduous or persistent through the flowering period. Petals 5–8, oblanceolate or spatulate, with a pronounced claw, rounded at the tip, yellow, 4½–6 mm long, usually slightly longer than the sepals in our material; the nectary minute, pouch-like, with a convex-margined scale. Stamens 10–12, short, the anthers extended a little past the nectaries. Carpels 6–20 in a globose head, with prominant hooked styles and stigmas, glabrous.

Achenes in a globose head; the shortly stipitate body about 3 mm long in our material; the seed distending the basal two thirds of the body, the upper part being filled with spongy tissue, the stylar beak prominent, hooked, 1–2 mm long. 2n = 16.

Ranunculus lapponicus has a circumboreal range; on this continent extending southward to British Columbia, Alberta, Ontario and the northeastern United States. A plant typically of the Boreal and northern Subalpine Forest regions, and the Arctic Tundra, it is found in bogs, black spruce swamp-forest, and open wet mossy tundra. In British Columbia it occurs in the Interior from about latitude 53° northward.

Section *Hecatonia*

Amphibious to aquatic. Sepals 5. Petals 5, yellow; the nectary surrounded by a collar-like scale or the latter reduced to a low crescent. Achenes small, with swollen tissue near base and margin; the beak very short. x = 8.

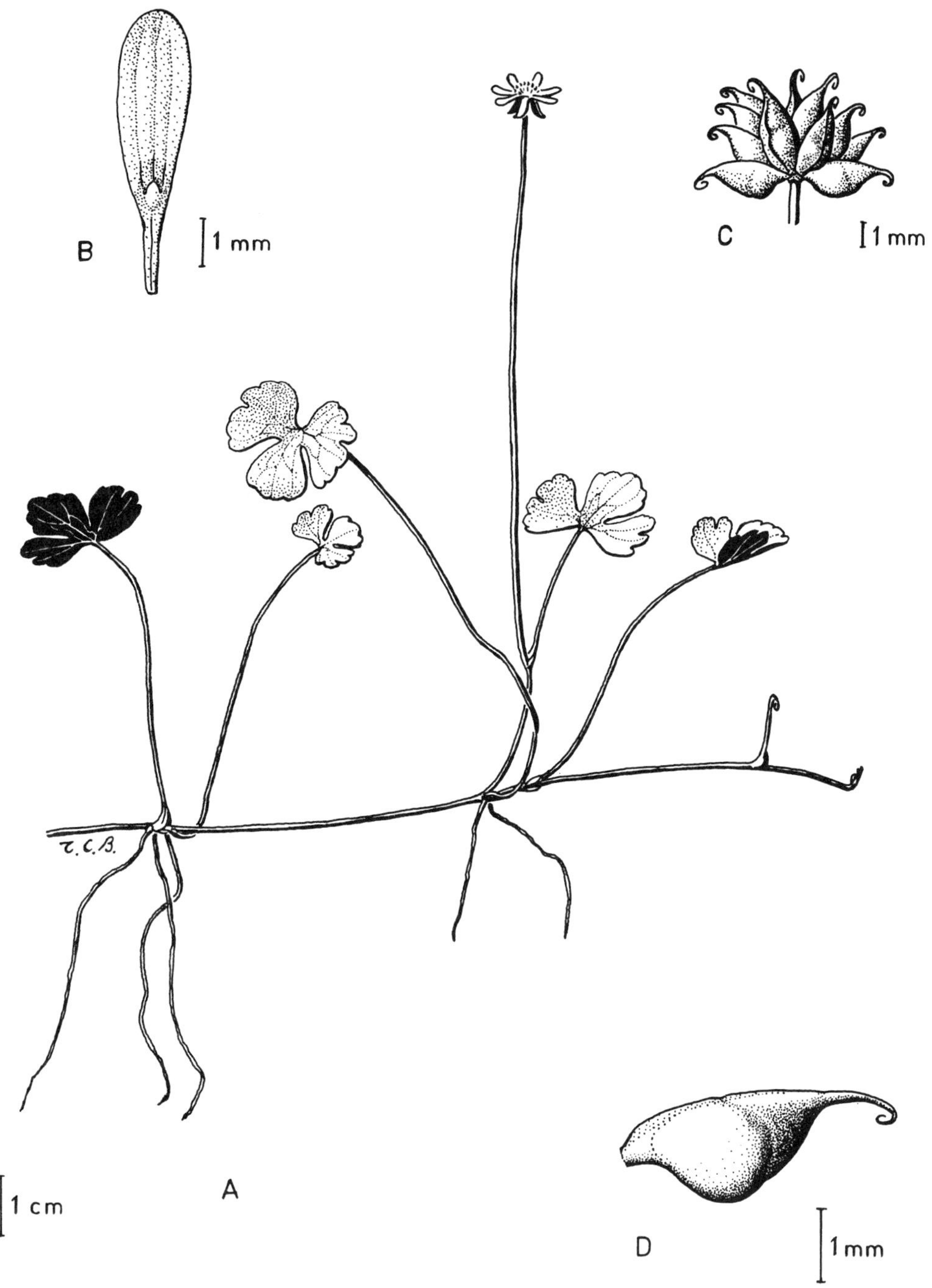

Figure 78. *Ranunculus lapponicus*

A. plant.   B. petal.   C. head of achenes.   D. achene.

Ranunculus gmelinii **DC.**   **Small Yellow Water-Crowfoot**
　　R. purshii **Richardson**
　　R. hyperboreus **Rottboell var.** *turquetillianus* **Polunin**
　　R. gmelinii **DC. ssp.** *purshii* **(Richardson) Hulten** [= *R. gmelinii* **var.** *hookeri* **(G. Don)**
　　　　L. Benson + var. *limosus* **(Nuttall) Hara]**

Aquatic to amphibious, hirsute to glabrous perennial with slender roots; the weak, hollow, slender or somewhat dilated stems trailing or floating, and rooting at those nodes in contact with the ground.

Leaves all along the stem, and very variable in form: pentagonal to reniform in outline, basically 3-lobed, with the lobes further 3-lobed; the submersed leaves deeply divided into delicate, linear, flat and ribbon-like, attenuate to filiform lobes, and up to 10 cm wide; floating and emersed leaves smaller and less deeply incised, the ultimate lobes acute to rounded. Stipular sheaths ciliate, and commonly hirsute, even when the rest of the plant is glabrous.

Flowers variable in size, terminal, and solitary, or sometimes the terminal one followed by flowers from the upper axils or in a bracteate cyme. Sepals 5, 2–6 mm long, spreading, commonly hirsute on the back, and deciduous early. Petals 5 or sometimes more, obovate, yellow, the basal half to two thirds more deeply coloured than the outer parts, 2.5–10 mm long; the nectary surrounded by a collar-like scale with a free margin sometimes lobed above, or the scale reduced and pocket-like. Stamens 10–45, the anthers from 0.7–1.2 mm long. Carpels 20–70 on an ovoid hairy receptacle.

Achene slightly compressed, the wall over the basal half of the body and about ⅔ of the margins somewhat thickened and paler than the rest; the surface often minutely pitted, at least in dried material; the body 1–1.5 mm long, with an ascending, flattened beak 0.3–0.5 mm long. 2n = 16, 32, 64.

KEY TO VARIETIES

1. Plant hirsute; leaves 5–20 mm wide, with little difference between submersed and aerial leaves; petals 2–6 mm long... **2**

   Plant glabrous to sparsely hirsute; leaves commonly larger, the submersed ones up to 10 cm wide, deeply dissected, biternate; flowers larger, the petals 4–10 mm long.. **var.** *hookeri* **3**

2. Leaves deeply lobed but small. Petals 2.5-4 mm long .... **var.** *gmelinii*

   Leaves shallowly to deeply lobed, the lobes sometimes merely notched. Petals 4–6 mm long... **var.** *limosus*

3. Plant prostrate or floating; flowers usually solitary and terminal....................... **forma** *hookeri*

   Plant decumbent to sprawling, with the flowering distal segments of stems ascending to erect; flowers 2 to many in a cyme with small, simple or shallowly lobed bracts.................
   .... **forma** *prolificus*

Distinction among these varieties is often obscured by the plastic habit of these plants, the effects of varying environment in regulating leaf size and degree of dissection being superimposed on those of diverse genetic makeup. Deeply submersed leaves are often surprisingly large, often triternate and further dissected, and very delicate; and are often poorly preserved or not collected. Aerial leaves on the same plant may be much smaller and relatively shallowly and broadly lobed.

Because of this great plasticity of response, Scott (1974) declined formally to recognize any subspecific groupings in this species: perhaps the most realistic approach, suggesting the use of caution in the acceptance of these formal varieties. These are presented here, as they are often recognizable, even though the precise reasons for their appearance may be obscure.

Variety *gmelinii*, the smallest of our varieties, with the smallest flowers, is circumboreal in distribution, and in North America is more northern in range than the other varieties. In British Columbia it is found north of 57°N. latitude.

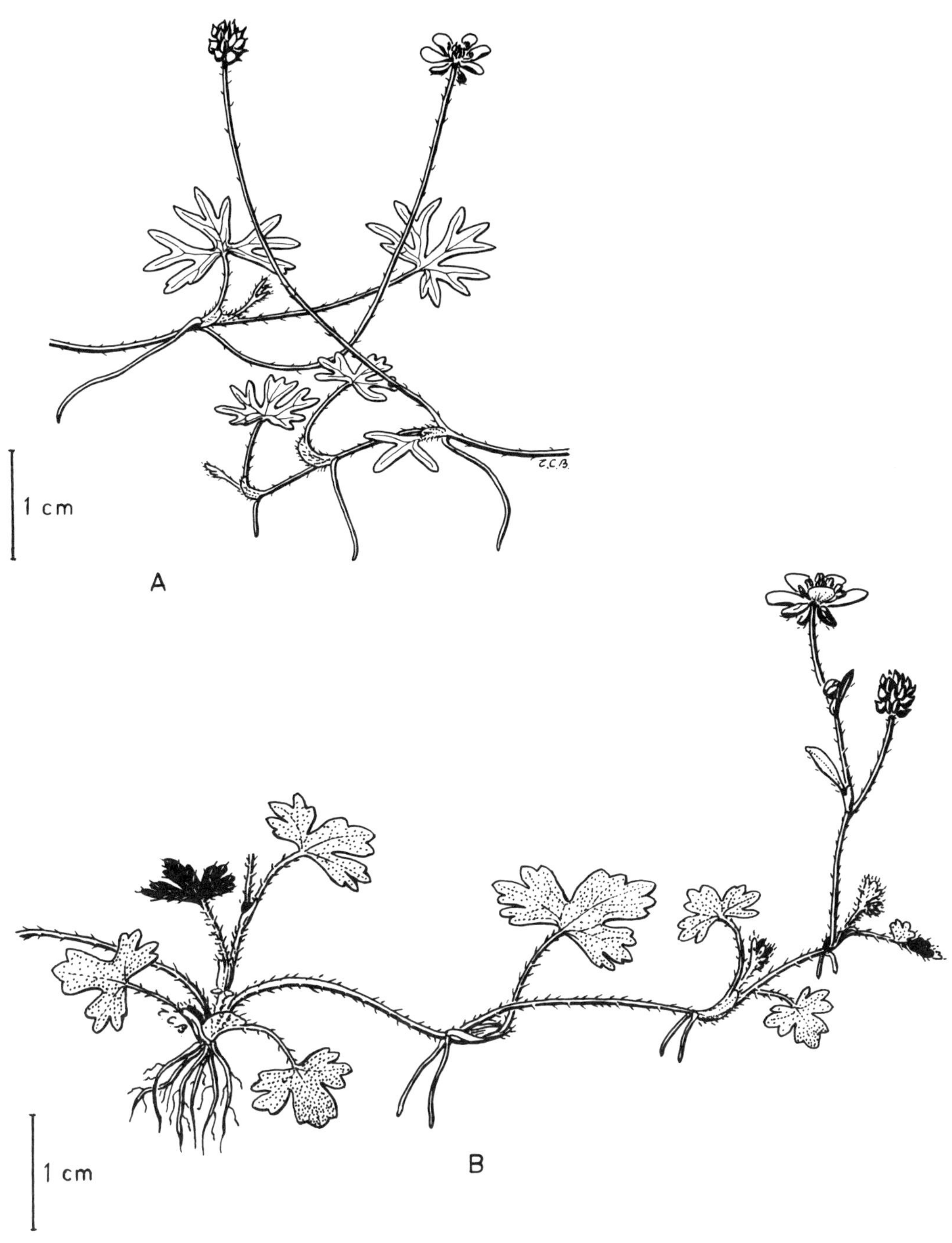

Figure 79. *Ranunculus gmelinii*

A. var. *gmelinii*.   B. var. *limosus*.

143

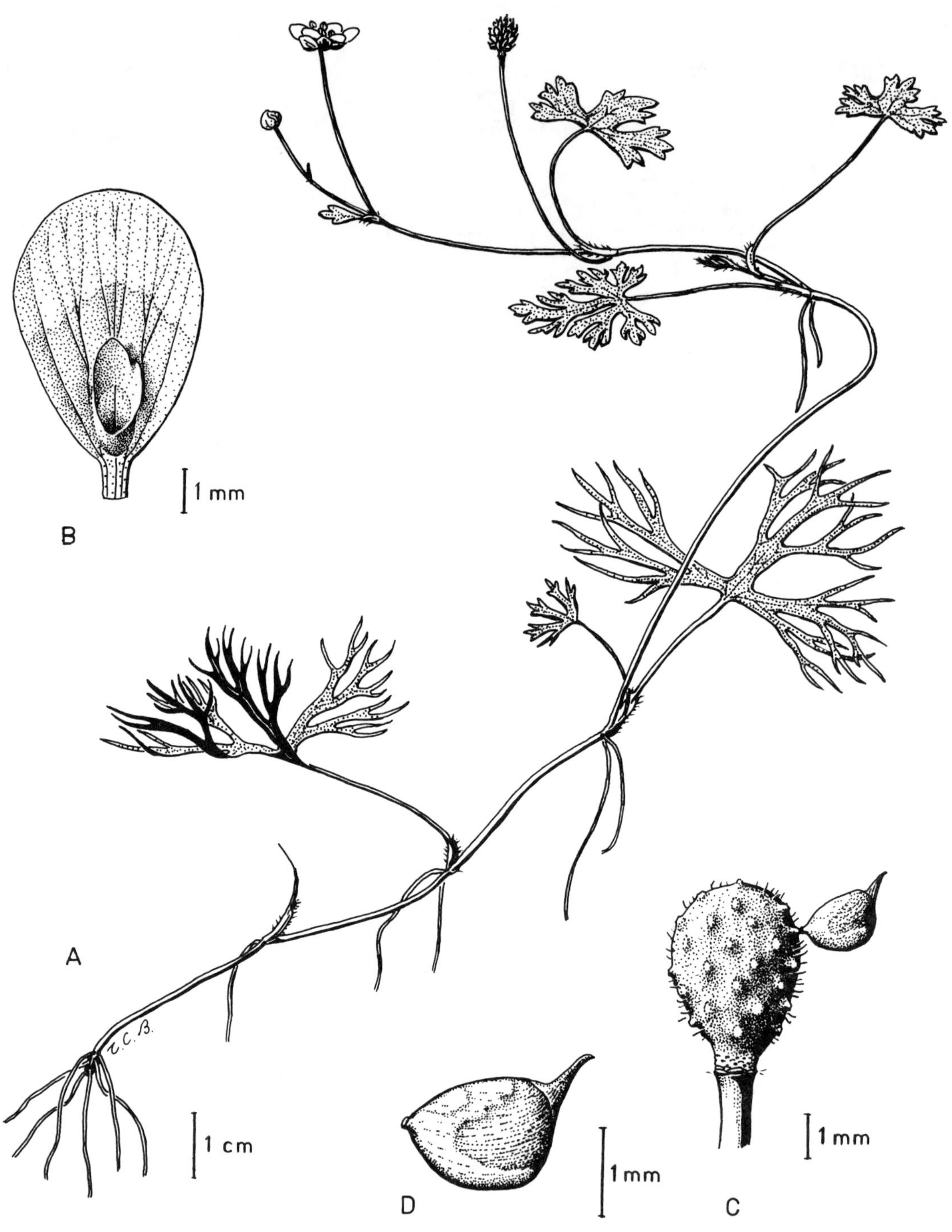

Figure 80. Ranunculus *gmelinii* var. *hookeri*

A. plant from aquatic habitat, with submersed and emersed leaves.
B. petal, showing nectary.
C. receptacle.
D. achene.

Variety *limosus* (Nuttall) Hara [*R. amphibius* James *sensu* James (1823) not L. Benson (1954): see note under *R. subrigidus*] occurs in the western United States, south to Utah and east to Colorado and the Dakotas, in the Canadian Prairie Provinces, and in central and southeastern British Columbia, and also in Asia.

Variety *hookeri* (G. Don) L. Benson is a larger plant, with larger flowers, than the other varieties. It is the common variety across North America. Extremely variable, it may be confused with *R. flabellaris*, from which it is poorly differentiated except by the achenes, and with *R. sceleratus*.

Forma *prolificus* (Fernald) T.C. Brayshaw *comb. nov.* [*R. purshii* Richardson var. *prolificus* Fernald, (Rhodora 19:135, 1917); *R. gmelinii* var. *prolificus* (Fernald) Hara] is included here, in var. *hookeri*. It differs from forma *hookeri* in being more floriferous: the terminal flower being accompanied by several more (up to 50) arising in the axils of small, entire or shallowly lobed bracts, forming a cyme. It does not form a geographically discrete population, except in the Magdalen Islands, Quebec, its type locality, but generally occurs as scattered individuals throughout the population of var. *hookeri*; being very scarce in British Columbia. This form can be mistaken for the strictly erect *R. sceleratus*; which however can be distinguished by its longer head of achenes, which have margins thickened all around and very short beaks.

Ranunculus flabellaris Rafinesque   **Yellow Water-Crowfoot**

    R. fluviatilis Bigelow *non* Willdenow
    R. multifidus Pursh *non* Forskal
    R. delphiniifolius Torrey *in* A. Eaton

Glabrous to sparsely hirsute, amphibious to aquatic perennial rather like a coarse version of *R. gmelinii*; with coarse, hollow, trailing or floating, branching stems rooting at the nodes when these are in contact with mud.

Leaves all along the stem. Submersed leaves finely and profusely 3–5 times ternately divided, the divisions narrow, flat, and somewhat ribbon-like basally, and with filiform ultimate segments, on short petioles or subsessile; the floating or emergent leaves with longer petioles, and varying from the deeply trifoliolate lower ones with long-petioluled leaflets and flat ultimate segments to deeply dissected simple upper leaves.

Flowers terminal and in the axils of bracts, relatively large; sepals 5; petals 5, sometimes more, sometimes lobed or divided, 7 to 15 mm long; with the nectary surrounded by an oblique collar that is lobed above, 1.5 mm long. Stamens 30–70. Receptacle ovoid, glabrous or slightly hairy.

Achenes 30–70 in a globose cluster, with a body 1.7–2.5 mm long, thickened around 2/3 to 3/4 of the margin, which is keeled distally; the faces sometimes with minute, blister-like swellings; the beak 1–2 mm long, flattened, and straight, or curved at the tip. 2n = 32.

Shallow water or muddy shores and marshes. From east of the Cascade Range to the Atlantic coast and southward to Louisiana. Recorded from southeastern British Columbia, but not seen since 1949.

The shape assumed by a developing leaf in this species is a plastic response of the leaf primordium directly to environmental influences. The gradual transition, from finely dissected submersed leaves to the palmately lobed, laminar, floating or emergent ones, has been found to be related to the temperature of the surrounding medium, and, to a lesser degree, to photoperiod (length of daylight) (Cook, 1968).

Ranunculus sceleratus L.   **Cursed Crowfoot**

Glabrous or sparsely pubescent, erect, branching, amphibious annual with clustered basal fleshy roots and soft hollow stems.

Basal leaves with long petioles, blades 2–6 cm long, reniform in outline, deeply 3-lobed with cuneate lobes commonly sinuately lobed with rounded tips. Stem leaves more deeply divided, with shorter petioles.

Flowers several to many, terminating the stem and branches. Sepals 5, spreading, 2–4.5 mm long, soon deciduous, or sometimes persistent. Petals 5, yellow, obovate, 2–6 mm long, the nectary

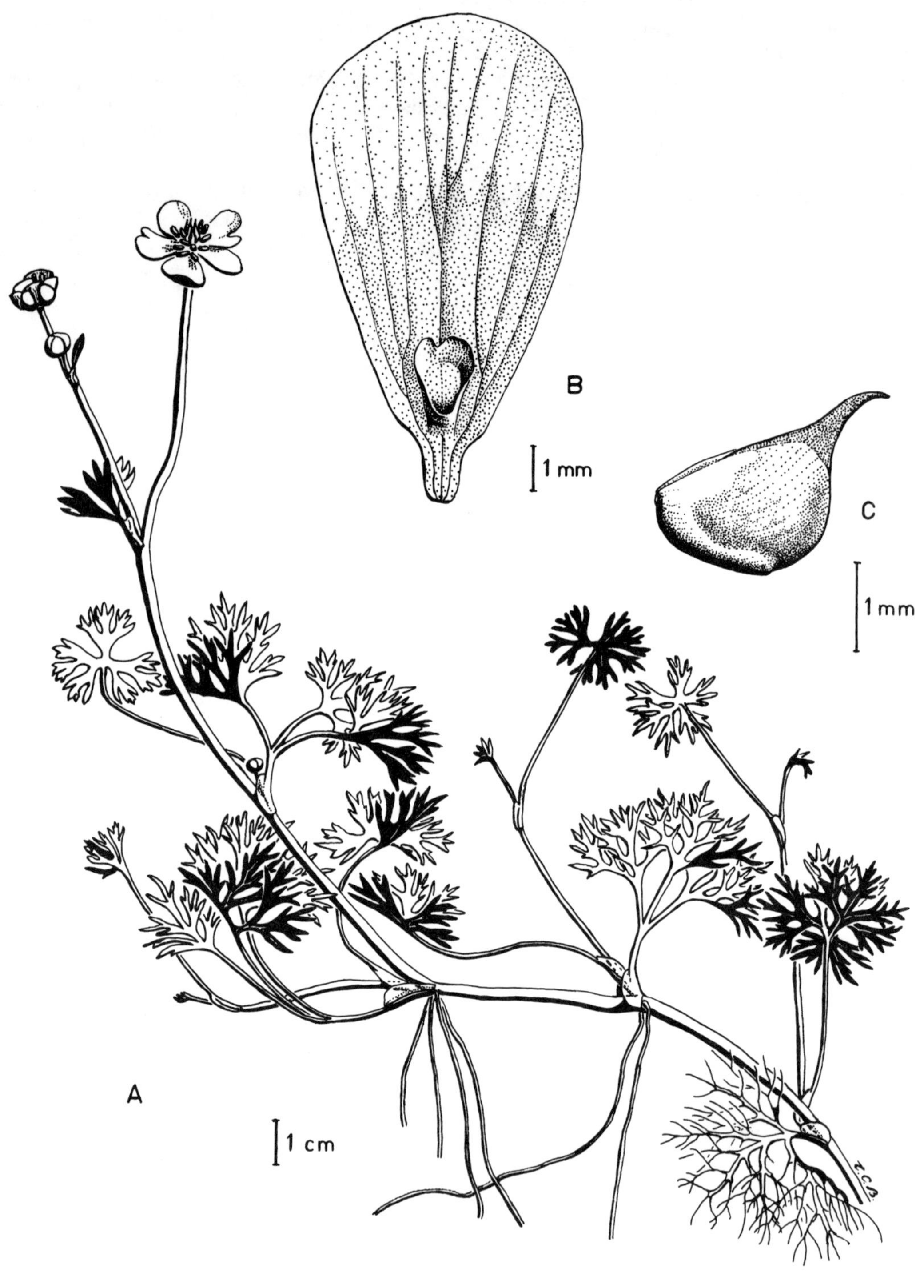

Figure 81. *Ranunculus flabellaris*

A. plant.   B. petal and nectary.   C. achene.

146

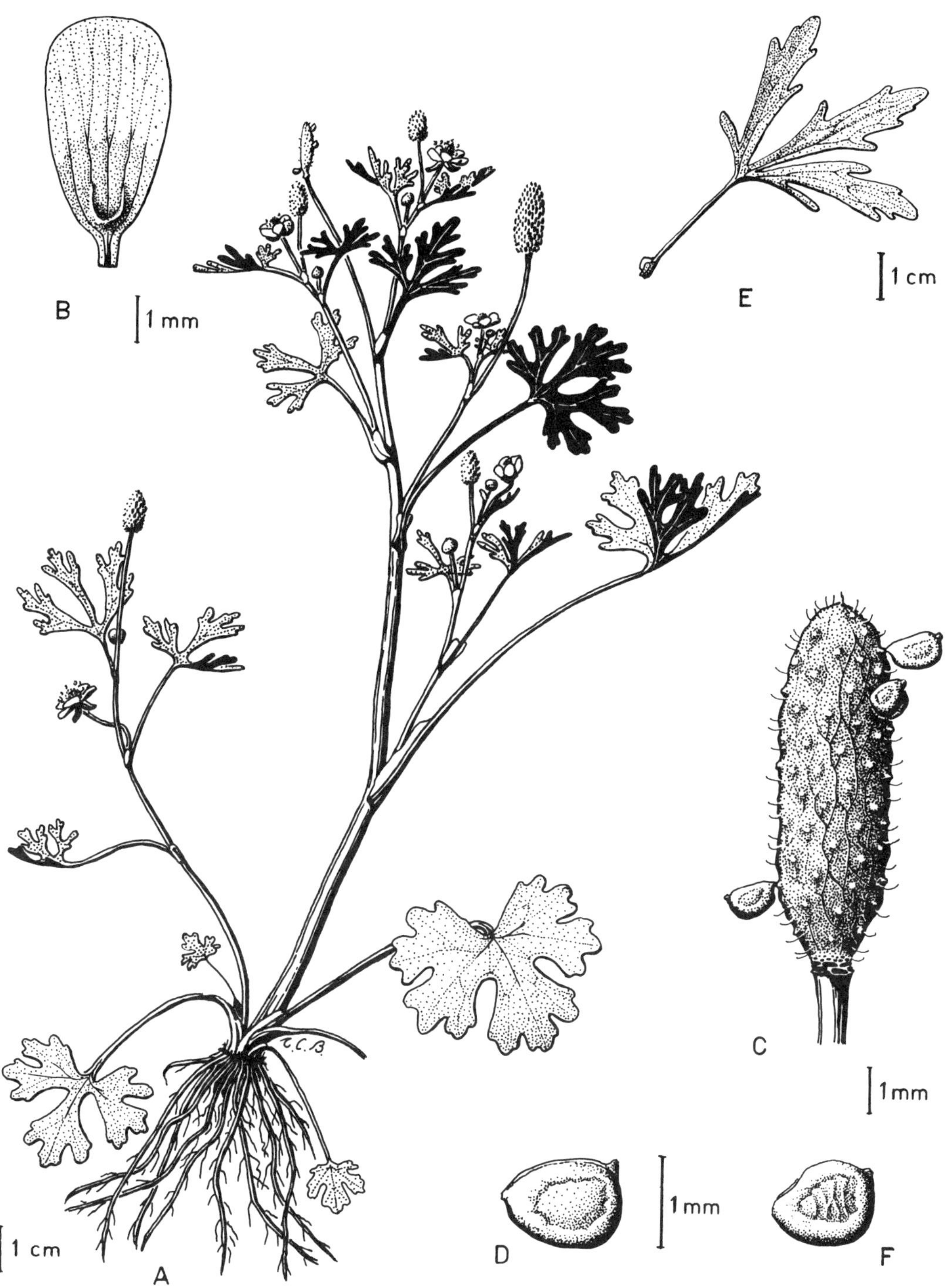

Figure 82. *Ranunculus sceleratus*

var. *multifidus*:
  A. plant.
  B. petal.
  C. receptacle.
  D. achene.
var. *sceleratus*:
  E. stem leaf.
  F. achene.

147

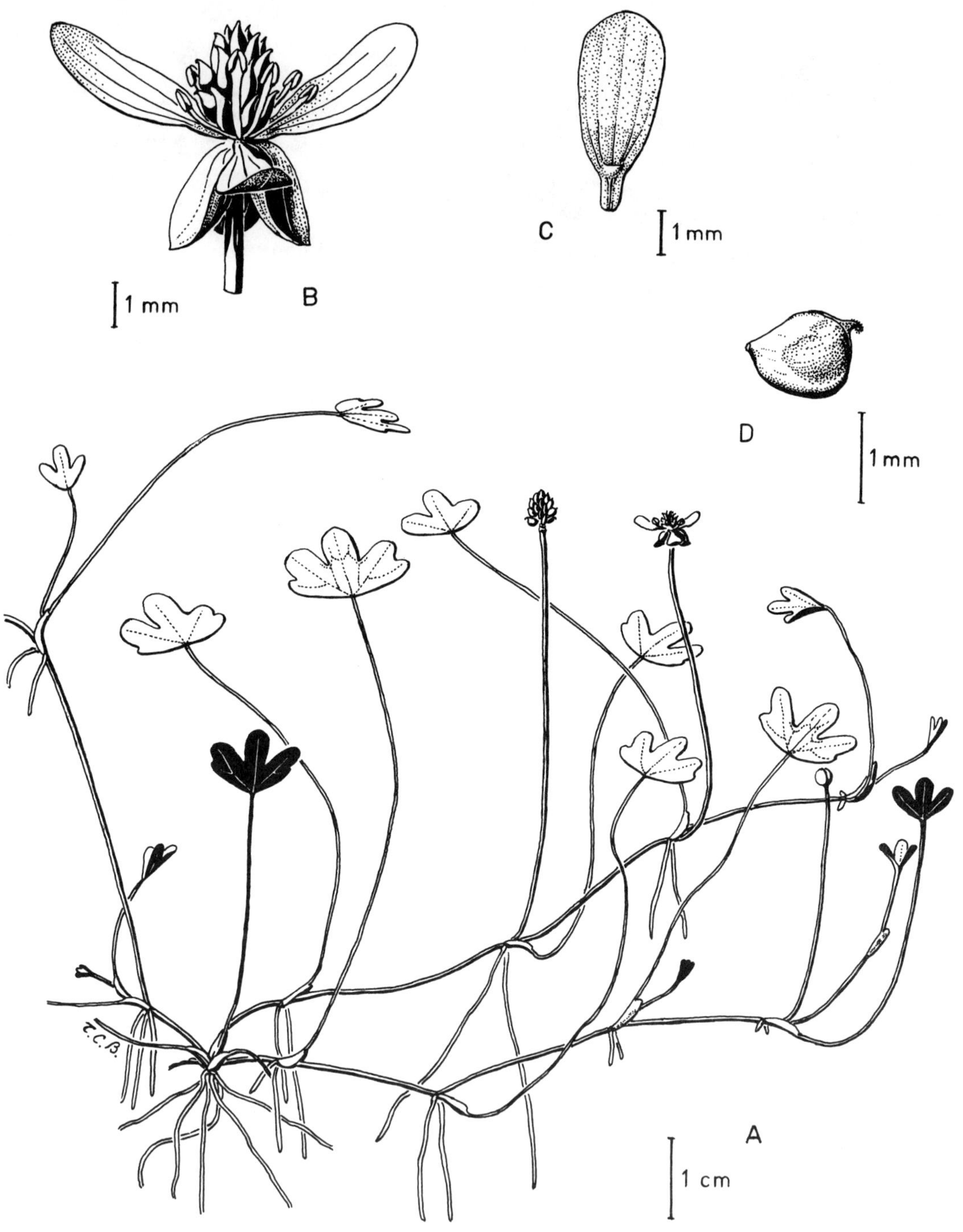

Figure 83. *Ranunculus hyperboreus*

A. plant.   B. flower.   C. petal.   D. achene.

pocket-like, with a short scale adnate along its lateral margin to the petal. Stamens 15–20. Receptacle ovoid, elongating to cylindrical and up to 14 mm long, usually sparsely pubescent, and bearing 50–200 or more carpels.

Achenes in a cylindrical head, rather flattened, about 1 mm long, the body with a thick margin all round, and sometimes transversely wrinkled or corrugated between the margins. Stylar beak almost lacking. 2n = 64 in North America, 32 in Europe.

KEY TO VARIETIES

1. Basal leaves with narrow sinuses between the lobes, the lobes shallowly lobed. Achene transversely corrugated or wrinkled between the margins. 2n = 32........................ **var.** *sceleratus*

   Basal leaves with wide, U-shaped sinuses, and the lobes secondarily deeply lobed. Achene smooth between the margins. 2n = 64.. **var.** *multifidus*

Ranunculus sceleratus typically inhabits wet, marshy or muddy sites. The species is circumboreal. Variety *multifidus* Nuttall is the common North American variety, and includes most of our material in British Columbia, where it is widespread. Variety *sceleratus* is the common native variety in Eurasia and may be native in eastern North America. It occurs as an introduced weed in some areas of the western United States, and at Osoyoos, British Columbia.

This species contains a high content of protoanemonin, especially at flowering time, and is considered to be one of our most toxic species in this genus (Kingsbury, 1964).

Ranunculus hyperboreus Rottboell

Arctic Buttercup
Northern Buttercup

Glabrous or almost glabrous, amphibious to aquatic perennial with slender prostrate or floating stems rooting at nodes in contact with the ground.

Leaves simple, wider than long, 2–11 mm long, 3-lobed, with a cuneate, truncate, or occasionally cordate base, the lobes oblique, broadly rounded, and sometimes again divided.

Flowers terminal, and sometimes also axillary. Sepals 3, 2–4 mm long. Petals 3, 2–5 mm long, obovate, yellow, deeper coloured on the basal half. Nectary pocket-like, its short, wide scale adnate at the sides to the petal surface. Stamens 10–20. Receptacle ovoid, glabrous.

Achenes 10–70, obovate, slightly flattened, smooth, slightly swollen on the basal part of the margin, the body 0.7–1.4 mm long, with a beak 0.1–0.3 mm long. 2n = 32.

Ranunculus hyperboreus is circumboreal in distribution, extending southward in North America along the Rocky Mountains to Colorado. This species typically inhabits shallow ponds, seeping wet mossy banks, and muddy shores, and may be found on brackish mud flats. In British Columbia it has been found from latitude 53° northward.

Section *Batrachium*

Aquatic. Stems weak, trailing, often rooting nodally. Submersed leaves penicillate — ternately divided and redivided into tufts of hair-like divisions. Floating leaves, if present, lamellar — broad and flat, 3- to 5-lobed, and long-petioled; the transition usually abrupt. Flowers terminal, on peduncles that often curve down into the water in fruit; appearing lateral as stem growth is prolonged further by growth from axillary buds. Sepals 5, glabrous in ours, and often bluish-tinged. Petals 5, white with a basal yellow spot; the nectary minute, with a crescentic or collar-like scale. Achenes commonly transversely corrugated, often hispid, with little or no beak, the style usually deciduous. x = 8. Plants non-poisonous.

This is a notoriously difficult section in which to try to distinguish species. Some species appear to intergrade with each other, and the distinctions are further blurred by variation in individual response to varying factors of the environment. This is reflected in extensive lists of synonyms of some of the species (Cook, 1966), of which only a few of the more commonly encountered names are given here.

Ranunculus aquatilis L.   **White Water-Crowfoot**
    R. heterophyllus **Weber**
    Batrachium aquatile **(L.) Dumortier**
    Ranunculus grayanus **Freyn**
    R. trichophyllus **Chaix** *sensu* **W. B. Drew with respect to North American plants;**
      including *R. trichophyllus* **var.** *hispidulus* **(E. R. Drew) W. B. Drew, which is** *R. aquatilis*
      L. var. *hispidulus* **E. R. Drew.**

Aquatic perennial or sometimes annual herb, with weak, sparingly branched, obliquely ascending or trailing, submersed stems, rooting from at least the lower nodes.

Leaves cauline, of three kinds–(A) Penicillate leaves–3-ternately to 5-ternately dissected leaves with filiform divisions, and petioles of various lengths (1–25 mm) but generally short (1–12 mm) in ours; permanently submersed. (B) Small, compact versions of type (A) with flattened ultimate divisions, produced on stranded, terrestrially growing stems. (C) Simple, lamellar leaves–with broad flat lobed blades, on petioles 15–40 mm long; commonly but not invariably floating; induced by the long days of summer. The transformation from type A to type C is usually abrupt, but transitional leaves are sometimes produced. Lamellar leaves are not always produced. The petioles of dissected submersed leaves vary in length in a systematic way; but the actual dimensions show much variation among individual plants. Petioles of the lowermost stem leaves are relatively long (6–15 mm), those of successively produced leaves being progressively shorter, until the uppermost are very short beyond the adherent sheath. This pattern of variation is repeated in leaves on axillary branches. Stipular sheaths oblong at vegetative nodes, but more broadly ovate to circular and dilated at flowering nodes. The sheaths adhere for ¾ to all of their lengths to the petioles, but soon become eroded and eventually almost disappear, leaving petioles that thus appear longer than when originally formed. Leaves, petioles and stipular sheaths are normally finely hispidulous, mainly on the dorsal surfaces, seldom quite glabrous.

Flowers single, terminal, on pedicels 7–50 mm long, that curve away from the direction of stem growth to bring the flowers to the water surface; and may appear opposite leaves of any type. Prolongation of stem growth then develops from a bud in the axil of the ultimate stem leaf. Sepals 5, ovate, glabrous, pale greenish with a bluish or purplish cast, 3–5 mm long, usually caducous. Petals 5, white, with yellow bases, 4.5–10 mm long, obovate, usually not contiguous in the open flower, at least in our material. Nectary variable, but commonly a minute circular yellow pit, with a subtending scale that may surround it as a circular collar, or, more often in ours, merely form a low, crescentic scale over the lower half of the pit. Stamens 14–22, the outer anthers sometimes dehiscing before the perianth opens. Receptacle ± globose, normally hairy. Carpels 15–32, hispid to glabrous.

Achenes obovoid, hispid to glabrous, the surface commonly, but not always, corrugated transversely, 1.3–2.4 mm long; the stigma and most of the style being shed when the achene is ripe, leaving a very short stub of a beak. At the fruiting stage the pedicel is divergent by 90° or more from the stem, or may be quite reflexed. 2n = 48.

KEY TO VARIETIES

1.   Pedicels 2.5–4 cm long in fruit: ½–⅔ as long as petiole of opposed lamellar leaf. Petiole of penicillate leaf (1–) 6 (–12) mm long. Achenes distinctly hispid with stout hairs, corrugated, deep green when ripe, broadly obovoid to semi-lunate, 1.3–1.6 mm long, with a subterminal or sublateral beak.. **var.** *aquatilis*

Pedicels shorter and stout: 0.7–2 cm long in fruit, often less than half as long as the petiole of the opposed lamellar leaf. Petiole of penicillate leaf (0–) 4 (–8) mm long. Achenes glabrous or thinly hispidulous with slender hairs; the surface weakly corrugated to smooth, pale yellowish brown when ripe, nearly symmetrically narrowly obovoid, the body 1.6–2.4 mm long, with a decidedly terminal beak.. **var.** *hispidulus*

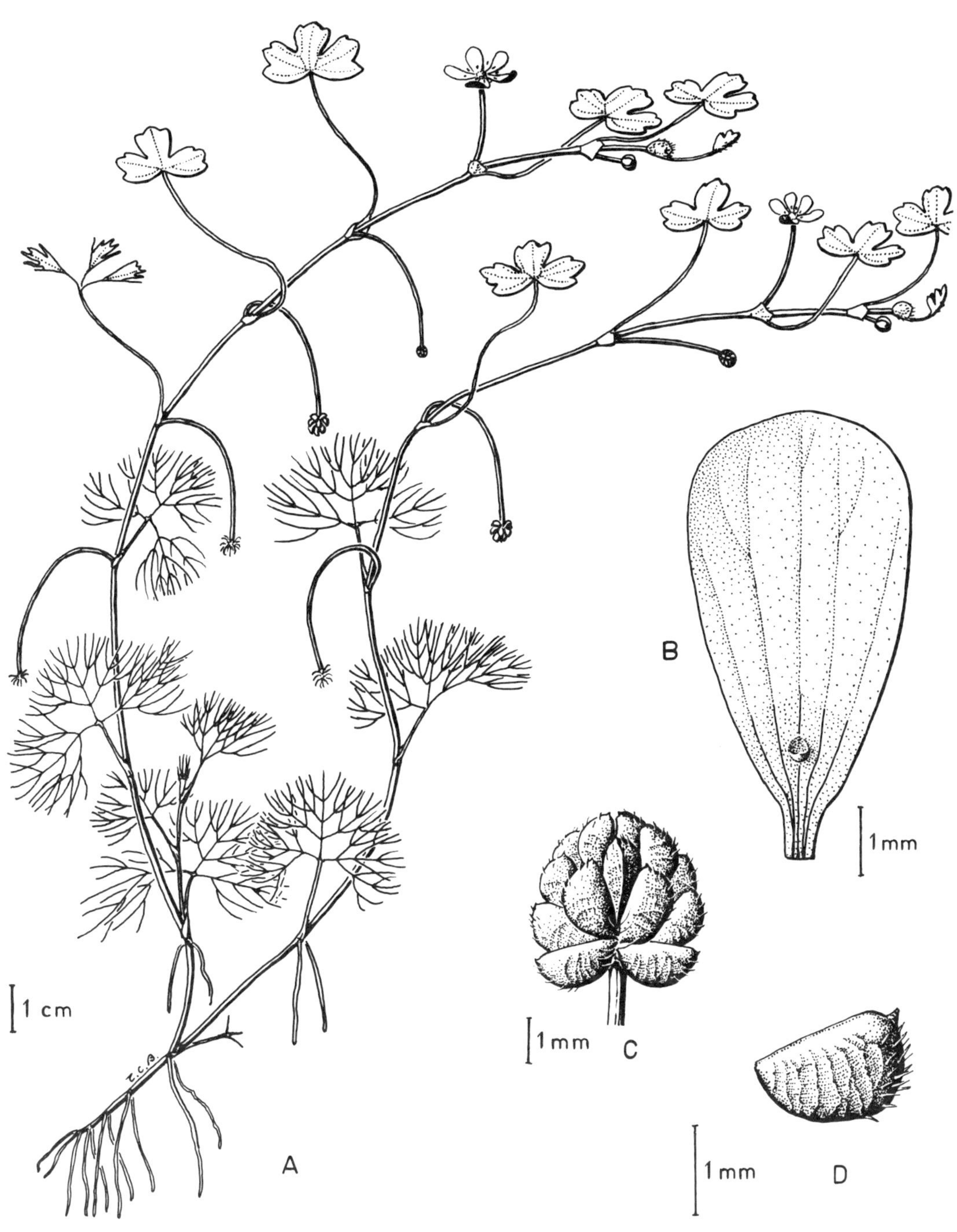

Figure 84. *Ranunculus aquatilis*

A. plant.   B. petal.   C. head of achenes.   D. achene.

Ranunculus aquatilis is found in shallow ponds and ditches, typically in disturbed situations. Elsewhere it has been found to be associated with eutrophic waters. In permanent water bodies it is a perennial, but where they dry out seasonally, it tends to behave as an annual.

This species is found over a large part of the temperate regions of the world: scattered across Eurasia, and in western North and South America; in tropical South America reaching high altitudes in the Andes. In British Columbia, a plant that is not clearly distinguishable from the European *R. aquatilis* var. *aquatilis* is found occasionally in the southern Interior, from the Coast Range eastward. The European material of *R. aquatilis* seen by the author differs from ours, on average, in being rather coarser; with stouter stems, larger submersed leaves, more deeply dissected lamellar leaves, and larger flowers, though the ranges of variation overlap almost completely.

Variety *hispidulus* E. R. Drew (*R. grayanus* Freyn, according to L.Benson, 1948) is found on the Coast, including Vancouver Island, where it is seen to have petioles proportionally shorter for all leaves than var. *aquatilis*. It has been reported from Lytton by Benson (1948) as the type specimen of *R. grayanus*, but that material has not been seen by the author.

The type specimen of var. *hispidulus*, which is at the University of California, Berkeley (UC), was collected at Jarnigan's, on the Mad River, Humboldt County, California. It was named and distinguished by its author on the basis of the fine short hairs on the leaves, petioles and stipules: a weak distinction for a variety of this species, in other varieties of which such hairs are also found. E. R. Drew's original description made no mention of flowers or fruit, although they are present on the type specimen. In this material, the pedicels are 1½ cm long in flower, 1½–2 cm long in fruit. Petals, 5½ × 3½ mm, and obovate. Achenes are 2–2.4 × 1.5 mm, weakly corrugated, yellowish brown when ripe, and thinly hispidulous, commonly with 6–8 fine short hairs which may be shed from some achenes as they ripen; the beak terminal to subterminal. The nectaries are not visible on the type specimen, but the topotype, collected subsequently at the same location, has crescentic nectary scales over very minute pits. The lamellar leaves have petioles relatively shorter than is usual in our material.

Material from Vancouver Island, except for the Alberni district, has carpels glabrous from the start. In other respects the plants are similar. Although the type material thus appears to be in a slight way intermediate between this variety as it appears on the coast of British Columbia and the var. *aquatilis* as it is found in the Interior, it is definitely closer in form to our coastal material.

Cook (1966) reports sterile hybrids between *R. aquatilis* and *R. trichophyllus*, as *R.* X *lutzii* A. Felix in Schaed., at sites where both parent species occur. He notes that these hybrids develop many more transitional leaves than is normal for *R. aquatilis*. Such plants may be looked for in this province.

The variation in leaf form in this and related species is of a kind that has given rise to endless confusion regarding the identities and limits of species and varieties; and a large number of species and varieties has been described based on the diversity of form. This confusion still persists; and the present treatment does not claim to be final.

The determination of leaf form is dependent first on photoperiod (Cook, 1966, 1968). In the short days of spring, only finely dissected, penicillate leaves are produced. Then a certain threshold of day-length is attained that operates a physiological switch that turns on the ability to change leaf shape. In longer days, a lack of water around a leaf primordium induces it to develop into a terrestrial dissected leaf. Otherwise, in the long days of summer (but possibly not facilitated by an identical threshold photoperiod) the plant in water switches to producing lamellar leaves, which may be floating or submersed when fully expanded, depending on the water level of the moment. The threshold photoperiod that determines this change in leaf form varies from individual to individual; and a few plants seem never to be able to produce lamellar leaves.

This species, by responding to photoperiod, a predetermined seasonal cue, can anticipate changes in its environment that are normally expected, but cannot be precisely predicted in time in any particular year.

Plants of this species that lack lamellar leaves and also lack flowers or fruit are indistinguishable from *R. trichophyllus* (Cook, 1966).

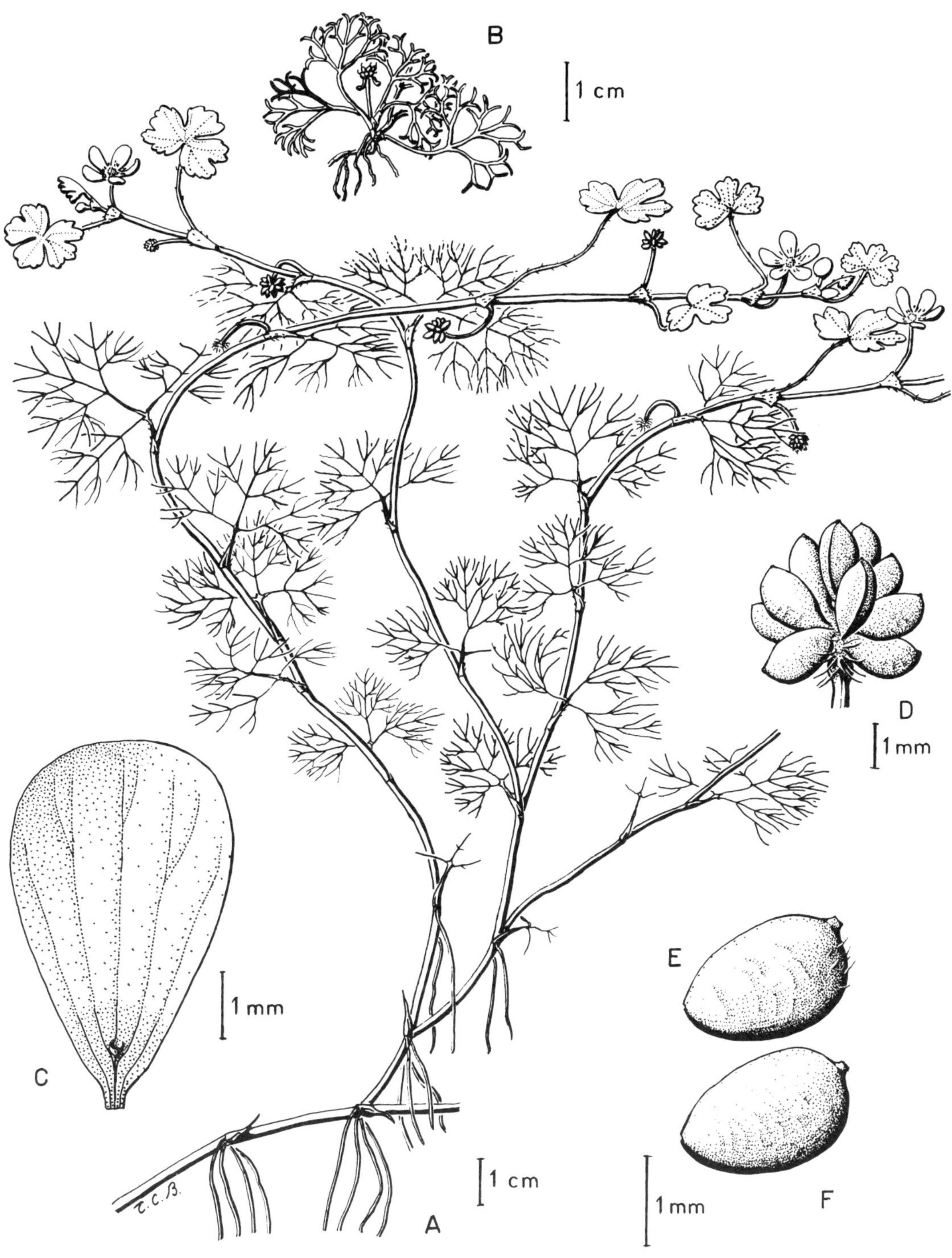

Figure 85. *Ranunculus aquatilis* var. *hispidulus*

A. flowering and fruiting aquatic plant.
B. terrestrially growing plant.
C. petal.

D. head of achenes.
E. achene from type specimen.
F. achene from Vancouver Island specimen.

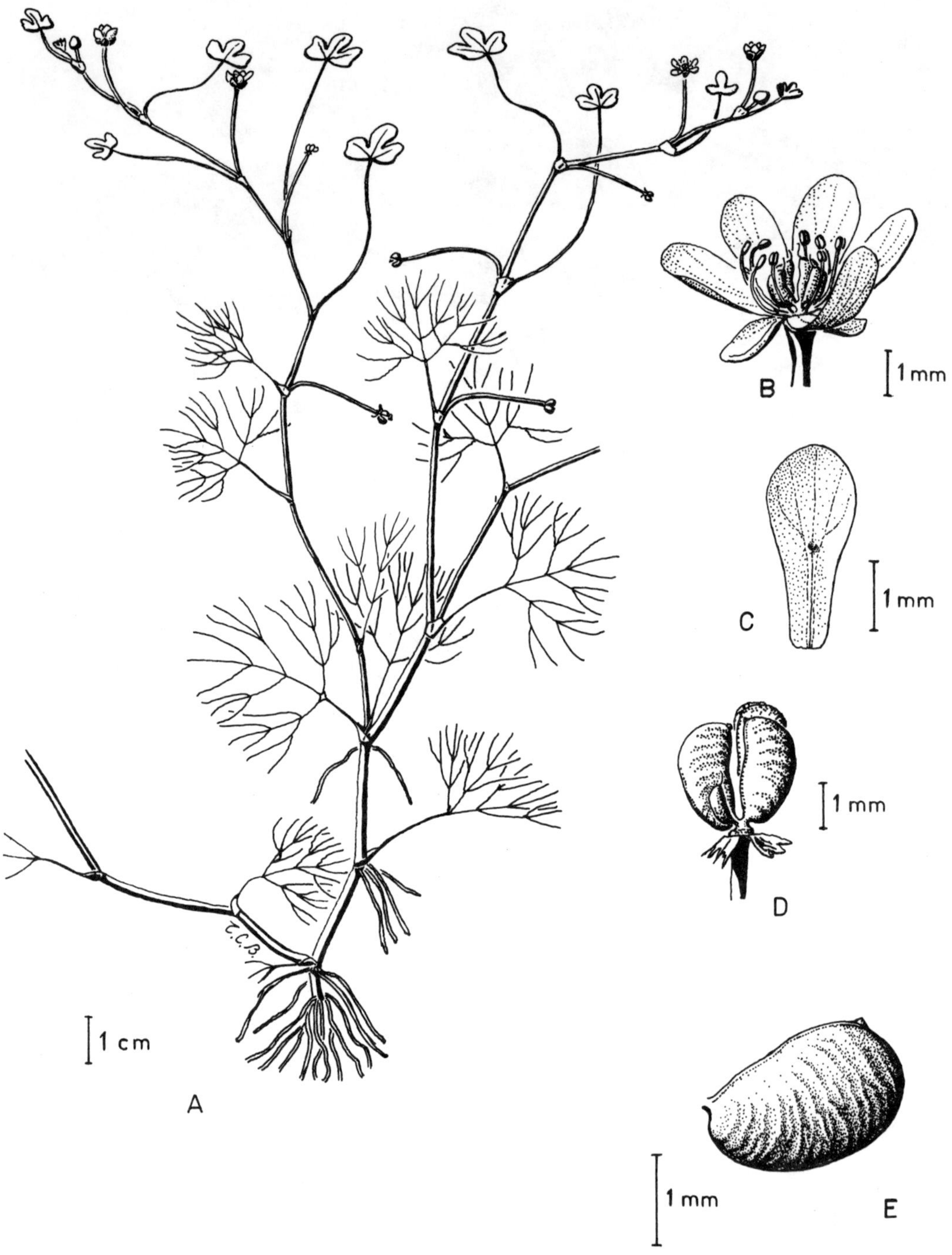

Figure 86. *Ranunculus lobbii*

A. flowering plant.
B. flower.
C. petal.

D. cluster of achenes.
E. achene.

Ranunculus lobbii (Hiern) Gray

Glabrous aquatic spring annual, the stems weak, slender, normally submersed. Earlier, lower leaves submersed, on petioles up to 2 cm long, finely bipalmately dissected into thread-like divisions. Later, upper leaves floating, with longer petioles and lamellar, lobed blades; the lobes simple, broadly rounded, the lateral lobes slightly notched or truncate.

Flowers terminal, on short (1–3 cm long) pedicels that are often straight, though sometimes turned down in fruit. Sepals 2–2.5 mm long, glabrous, sometimes persistent. Petals 3– 6 mm long, with minute crescentic nectary scales nearly midway up the petals. Stamens 10–16. Carpels 2–7, glabrous. Receptacle low, glabrous.

Achenes large for this section (2–2.5 mm long), obliquely obovoid, plump, finely corrugated or wrinkled, beakless or with a subterminal stub; the body sometimes slightly keeled.

Ranunculus lobbii has been found from southern Vancouver Island southward along the coast to California, in shallow vernal pools and pastures that flood in winter. The most recent record at the Royal British Columbia Museum (V) is dated 1948. Recent searches for it in localities where it has previously been reported have failed to reveal it. This species may now be extinct in this country.

Ranunculus trichophyllus Chaix *in* Villars   **Thread-Leafed Buttercup**
R. capillaceus Thuiller
R. aquatilis L. var. *capillaceus* (Thuiller) DC.

Flaccid-stemmed aquatic perennial, normally fully submersed except for the open flowers. Stem slender, weak, up to a metre long, rooting from the lower nodes.

Leaves all alike: dissected-filiform (penicillate); typically all petioled above their dilated stipular sheaths, sessile leaves rare; the blades twice to thrice palmatifid. Stems, sheaths and leaves commonly shortly hispid.

Flowers on pedicels 4–8 cm long, held erect and emergent at flowering time, but subsequently the pedicels curving downward, drawing the ripening fruits beneath the water surface. Sepals 5, glabrous, purplish-tinged. Petals 5, white with a basal yellow spot, 3–9 mm long in North American material, in contrast to the 3–5 mm long petals of European plants. Nectary variously crescentic to circular or even pear-shaped in outline. Stamens 5–15. Receptacle globose, not elongating in fruit, normally hispid, but occasionally glabrous or almost so in var. *calvescens* W. B. Drew. Carpels 12–32, obovoid, normally hispid except in var. *calvescens*.

Achenes obovoid to semi-lunate, normally hispid, though sometimes the hairs present at flowering time are shed from the ripening achenes; the body transversely corrugated, and commonly green when ripe and sometimes bluish-tinged, 1.6–2.0 mm long; the beak very short (½ mm or less long), subterminal. 2n = 32. Reports of 16 and 48 chromosomes are suspect. The plants used may have been misidentified specimens of *R. subrigidus* or *R. aquatilis*.

KEY TO VARIETIES

1. Stem rooting from the lowest nodes, slender to stoutish. Petals 4–9 mm long. Stamens usually 7–15. Carpels 10–32 .... **2**

  Stem rooting from all nodes, thread-like, short and weak. Petals 3–4.5 mm long. Stamens 5–10. Carpels 8–20, usually hairy when immature but the hairs often shed as achenes mature .... **var. *eradicatus***

2. Receptacle and carpels glabrous .... **var. *calvescens***

  Receptacle and carpels hispid .... **var. *trichophyllus***

Ranunculus trichophyllus is a widespread species in Eurasia and North America, and is found all over British Columbia. It is found in shallow, usually still water of lakes and ponds. The common variety here is var. *trichophyllus*.

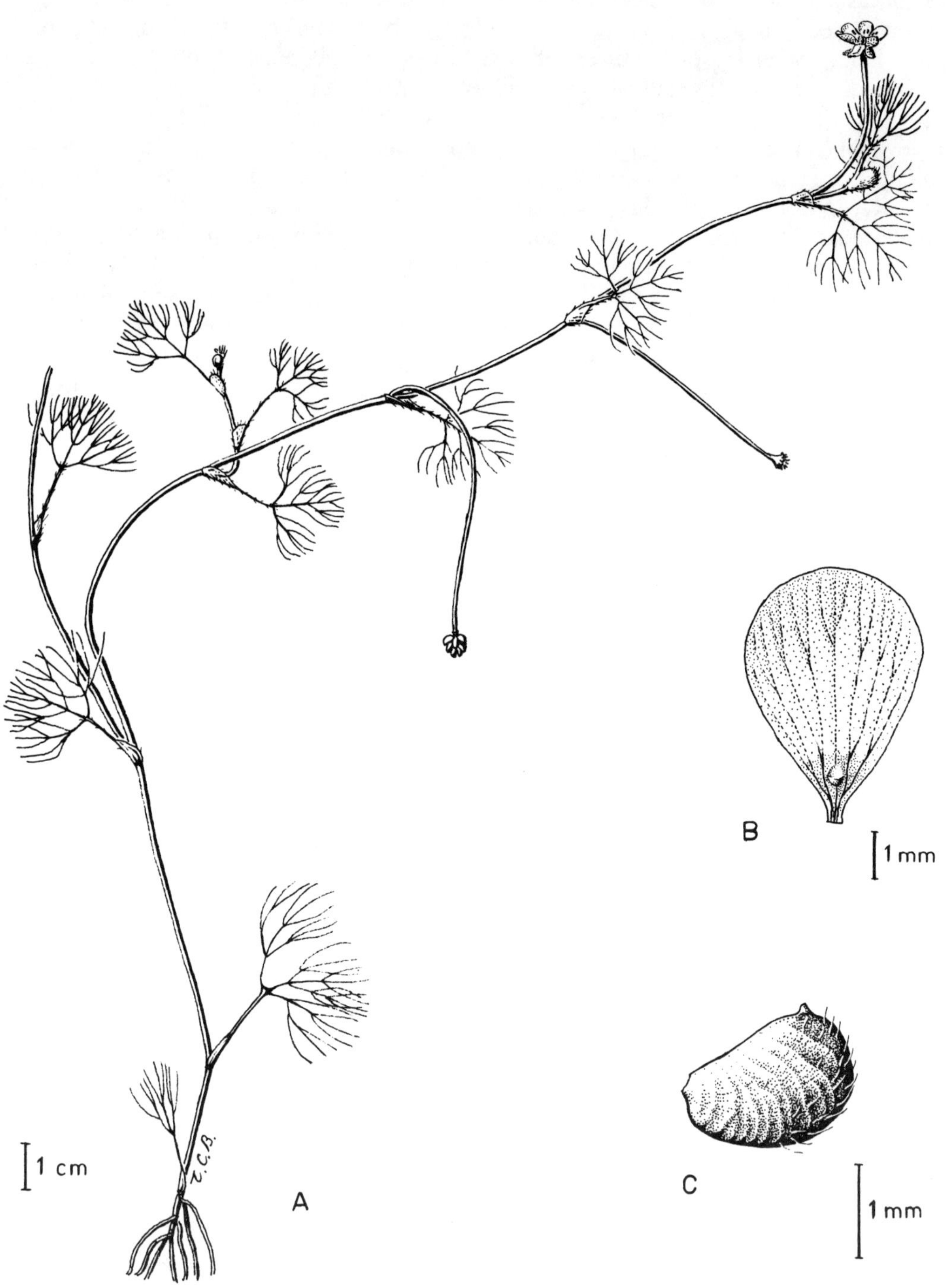

Figure 87. *Ranunculus trichophyllus*

A. flowering and fruiting plant.   B. petal.   C. achene.

156

Variety *calvescens* W. B. Drew occurs scattered through the North American range of this species, including some localities in British Columbia. It may be found growing with the typical variety; and plants of intermediate character, having very sparse hairs on the receptacle and glabrescent achenes, are also found.

Variety *eradicatus* (Laestadius) W. B. Drew [*R. confervoides* Fries, *R. lutulentus* Perrier & Songeon, *R. trichophyllus* Chaix subsp. *lutulentus* (Perrier & Songeon) Vierhapper]. The flowers commonly self-pollinate beneath the water surface (Cook, 1966). This is a far northern, circum-boreal, and in Europe, alpine, plant of shallow to deep water. It was postulated by Laestadius to be a product of injury or habitat destruction by ice. However, Cook (1966) found that it usually retains its distinguishing characteristics in cultivation. On the other hand, he also found that an *eradicatus*-like form could be induced in var. *trichophyllus* by cultivation in cold (6°–8°C) water (Cook, 1966). This plant is known in the Northwest Territories; and though not yet found in British Columbia, it may be looked for in the northernmost parts of this province, and especially at high altitudes.

Ranunculus trichophyllus is confusingly intergradient with its relatives: with *R. subrigidus* on one hand (see note under that species), and with *R. aquatilis* on the other hand. On morphological grounds alone, in this region, sterile individuals of *R. aquatilis* (except var. *hispidulus*) that have failed for any reason to display lamellar leaves are indistinguishable from *R. trichophyllus*. Cook (1966) reported a similar difficulty with European material.

The few chromosome counts that have been made on North American material of *R. trichophyllus* suggest that, like the European population, it is tetraploid (i.e. 2n = 4x) with respect to the basic number (x = 8) for this section; and that there is genetic discontinuity between it and *R. aquatilis*, which is hexaploid (2n = 6x). Otherwise, there are good grounds for accepting the varietal status assigned to it by De Candolle (1824) and followed in their floras of this region by Hitchcock and Cronquist (1964, 1973) as var. *capillaceus* (Thuiller) DC.

Ranunculus subrigidus W. B. Drew
R. circinatus Sibthorp var. *subrigidus* (W. B. Drew) L. Benson
R. codyanus B. Boivin
R. amphibius sensu L. Benson and lectotype designated by L. Benson, not James

Submersed aquatic perennial; only the flowers projecting above the water surface. Stems slender and weak.

Leaves all submersed and penicillate: bipalmately and further divided into thread-like divisions, forming rather globular tufts. Leaves on the upper part of the stem sessile on their stipular sheaths, but the lower leaves and those on branches often with petioles 1–5 mm long; stiff or flaccid.

Flowers terminal, on pedicels 3–7 cm long. Sepals 5, glabrous, ovate, commonly purplish-tinged. Petals white with a basal yellow blotch, 4–6 mm long, broadly obovate, contiguous in the living flower. Nectary more or less surrounded by a collar-like scale, pear-shaped in outline. Stamens 10–20. Receptacle obovoid, hispid, commonly but not always elongating and becoming cylindrical as the fruit matures. Carpels 30–80, hispid.

Achenes (22–) 30–80, normally hispid, corrugated transversely, 1–1.3 (–1.5) mm long, in a globose to rather ovoid head. Fruiting pedicels curving downward, drawing the ripening achenes beneath the water surface. 2n = 16.

Ranunculus subrigidus is widespread across Canada, and southward through the western United States to Mexico; and is found in lakes and ponds on the mainland of British Columbia east of the Coast and Cascade Ranges.

This species is often difficult to distinguish from its near relatives. It has been variously reported as *R. aquatilis* or *R. circinatus* (the latter a European species) when sterile, and as *R. sphaerospermus* Boisser & Blanche (an Asiatic and southeastern European species, also with elongating receptacles and many small achenes) when fertile. In achene size and petiole development it is not sharply separated from *R. trichophyllus*, which typically has achenes 1.6–2.0 mm long and all leaves petioled. Some specimens with sessile upper leaves and achenes 1.6–1.7 mm long have been collected at scattered localities on the mainland of British Columbia.

Lyman Benson (1948) assigned this species to varietal rank in *R. circinatus* Sibthorp. In 1954 he pointed out that the recently rediscovered type specimen of *R. amphibius* James, at the New York Botanical Garden (NY), belongs to the same species as that of *R. subrigidus*; and noted that at the level of a species, the name *R. amphibius* has priority over *R. subrigidus*, having been published over a century earlier. The author has examined the type specimens of both *R. subrigidus* and *R. amphibius*, and concurs with Benson's (1954) statement regarding the identity of the *R. amphibius* type specimen and the priority of *R. amphibius* as a published name. The matter, however, is not as clear as it might at first appear.

The specimen at the New York Botanical Garden labelled "Ranunculus amphibius n s?" (in James' handwriting?) was designated the holotype of this species by L. Benson in 1952, and thus of his *R. circinatus* Sibth. var. *subrigidus* (W. B. Drew) L. Benson (Benson, 1954). As types are currently defined, this would be termed a lectotype. James designated no specimen; and the above specimen bears no data on the collector, date or location of its collection.

Closer examination reveals the striking discordance between the characteristics of the above specimen and the description of *R. amphibius* published by James (1823). In no way can they be the same species.

The specimen is an undoubted example of the species known as *R. subrigidus* W. B. Drew; with sessile, ternate and multifid leaves like hair-tufts, and a flower that however has now lost its original colour. However, James (1823) described *R. amphibius* as "Slender, floating or decumbent, leaves reniform, four or five lobed, divisions cuneate, oblong, margin crenate, petioles long and alternate. The submersed leaves are, in every respect, similar to the floating ones." Although flowers were present, James remarked only on their size: "larger than that of *R. fluviatilis* (= *R. flabellaris* Raf.) with which it grows", and not on what should have been an obvious difference in their colour; which is white in *R. subrigidus*, and yellow in *R. flabellaris*. From this, one can infer that the flowers of the plant James was describing were yellow, not white. The description appears to be of some variety of *R. gmelinii* DC.: possibly of a large flowered form of *R. gmelinii* var. *limosus* (Nutt.) Hara.

In the 1983 edition of the International Code of Botanical Nomenclature (Voss *et al.*, 1983), the *Guide for the determination of types*: paragraph T4(f) includes this condition on the choice of a lectotype: "The first choice of a lectotype must be followed . . . unless it can be shown that the choice was in serious conflict with the protologue"; a clause that did not occur in the 1952 edition (Lanjouw *et al.*, 1952), which Benson would have seen. Such is obviously the situation in this case and raises serious doubt as to whether this specimen is an acceptable type of *R. amphibius*, or, under that name, of *R. subrigidus*.

Regarding the relationship between *R. circinatus* and *R. subrigidus*; it is noted that the main characteristic used by Cook (1966) to distinguish *R. circinatus* from its relatives is the planar attitude of the leaf divisions: all lying in one plane instead of forming conical or globular tufts. This may work well with living material in the water, but is scarcely distinguishable in dried, pressed specimens, where all leaves are in one plane, regardless of species. The author has not seen this characteristic in living plants of *R. subrigidus* in North America; as its leaves tend to form globular tufts.

While there is a general resemblance between *R. circinatus* of Europe, and *R. subrigidus*, including having the same chromosome number, there are a few differences. Apart from the difference in leaf attitude, noted above, *R. circinatus* has lunate (crescentic) nectaries; and the admittedly few specimens seen by the author had non-elongating receptacles bearing around 30 very sparsely hispid, uncorrugated achenes, on non-recurving pedicels.

In view of these differences, it is felt preferable to retain *R. subrigidus* as a distinct species. The illustration (Figure 88) is based mainly on the type specimen of *R. subrigidus*.

Section *Halodes*

Stoloniferous. Leaves apically shallowly lobed. Petals yellow; the nectary scale attached at the sides. Achene longitudinally veined and ribbed; with a very short beak. x = 8.

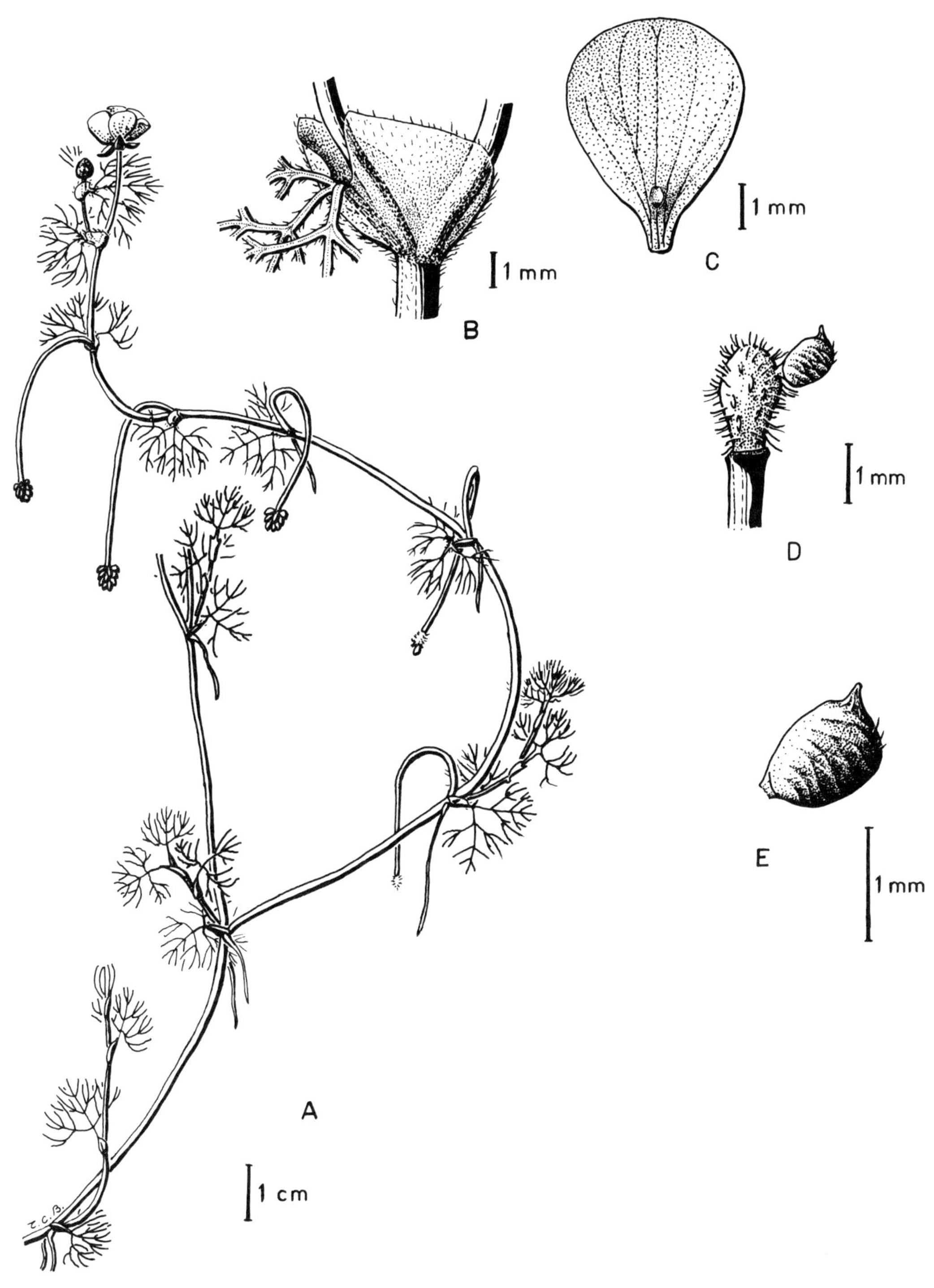

Figure 88. *Ranunculus subrigidus*

A. flowering and fruiting shoot.
B. node; showing stipular sheath and base of sessile leaf.
C. petal.
D. receptacle.
E. achene.
A, C, D, and E based on type specimen of *R. subrigidus*.

Ranunculus cymbalaria **Pursh**   **Seashore Buttercup**
    Halerpestes cymbalaria **(Pursh) Greene**

Stoloniferous, low, puberulent or glabrous perennial, rooting at the nodes in contact with the ground; perennating by axillary shoots arising at the rooted nodes.

Basal leaves long-petioled, broadly ovate, the base rounded to subcordate, the margins with few shallow rounded lobes. Stem leaves sessile, bract-like, usually entire.

Flowers solitary on erect scapes or few in erect, cymose bracteate inflorescences. Sepals normally 5, greenish yellow, often pubescent, 2–4 mm long. Petals usually 5, pale yellow, oblanceolate or spatula-like to obovate, from slightly shorter than to about twice as long as the sepals. Nectary pocket-like, with a narrow crescentic scale. Stamens about 15–30. Receptacle thinly hairy to glabrous, elongating in fruit. Carpels 15 to about 200.

Achenes in a cylindrical spike on the slender smooth receptacle; obovoid, with a projecting, tapering beak; the body 1–1.5 mm long, the beak about 0.5 mm long; the sides marked by several longitudinal raised veins forming ribs. 2n = 16.

KEY TO VARIETIES

1.   Stamens less than 25. Stems slender (0.5 mm thick). Achenes strongly ribbed ....
    .... **var.** *cymbalaria*
    Stamens 25 or more. Stems stouter (up to 1 mm thick). Achenes more weakly ribbed ......
    .... **var.** *saximontanus*

These varieties are completely intergradient in our range, and much of the material collected in British Columbia is intermediate in form. Their differing characteristics undoubtedly are largely environmentally determined.

Variety *cymbalaria* is widespread in Asia as well as North America, and is present as an introduced plant in Europe. Variety *saximontanus* Fernald is restricted to western North America.

Ranunculus cymbalaria shows a preference for, but is not restricted to, moist sites such as shores of ponds, especially where the available soil moisture is saline or alkaline. It is common around the shores of alkaline ponds in the interior of British Columbia, but has also been found in dry sagebrush communities, and on tidal flats at the coast.

Section *Arcteranthus*

Glabrous, scapose perennials. Leaves reniform to circular, deeply divided and redivided into round-tipped lobes, rather fleshy. Bracts scale-like if present. Sepals 5. Petals 7–15, yellow; the nectary scale divided, V-shaped. Achenes longitudinally veined and ribbed, with long hooked beaks. x = 8.

Ranunculus cooleyae **Vasey & Rose**   **Cooley's Buttercup**
    Kumliena cooleyae **(Vasey & Rose) Greene**
    Arcteranthus cooleyae **(Vasey & Rose) Greene**

Small, glabrous, scapose perennial, with a small buried caudex wrapped in the ample sheathing petiole bases, and long fibrous roots. Stem 2–14 cm tall in flower, and up to 23 cm tall in fruit.

Leaves basal; the long petioles with prominently sheathing bases; the blades circular or kidney-shaped in outline, deeply dissected into three main divisions, each of which is again dissected ⅓ to ⅔ of the way to the base into several lobes, the rounded tips of which bear each a minute projecting vein tip, glabrous, rather fleshy, deep green above, and rather glaucous beneath.

Flower solitary, terminating a scape that sometimes bears a minute bract on its upper half. Sepals 5, narrowly elliptic or obovate, 7–11 mm long, greenish yellow, glabrous, spreading to reflexed, persistent through the flowering stage. Petals 7–15, oblanceolate to spatulate, with prominent claws and rounded tips, yellow, from shorter than to rather longer than the sepals; the nectary scale divided and narrowly V-shaped, its lateral margins adnate to the petal surface. Stamens 50–80. Carpels 30–90 in a hemispheric head, with prominent hooked styles. Receptacle globose, glabrous.

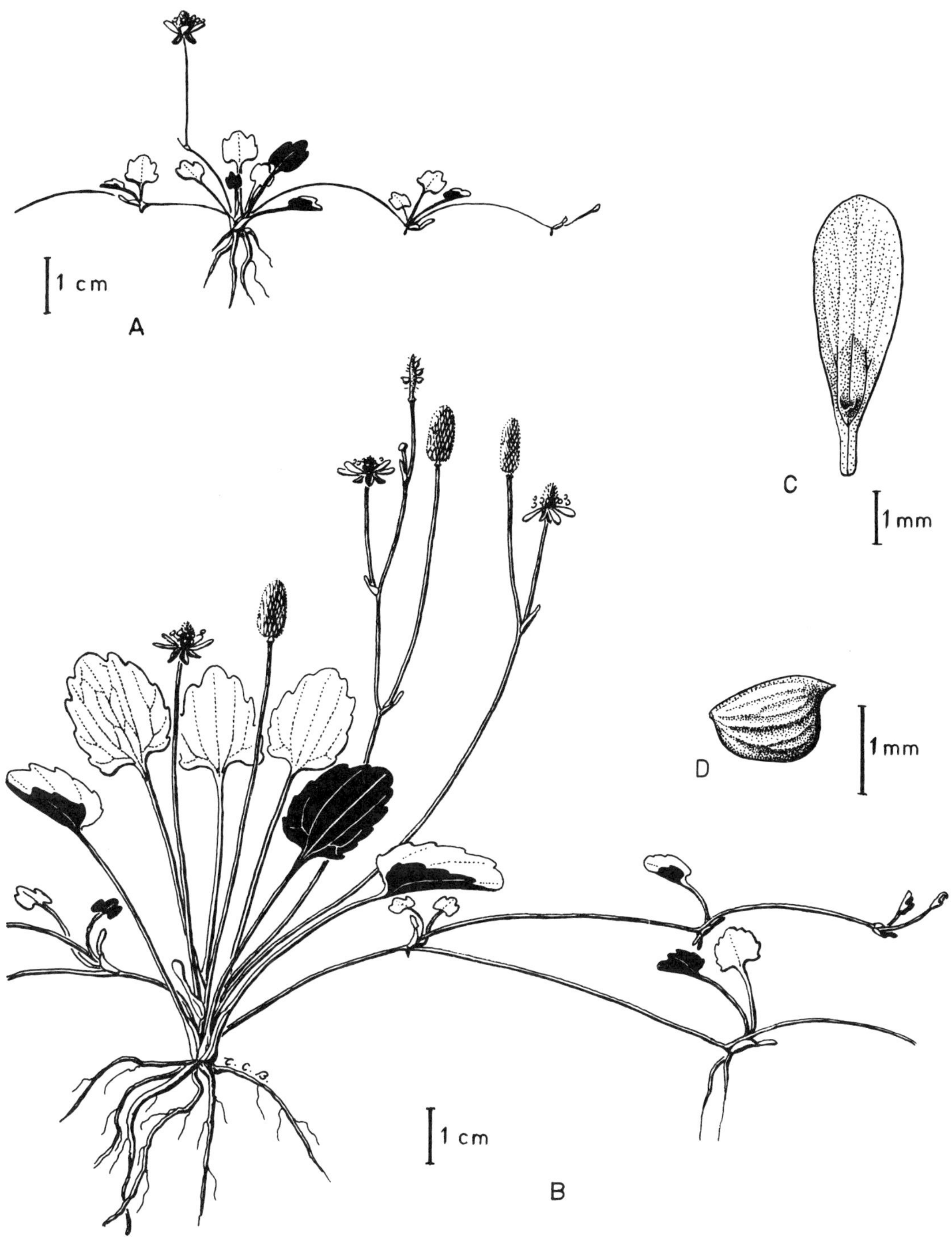

Figure 89. *Ranunculus cymbalaria*

A. plant of var. *cymbalaria*.
B. plant of var. *saximontana*.

C. petal.
D. achene.

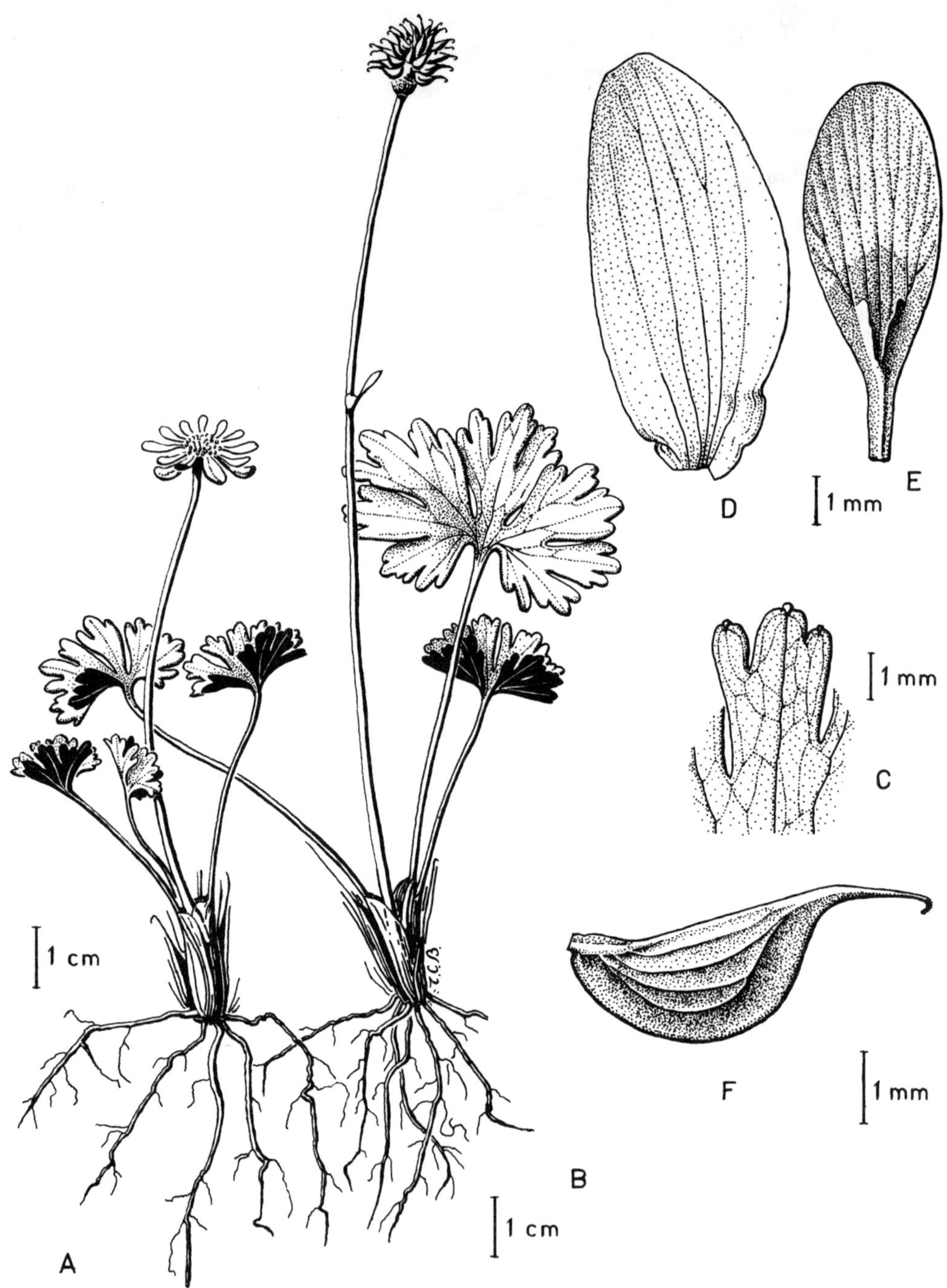

Figure 90. *Ranunculus cooleyae*

A. flowering plant.
B. fruiting plant.
C. tip of leaf lobe.

D. sepal.
E. petal.
F. achene.

Achenes in a globose head, glabrous; the achene body semi-lunate, strongly keeled, conspicuously longitudinally veined, membranous-walled, 2–3 mm long by about half as deep; tapering into a prominent stylar beak 1.2–1.5 mm long that is sharply hooked at the tip. 2n = 16.

This is a species primarily of the coastal mountain ranges, from Alaska to Washington State, with a few outlying colonies further inland. It is found in alpine tundra, around late-melting snow banks.

Benson (1948) reported a specimen from Mt. Sir Sandford [Butters & Holway 676 GH (= Gray Herbarium of Harvard University)]. Mount Sir Sandford is in the Selkirk Range; and this is the only reported occurrence of this species in that range, which is well outside the currently confirmed range of *R. cooleyae*. Unfortunately the specimen cannot now be located; so it would be desirable to seek again for this species in the northern Selkirk Range.

Section *Ceratocephalus*

Tomentose scapose annuals. Leaves 3-parted; the divisions linear, acute. Achenes tomentose, with inflated lateral chambers and long, prominent beaks. x = 7.

Ranunculus testiculatus **Crantz**　　　　　　　　　　　**Hornseed Buttercup**
Ceratocephalus testiculatus (Crantz) **Roth**

Dwarf, scapose, greyish-floccose or tomentose annual, 2–10 cm tall, with a short taproot-like hypocotyl giving rise to a cluster of fibrous roots.

Leaves all basal, slender-petioled; the petiole base narrowly dilated; the blade 1–3 cm long, divided into three linear divisions which may be again divided, pale green. Small, narrowly linear bracts are occasionally present on the scapes.

Flower solitary on each scape. Sepals 5, lanceolate to ovate, acute-tipped, 2½–5 mm long, tomentose on the outer surface, persistent in fruit. Petals 5, obovate or spatulate, 5–6 mm long, distinctly clawed, light yellow, drying white; the nectary scale truncate to convex or lanceolate and free at the sides. Carpels 20–70 in an ovoid to cylindrical head; tomentose, 3-chambered, two of the chambers empty. Receptacle slenderly cylindrical. Flowering in April, fruit ripe in May.

Achenes in an ovoid to cylindrical spike 1–3 cm long by up to 1 cm thick. The achene body broader than deep, owing to the dilation of the empty lateral chambers, keeled dorsally, and drawn out into a strong, straight or gently upcurved beak 3– 4 mm long, that is triangular in cross section at the base, tomentose, and tapering to a sharp glabrescent point. 2n = 14.

Ranunculus testiculatus is a native of southeastern Europe that has been present in the inland regions of the northwestern United States for many years. It is found from Washington to Idaho and Colorado. In British Columbia it is known from the Thompson and Okanagan Valleys. It occurs in disturbed semi-arid environments such as grassland and sagebrush steppe.

THALICTRUM L.

Meadow-Rue

Perennial rhizomatous herbs. Each segment of the horizontal, scaly rhizome terminates in an erect shoot. Branches and prolongation of the rhizome develop from buds in the axils of the scale-like leaves. Aerial leaves are reminiscent of those of *Aquilegia*: doubly or triply compound, with rounded-lobed or crenate leaflets.

Flowers apetalous, in panicles or racemes. Sepals 3–5, greenish, purplish, or pale yellowish to white. Stamens many, exserted on usually slender filaments. Carpels 2–15, with linear or lanceolate stigmas; the carpel margins sometimes not sealed at flowering time. Achene with conspicuous longitudinal veins, with a single apically attached seed in each achene, and a terminal linear stigmatic beak. x = 7.

Thalictrum includes a hundred or more species of north temperate latitudes; generally in woodlands, but sometimes in alpine or other open habitats.

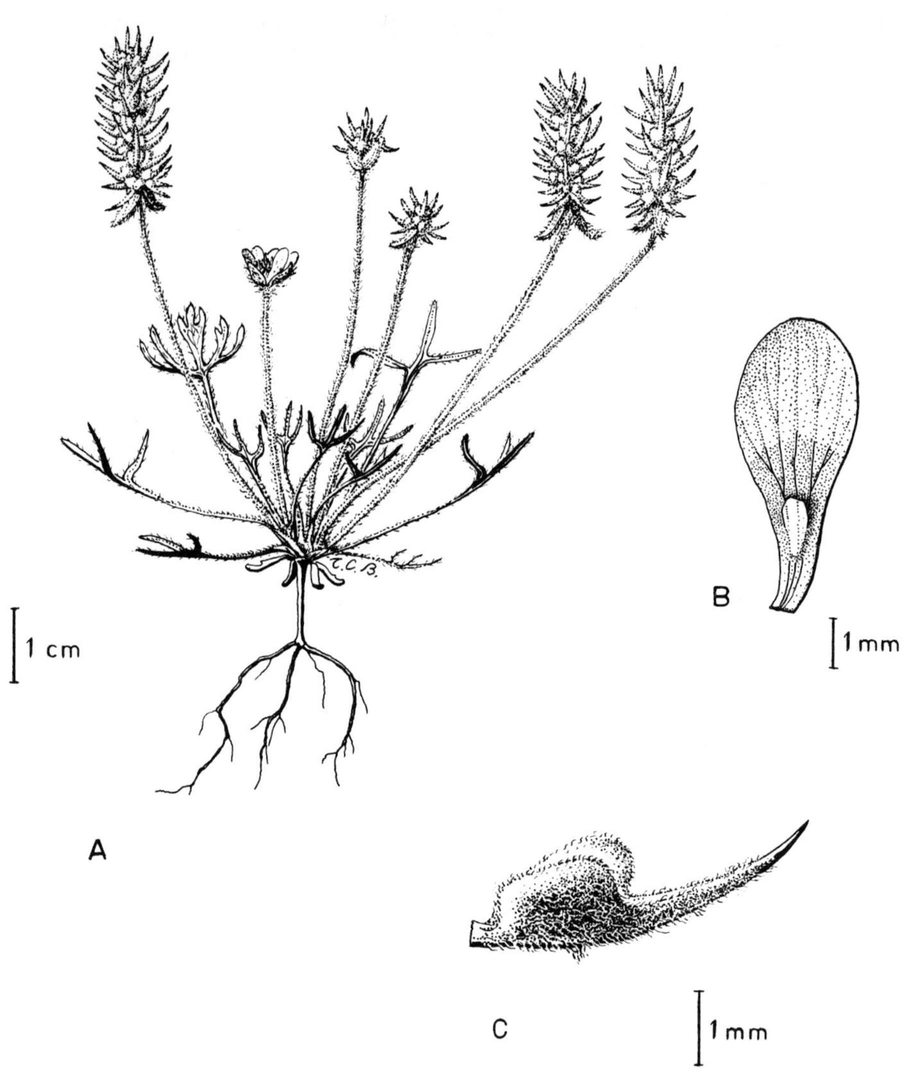

Figure 91. *Ranunculus testiculatus*

A. plant.   *B. petal.*   *C. achene.*

The flowers of *Thalictrum* are generally pollinated by wind. Though no nectar is produced in this genus, some species have white or variously coloured sepals and stamens, and appear adapted to attracting insects for pollination. All our species, however, with their small, dull flowers, the anthers dangling on threadlike, flexible filaments, appear to be adapted for wind-pollination.

KEY TO SPECIES

1. Scapose; seldom with one stem leaf; 5–20 cm tall. Flowers perfect, in a raceme. Plant of alpine habitats .... ***T. alpinum***

   Stems leafy, taller, inflorescence usually paniculate. Plants of woodlands and glades at lower elevations **2**

2. Flowers perfect. Sepals pale, whitish or greenish. Anthers less than 1 mm long. Fruit with anastomosing veins. Panicle leafy, with few flowers .... ***T. sparsiflorum***

   Flowers dioicous. Sepals green or dull purplish. Anthers more than 1 mm long. Fruit with mostly simple veins. Panicle with more flowers and reduced bracts **3**

3. Leaflets coarse, entire or shallowly 3-lobed, minutely pilose beneath; the margins revolute. Achenes about 4 mm long, with a 3 mm beak .... ***T. dasycarpum***

   Leaflets smaller, or more lobed into rounded lobes. Achenes longer, or if not, leaflet margins revolute **4**

4. Leaflets with revolute margins and veins raised beneath, 1–2 cm long. Achenes subsessile, ascending to erect, the body rather plump, 3–5 mm long by about half as wide; the beak 1–2.5 (–3) mm long. Fruiting pedicels of various lengths, some no longer than the achenes
   .... ***T. venulosum***

   Leaflets with flat, not revolute, margins, and veins little if at all raised beneath, 1.5–3 cm long. Achenes shortly stipitate, reflexed or widely spreading; the body compressed, 4–10 mm long and less than half as wide; the beak (2.5–) 3–4.5 mm long. Fruiting pedicels relatively uniform in length, all longer than the achenes .... ***T. occidentale***

Thalictrum alpinum L. var. *hebetum* Boivin   Alpine Meadow-Rue

Low (5–25 cm high) glabrous, slenderly rhizomatous perennial herb of rather fern-like aspect; the leaves usually all basal, but sometimes with one stem leaf.

Leaves usually bipinnate, the leaflets ½–1 cm long, rounded- to cuneate-based, each with usually 3 shallow rounded lobes, revolute margins, veins impressed above and raised beneath; and, in ours, moderately to strongly glaucous surfaces, especially beneath.

Inflorescence a simple, bracted raceme, sometimes with a small secondary raceme from the axil of the lowest bract. Pedicels ascending at flowering time, later arching over. Flowers perfect. Sepals 4 (–5), pale, caducous. Stamens several (8–15) with filiform flexible filaments 1.5–3 mm long, and linear anthers 1–2 mm long, the connective of each with a distinct apical projection 0.1–0.3 mm long. Carpels 2–4, with lanceolate stigmas 1 mm or less long.

Achenes 2–4, recurved, sessile or shortly stipitate, 2 ½–3 mm long, longitudinally ribbed by the raised veins, with recurved lanceolate stigmas 1 mm or less long, that may be shed by the time the achenes are ripe. 2n = 14.

Thalictrum alpinum is circumboreal in distribution, with southern extensions and outlying populations in the high mountains of North America and Eurasia. It inhabits arctic and high alpine tundra.

Our variety *hebetum* Boivin differs from the typical var. *alpinum* of the Old World by the dull, glaucous appearance of the foliage and stem; those of the var. *alpinum* being of a deeper, shiny green colour (Boivin; 1944: p. 356).

Thalictrum dasycarpum Fischer, Meyer & Ave-Lallemant *in* Fischer & Meyer

Normally dioicous, thinly puberulent to glabrous, perennial herb with a simple or branched, hollow stem 6–18 dm tall.

Leaves mainly on the stem, at least on flowering shoots. Lower leaves petioled, 3–5 times ternate, or biternate and the divisions then pinnately divided with 5 leaflets; the leaflets obovate, rounded at base, acutely or obtusely 3- to 5-lobed at apex, or sometimes unlobed, short-pubescent to glabrous, rather leathery in texture, glaucous beneath, often darkening on drying; the margin often minutely revolute. Upper leaves sessile, further up becoming smaller, less divided versions of the same general pattern as the lower leaves.

Flowers in panicles terminating the stem and branches, with bracts like very reduced leaves. Sepals greenish white. Stamens 18–25; the anthers 1.5–3 mm long, linear and apiculate, on slightly longer filaments. Carpels around 10. Occasionally plants are found with perfect flowers.

Achenes shortly stipitate; the body slenderly obovoid to ovoid, 4–4.5 mm long, its veins forming prominent longitudinal ribs, glabrous or sparsely puberulent; the beak, formed mainly of the persistent linear stigma, about 3 mm long. $2n = 14, 168$.

KEY TO VARIETIES

1. Anthers 1.8–2.2 mm long. Leaflets leathery, revolute. Stigma 2–3 mm long **var. *dasycarpum***
    Anthers 2.2–3.2 mm long. Leaflets membranous. Stigma 2.5–5 mm long................................
    .... **var. *hypoglaucum***

Thalictrum dasycarpum ranges from eastern British Columbia eastward to Ontario, and southward through the Rocky Mountain States to Arizona, Texas and Louisiana. It occurs in open woods and alpine meadows. In British Columbia, where it is uncommon, it is found in the Columbia Forest region. The material seen in this province shows an intergradation between var. *dasycarpum* and var. *hypoglaucum* (Rydberg) Boivin.

Thalictrum occidentale **Gray**   **Western Meadow-Rue**

Erect, dioicous perennial herb with a creeping scaly rhizome, one or two basal leaves, and a leafy stem 2–10 dm tall, glandular-puberulent to glabrous.

Strictly basal leaves commonly absent from flowering shoots, having been produced the year before. Stem leaves with flaring stipular sheaths, petioled, 3–4 times ternate or pinnate; the leaflets broadly obovate to orbicular, rounded to cordate at base; the veins usually not conspicuous, the margin usually flat.

Inflorescence a terminal panicle, the lower bracts of which are like reduced leaves, the upper ones small, simple, and similar in size to their stipular sheaths. The panicle is small, compact and dense when flowering starts, but expands considerably, with elongation of the peduncles and pedicels, as flowering continues. Mature pedicels are generally of more or less equal lengths; and at fruiting time are longer than the achenes. Sepals commonly 4, whitish to pale greenish, sometimes tinged with purple. Stamens 15–30, the anthers linear, often purplish, 1.5–4 mm long, dangling on the ends of filaments that elongate to 4–8 mm long. Carpels 7–16, glandular-puberulent, becoming glabrous; the stigmas linear, purple, 3–6 mm long (or often rather shorter on Vancouver Island, and in the northern var. *breitungii*).

Achene glabrous, with a very short stipe, and a somewhat flattened, simple-veined body 4–10 mm long, tipped by the short style and linear stigma.

A very variable species, in which several varieties have been distinguished. When in fruit, varieties may be diagnosed by the following key:

KEY TO VARIETIES

1. Achene body 5–9 mm long. Achenes spreading... **2**
    Achene body 3–5 mm long.. **3**

2. Stigmas 5–6 mm long... **var. *macounii***
    Stigmas (2.5–) 3–5 mm long... **var. *occidentale***

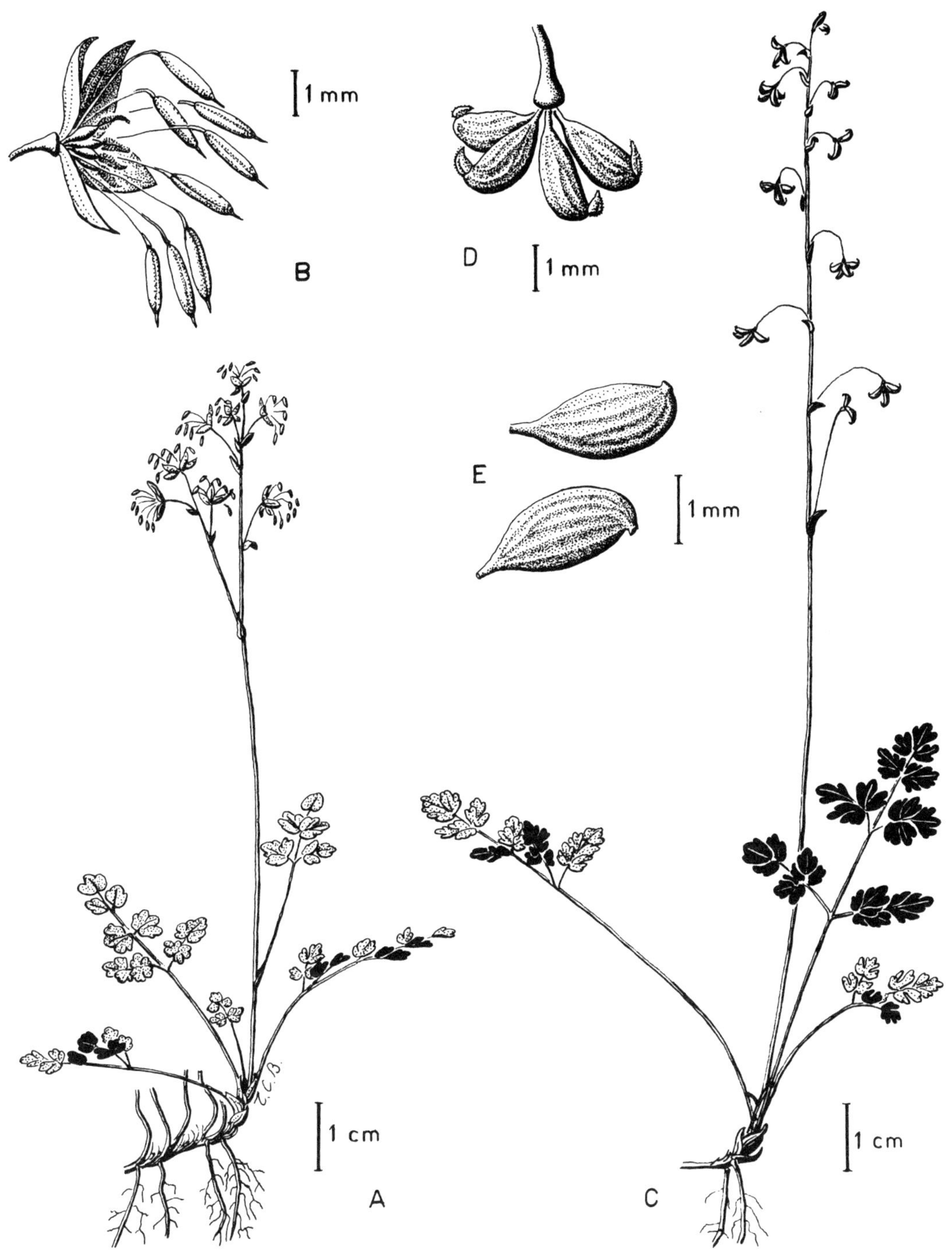

Figure 92. *Thalictrum alpinum*

A. flowering plant.   B. flower.   C. fruiting plant.   D. cluster of achenes.   E. achenes.

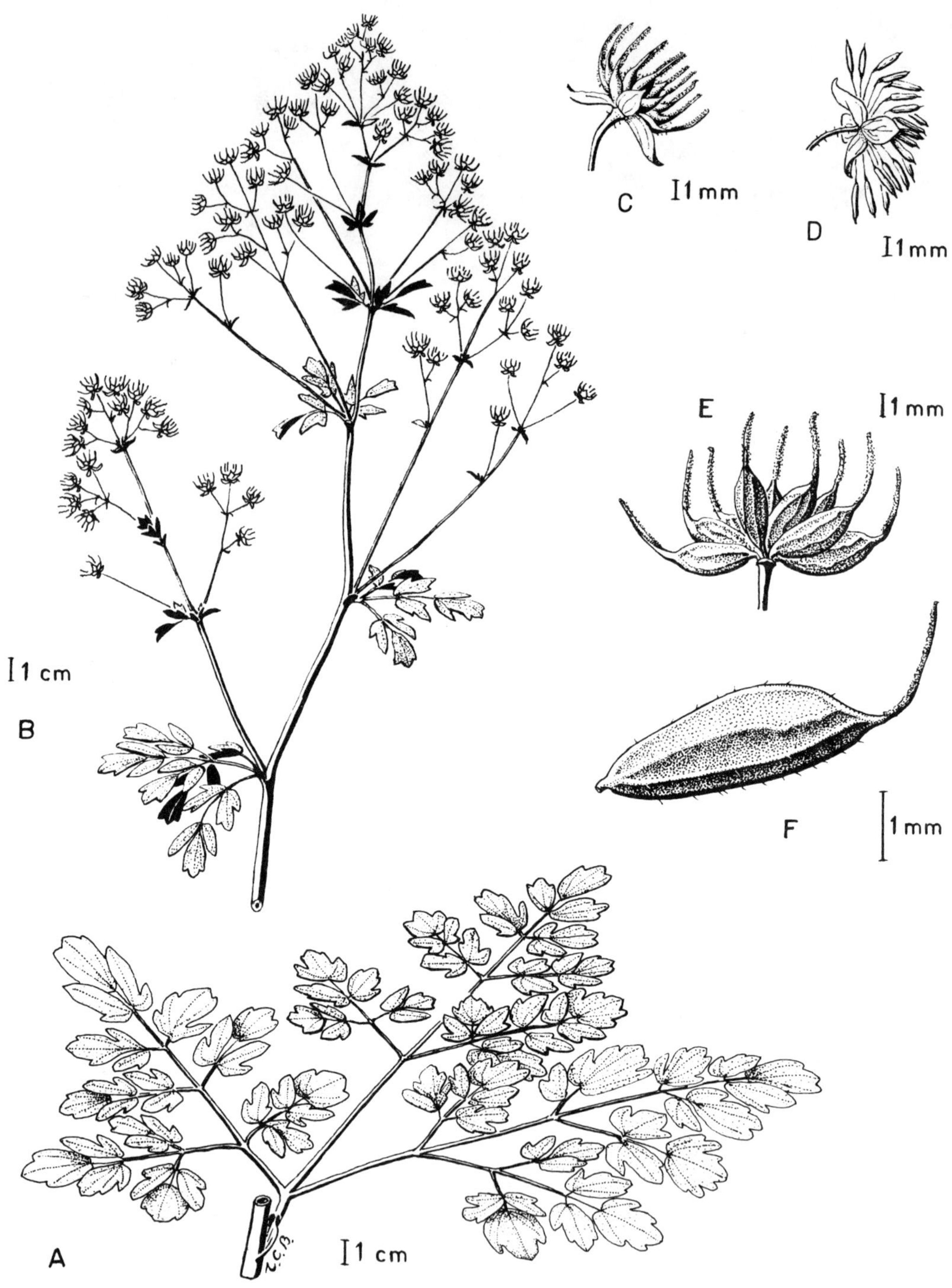

Figure 93. *Thalictrum dasycarpum*

A. mid-stem leaf.
B. upper part of pistillate plant and panicle.
C. pistillate flower.

D. staminate flower.
E. head of achenes.
F. achene.

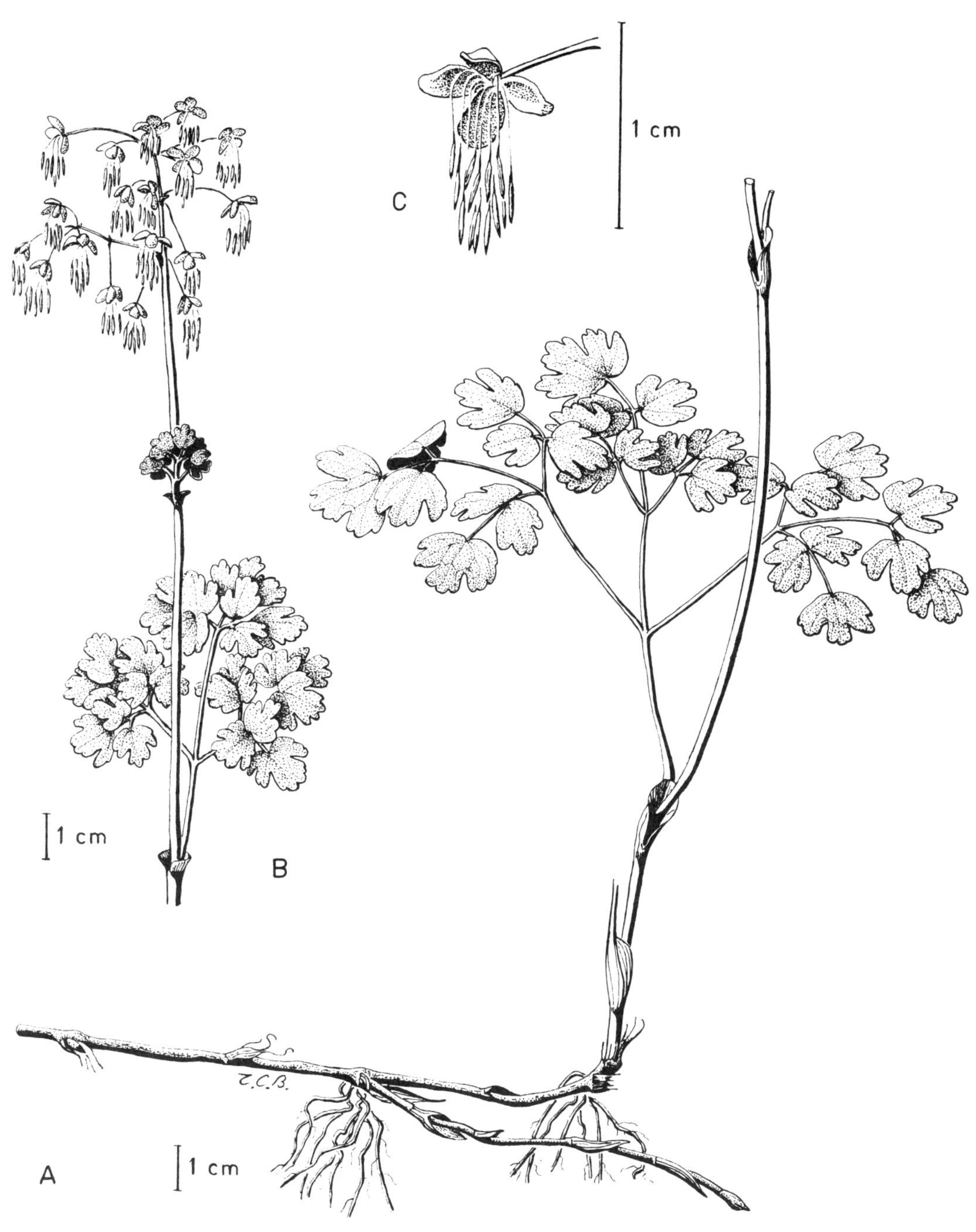

Figure 94. *Thalictrum occidentale*

A. rhizome and base of flowering shoot.
B. top of staminate flowering plant; flowering well advanced.
C. staminate flower.

3. Achenes ascending .... **var. *breitungii***
   Achenes reflexed .... **var. *palousense***

Our commonest and most widespread species of *Thalictrum* in British Columbia, *T. occidentale* ranges from southeastern Alaska to northern California, and eastward to Alberta, Montana and Colorado. In British Columbia, this is the only species of this genus found west of the Coast and Cascade Ranges in non-alpine habitats.

Typically a plant of mixed or deciduous woodlands, it may be found in grassy open areas and even at alpine levels.

Though in their typical forms *T. occidentale* and *T. venulosum* are clearly distinct, at least when mature pistillate plants are examined, difficulty is sometimes encountered, particularly in northern areas, where the stigmas of the former are relatively short. Differences in pedicel length are less clearly marked in staminate than in pistillate material; and leaf characters may then be the most reliable means for distinguishing between these species.

Of the material of this species from British Columbia that can be assigned to any specific variety, the greatest part falls into var. *occidentale*; but a few specimens are assignable to each of the following varieties.

**Var. *breitungii* (B. Boivin) T. C. Brayshaw *stat. nov.* (*T. breitungii* B. Boivin, 1948), is distinguished from other varieties of *T. occidentale* by its ascending achenes with short, compressed achene bodies 3–5 mm long, and stigmas only 2–4 mm long. In these respects it appears rather intermediate between this species and *T. venulosum*. It is found in scattered locations in northern British Columbia [Portage Mountain, Blanchard and Tatshenshini Rivers, Ghost Mountain (57°35′N., 128°47′W.)].

Var. *macounii* B. Boivin occurs in widely scattered localities in southern British Columbia, including Vancouver Island.

Var. *palousense* St. John is uncommon on the southern mainland of the province.

Thalictrum sparsiflorum **Turczaninow** *ex* **Fischer & Meyer var. *richardsonii*
(Gray) Boivin**   **Few-Flowered Meadow-Rue**

Leafy-stemmed plant up to a metre tall. Leaves biternate to triternate, the leaflets 6–20 mm long, divided into small rounded ascending lobes; the lower surface often glandular, somewhat glaucous. Plant sweet-scented on drying.

Inflorescence few-flowered and notably leafy; the bracts leaf-like, though smaller than the leaves proper.

Flowers perfect, or rarely unisexual, small. Sepals pale, scarcely greenish, 2.5–3 mm long. Stamens 7–20, white; anthers 0.4–0.8 mm long, on filaments that elongate to about 5 mm long. Carpels 5–20, glabrous, glandular, or thinly puberulent, about 2 mm long, with linear stigmas about 1–1.5 mm long.

Achenes erect to spreading, typically glabrous, on pedicels that develop apical crooks; the achene body 4–6 mm long by 2.3–3 mm wide, compressed, thin-walled, with a straight or convex dorsal edge, and 3 or 4 anastomosing veins. 2n = 14.

Thalictrum sparsiflorum is found in woods and thickets. The species ranges across Siberia and western North America; on this continent extending from Alaska southward via the Rocky Mountains to Colorado and California. Variety *richardsonii* occurs in British Columbia, Alberta, Yukon, Alaska, and Kamtschatka. In British Columbia it is found mainly in the Boreal and northern Subalpine Forest regions. Variety *sparsiflorum*, with the glabrous achene body 2 mm wide, and with a concave dorsal edge, is confined to Siberia. Variety *saximontana* Boivin, with pubescent achenes on short stipes 0.3–1 mm long, occurs in the Rocky Mountain States and Oregon (Boivin, 1944; 369–373).

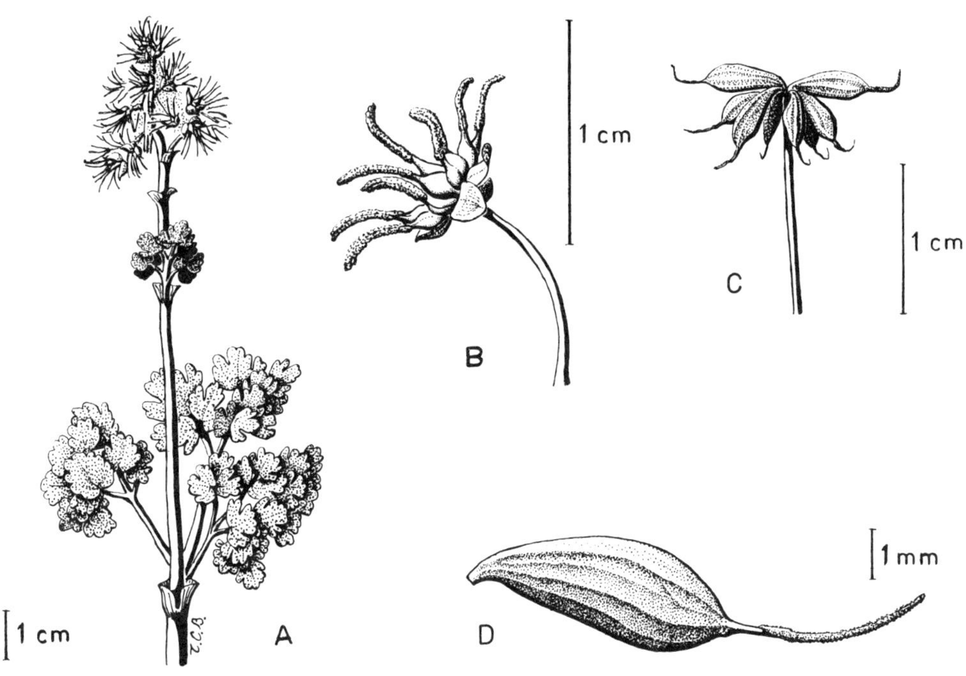

Figure 95. *Thalictrum occidentale*

A. top of pistillate flowering plant at beginning of flowering.
B. pistillate flower.
C. head of achenes.
D. achene.

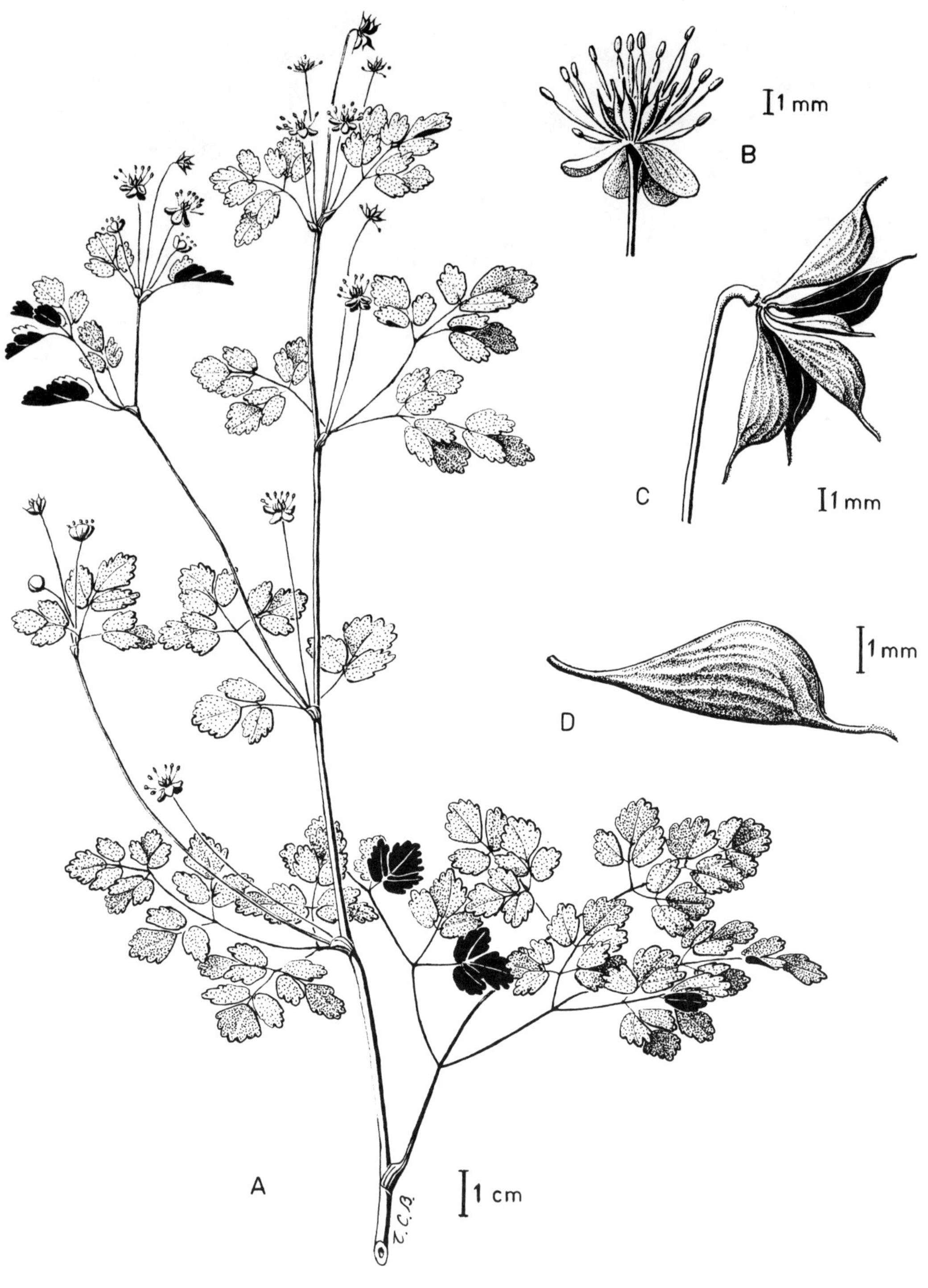

Figure 96. *Thalictrum sparsiflorum*

A. upper part of flowering plant.
B. flower.
C. head of achenes.
D. achene.

Thalictrum venulosum **Trelease**
  T. **columbianum Rydberg**
  T. **occidentale Gray var.** *columbianum* **(Rydberg) St. John**
  T. **turneri Boivin**

Dioicous herbaceous perennial resembling *T. occidentale* in habit, 3–7 dm tall, glabrous or sparsely glandular or puberulent.

Leaves mostly cauline; the basal leaves often absent from flowering shoots. Basal and lower stem leaves 3–4 times ternate or pinnate, and long-petioled; the leaflets nearly orbicular to broadly obovate, more or less cordate to rounded at base, ¾–2 cm long, glaucous beneath, with blunt to rounded lobes or teeth, and generally revolute margins; the veins in a fine but conspicuous net, the main veins prominent and raised beneath. Upper leaves sessile, progressively smaller, and with fewer leaflets, in a gradual transition to the bracts of the inflorescence.

Flowers in a rather slender terminal panicle; the pistillate flowers on pedicels of diverse lengths, the shorter pedicels often shorter than the achenes; this diversity not so pronounced in staminate inflorescences. Sepals 4 or rarely 5, pale greenish white, 2–5 mm long, broadly ovate. Stamens 10–20; the anthers 2–3.5 mm long, linear and apiculate, on slightly longer filaments. Carpels 7–15; the linear to lanceolate stigmas up to 3 mm long and often basally broadened.

Achenes glabrous, subsessile, more or less ascending to erect; the body plump, 3–5 mm long, with simple longitudinal raised veins; the stigmatic beak 1–3 mm long. 2n = 42.

Thalictrum venulosum is a plant of open woods and grassy clearings. It ranges from the interior of British Columbia, and possibly Yukon, across the Prairie Provinces to Ontario, and perhaps Quebec, and southward to eastern Oregon and Minnesota.

On the average, *T. venulosum* is a finer-structured, more slender plant than *T. occidentale*; however, this is not a wholly reliable criterion for distinguishing between these species (see note under *T. occidentale*).

TRAUTVETTERIA Fischer & Meyer

False Bugbane

Rhizomatous perennial herbs with palmately lobed and veined leaves, and erect stems with terminal flattish corymbs of radially symmetrical, apetalous flowers.

Sepals 3–5, deciduous as the flower opens, many stamens with inconspicuous anthers, and several free, basally single-ovuled carpels with terminal hooked stigmas. Fruit a head of longitudinally ribbed achenes with hooked beaks derived from the styles and stigmas. x = 8.

A small genus of North America and eastern Asia; containing one, two, or three species, according to interpretation.

Trautvetteria caroliniensis **(Walter) Vail var.** *occidentalis* **(Gray) C.L. Hitchcock**
  T. **grandis Nuttall** *in* **Torrey & Gray**
Stems scattered on spreading rhizomes, producing colonies. Stems ½ to 1 metre tall, simple or sparingly branched, glabrous below but thinly crisp-puberulent in the inflorescences.

Basal leaves 1–3 dm wide, palmately cleft into 5–11 toothed lobes, the veins impressed. Stem leaves smaller and less divided.

Inflorescence a terminal flattish corymb at the apex of the stem or of a branch. Pedicels, except the lowest, bractless. Flowers opening simultaneously in each corymb, the corymb terminating the main stem flowering before any corymbs terminating axillary branches. Each flower about 2 cm wide when fully open, with 50–70 white stamens 5–10 mm long, with upwardly dilated filaments and inconspicuous anthers. Carpels 12–18 on a rather columnar receptacle.

Achenes 3–4 mm long, thin-walled and bladdery, with two prominent longitudinal veins on each side, and with a hooked terminal beak. 2n = 16.

In spite of the lack of petals, the densely tufted white stamens make the inflorescence conspicuous and ornamental. The flowers are visited by bees.

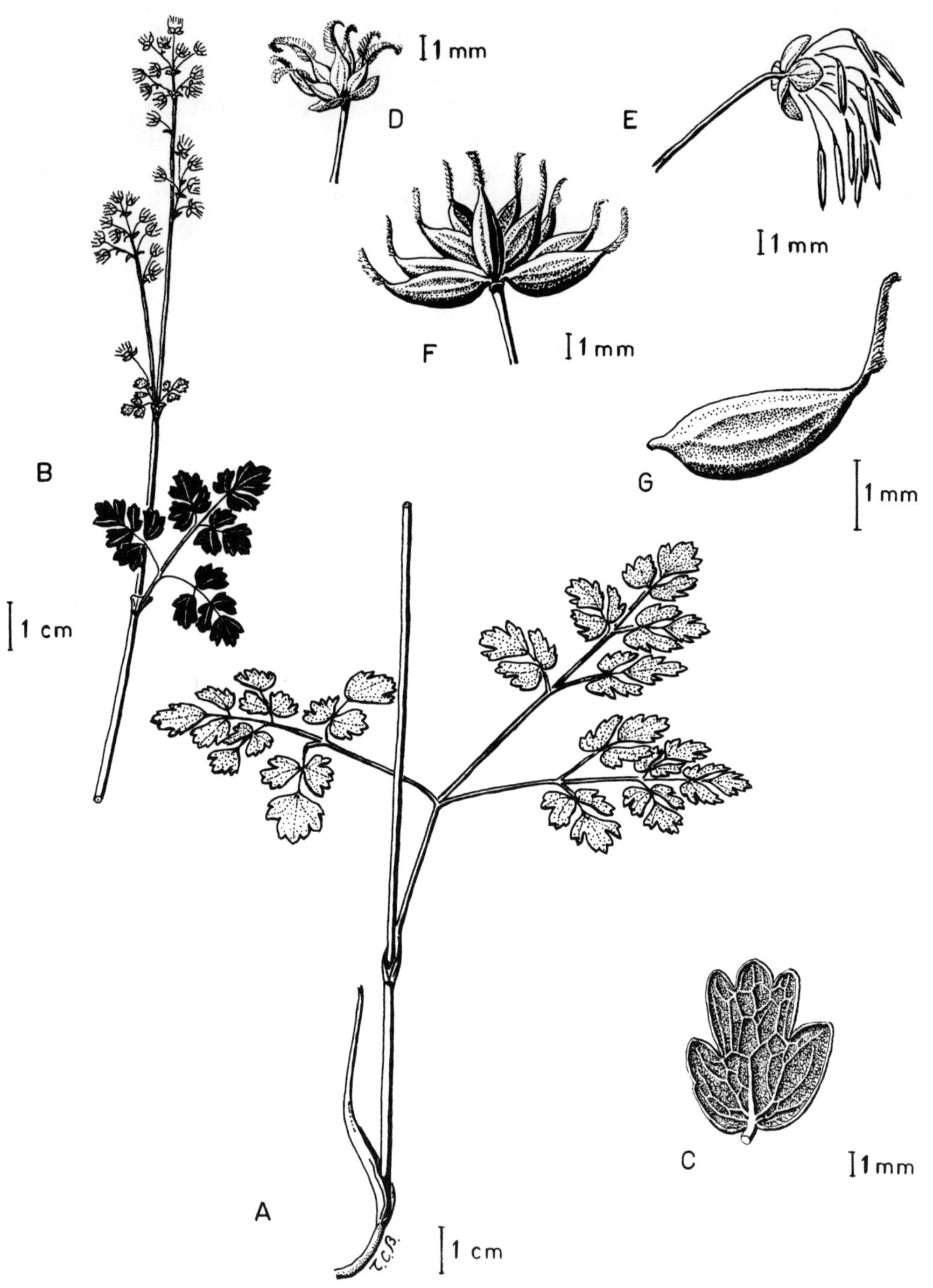

Figure 97. *Thalictrum venulosum*

A. base of plant, and leaf.
B. top of pistillate plant, and panicle.
C. leaflet, showing underside.
D. pistillate flower.
E. staminate flower.
F. head of achenes.
G. achene.

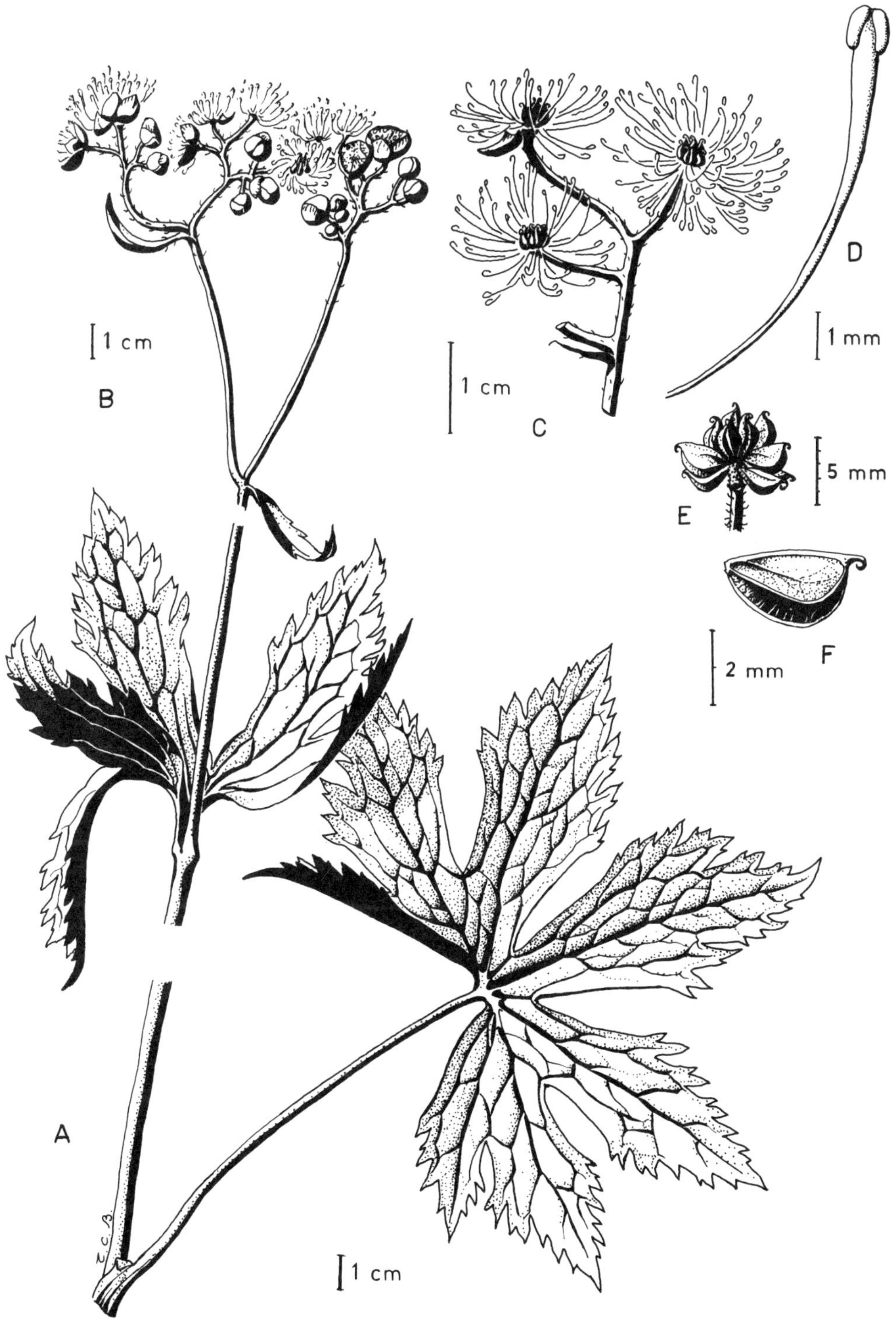

Figure 98. *Trautvetteria caroliniensis*

A. lower stem leaf.
B. top of flowering plant.
C. flowers.
D. stamen.
E. head of achenes.
F. achene.

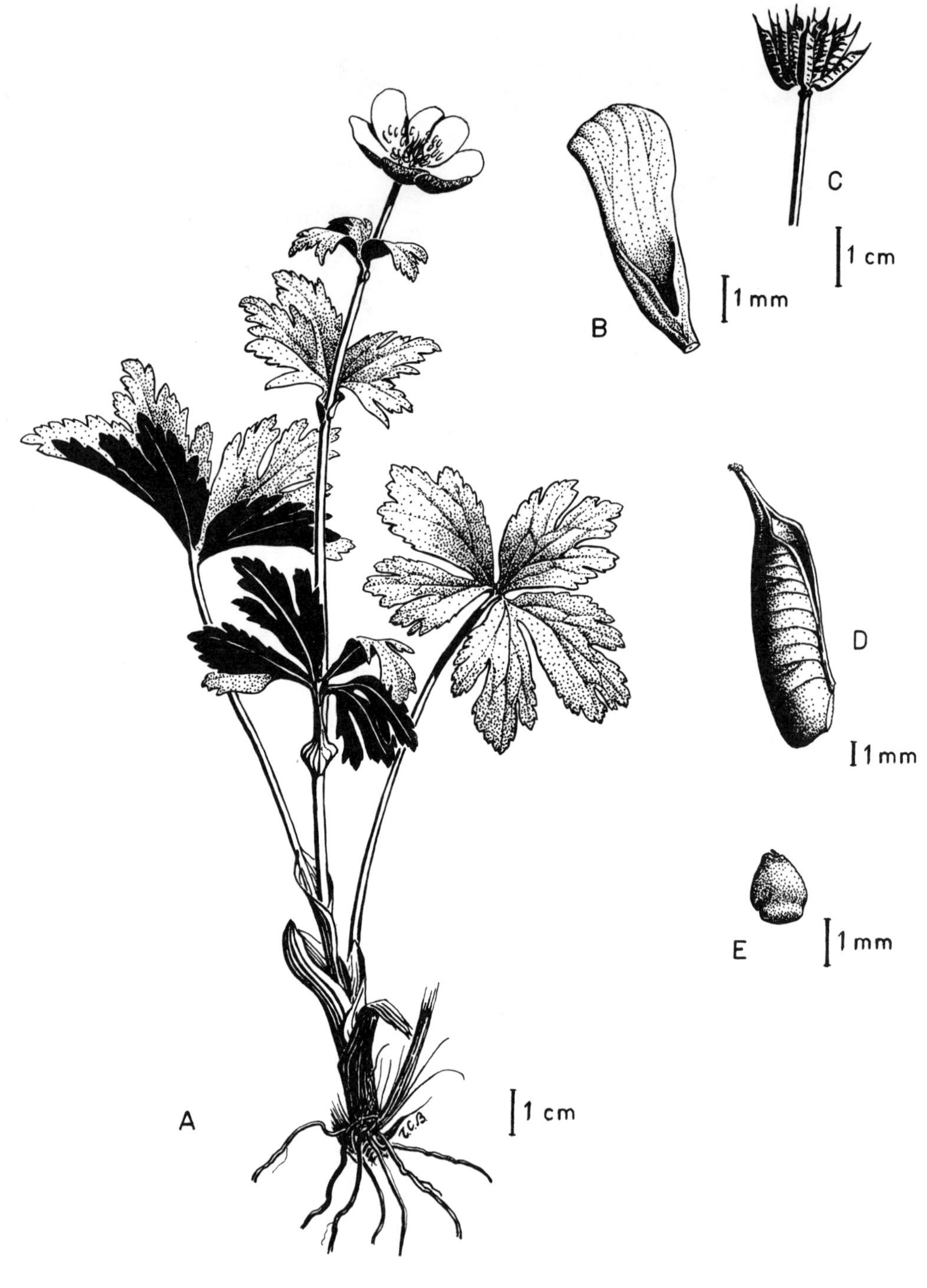

Figure 99. *Trollius laxus* var. *albiflorus*

A. flowering plant. B. petal. C. head of follicles. D. follicle. E. seed.

This species ranges from southern British Columbia to California; inhabiting shady, moist forest habitats in the Coast Forest, and, more rarely, in the Columbia and Subalpine Forest regions. Variety *caroliniensis*, distinguished by the firmer leaves, and the stamens averaging slightly longer, occurs in the eastern United States.

TROLLIUS L.

Globe Flower

Perennial herbs with erect, buried caudices and fibrous roots. Leaves alternate and spirally arranged, palmately lobed to compound.

Flowers one or few, terminal. Sepals 5–12, petaloid, showy, coloured, spirally arranged. Petals several, usually small and inconspicuous, and functioning only as nectaries, with nectar glands in pockets near their bases. Stamens and carpels numerous; the latter sessile on a short receptacle, free, multi-ovuled; forming a head of follicles in fruit. $x = 8$.

A genus of about 12 species of north temperate latitudes; one species in North America. All species are poisonous.

Trollius laxus Salisbury var. *albiflorus* Gray
T. albiflorus (Gray) Rydberg

Glabrous perennial with one to several stems, 10–50 cm high at flowering time; but often becoming taller by subsequent stem elongation. Leaf and petiole bases with ample stipular sheaths. Basal leaves with long petioles, stem leaves short-petioled or sessile on their sheaths, palmately lobed or compound.

Flower terminal and usually solitary; but occasionally, additional flowers develop from buds in the upper leaf axils. Sepals 5–12, white, 1–2 cm long, obovate. Petals 7–14, 2–6 mm long, oblong or narrowly obovate, yellowish, concealed under the numerous, much longer stamens; the nectar glands in narrow, near-basal pockets. Carpels 5–30, averaging around 13.

Follicles 7–13 mm long, distinctly transversely veined, and sometimes papillose, tipped by a short stylar beak and stigma; and opening from the apex along the ventral suture. Seeds ovoid to subglobose. $2n = 32$.

Trollius laxus var. *albiflorus* ranges from central British Columbia to Colorado in the mountains; commonly in moist to wet sites in the Subalpine Forest region. Flowering starts soon after the snow melts, and before the leaves, especially the basal leaves, have fully expanded; the flowers at first often being quite close to the ground. The stems and pedicels elongate as the flowers age and the fruits mature.

Variety *laxus*, of the central and eastern United States, differs from our variety in having yellowish to greenish sepals.

The Family BERBERIDACEAE

Barberry Family

Shrubs or herbs with compound or simple leaves. Flowers with multiple whorls of perianth segments, rarely perianth absent. Stamens opposite to and basally adnate to the petals, and usually of the same number; the anthers opening by uplifting valves. Carpel one, containing one to several ovules, and with a disc-like, usually sessile stigma. Fruit a berry, capsule or achene. Pollen grains with 3 to many grooves.

A family primarily but not exclusively of temperate climates of Eurasia, North Africa and the Americas; containing about 600 species in 10 genera.

KEY TO GENERA

1. Shrubs. Flower with multiple perianth whorls... ***Berberis***
   Herbs with basal trifoliolate leaves. Flower without perianth... ***Achlys***

ACHLYS DC.

Vanilla-Leaf

Rhizomatous perennial herbs without leafy aerial stems. Leaves trifoliolate. Flowers in a scapose spike, without sepals or petals. Stamens 8–20, each with 2 anther cells opening by uplifting valves. Carpel one with a sessile stigma. Fruit a single-seeded achene. x = 6.

Achlys consists of one or two species, according to interpretation (see note under *A. triphylla*), in western North America and Japan.

Achlys triphylla (J.E. Smith) DC.   **Vanilla-Leaf**
    Including *Achlys californica* Fukuda & Baker = *A. triphylla* ssp. *californica* (Fukuda & Baker) T.C. Brayshaw *stat. nov.*
    ***Leontice triphylla* J.E. Smith**

Scapose perennial with a slender rhizome rooting at the nodes; and glabrous except for the fruit.

Leaflets coarsely sinuately dentate, the lateral ones very asymmetrical, often in a nearly vertical attitude; emitting a vanilla-like odour on drying.

Flower spike white, 2–4 cm long by 1 cm thick in flower; elongating to twice as long or more in fruit. The lower flowers sometimes with minute bracts. Stamens white, the filaments widening upward. Anther opening partly by spreading of the lateral valve and partly by uplifting of the inner valve. Stamens maturing and elongating in succession in each flower; up to 6 mm long when extended.

Carpel straight at flowering time but curving back and becoming very asymmetrical at fruiting time.

Achene dark grey-brown, leathery, of a peculiar, thickly crescentic form; containing one basally attached seed. 2n = 12, 24. Attached externally, to the inner side of the 'crescent' is a fleshy appendage of no established function. Other small fruits and seeds provided with such appendages, including the seeds of the related *Vancouveria hexandra*, are dispersed by insects such as ants, or wasps in the case of *Vancouveria* (Pellmyr, 1985). This method of dispersal has not been recorded for *Achlys*, but may be watched for.

This plant inhabits shady woodlands in the Coast Forest region, from British Columbia to California, and also occurs in Japan (Fukuda, 1967).

Fukuda (1967) noted that the North American population of *Achlys triphylla* consists of two sub-populations with different chromosome numbers. One, which includes the type specimen, is diploid (2n = 12), has terminal leaflets averaging 1958 sq. mm. in size, commonly with 3 (range: 3–8) teeth, and stoma sizes averaging 50 micrometers. It tends to occupy well-drained, often open, upland sites. The other sub-population is tetraploid (2n = 24), has terminal leaflets averaging 6010 sq. mm. in size, with on average, 7 (range: 3–12) teeth, and stoma sizes averaging 62.5 micrometers, and tends to occupy deeply shaded forests on low bottomland sites.

Achlys japonica Maximowicz, of Japan, is diploid and so like the diploid North American *A. triphylla* that Fukuda proposed treating it as *A. triphylla* ssp. *japonica* (Maximowicz) Fukuda. The tetraploid population, on the other hand, is treated as a distinct species: *A. californica* Fukuda & Baker.

The distinction between the diploid and tetraploid forms, while undoubtedly real in that it is based on a genetic character difference that would confer a degree of genetic incompatibility (Fukuda reported few triploids), is not easy to discern through its gross morphological expression. The ranges in number of terminal leaflet teeth at the two ploidy levels overlap completely. Examination of material at the Royal British Columbia Museum has disclosed that the leaflet area, which is susceptible to influence by the environment and by the age and health of the plant, does not vary directly with tooth number except as a very general trend.

This leaves stomatal size and counts of the chromosomes themselves as the only sure means of distinguishing between the diploid and tetraploid populations. These observations require the use of high-powered microscopes and techniques that are not practical for the average naturalist.

The author feels therefore that it would be more practical to retain the tetraploid population within the species *A. triphylla*; giving it formal recognition as a subspecies: **Achlys triphylla (J. E. Smith) DC. ssp.** *californica* **(Fukuda & Baker) T. C. Brayshaw** *stat. nov.* (*A. californica* Fukuda & Baker *in* Fukuda, Ichiro; 1967. Taxon 16:315).

BERBERIS L.

Including *Mahonia* Nuttall

Barberry, Oregon-Grape

Shrubs with yellow wood and bark, sometimes with creeping subterranean rhizomes.

Leaves alternate or in alternate fascicles, commonly some of them modified as spines; simple or compound (sometimes with a single leaflet), deciduous or evergreen and stiff, entire or spinulose-margined.

Flowers usually in axillary racemes, yellow or sometimes orange. Perianth in five or six whorls of three members each. The outermost one or two whorls, often interpreted as bractlets, are smaller than the others, and deciduous early. The next two whorls, of sepals, enclose six petals in two whorls; each petal bearing a pair of bare nectar glands near the base, and apically notched or entire. Stamens six, opposite to, and basally adnate to, petals; the anthers opening by uplifting valves. The ripe stamens are sensitive to touch; and upon stimulation by a pollinating insect, turn forward to apply the pollen to it. Carpel one, with a sessile, peltate stigma. Fruit a berry containing one to several seeds.

Taken in a broad sense, including *Mahonia*, *Berberis* has about 500 species, and occurs mainly in warm temperate climates of all the vegetated continents except Australia.

The species with unarmed stems and pinnately compound evergreen leaves are commonly placed in a separate genus: *Mahonia* Nuttall, which includes our western North American native species and many more in Asia.

Many species, both of *Berberis* proper and *Mahonia*, are very ornamental, and are widely grown in gardens. Deciduous species of *Berberis* are alternate hosts of the Wheat Rust disease, and their sale and cultivation are prohibited. Fortunately, our native species are not affected by this disease.

KEY TO SPECIES

1. Leaves simple, deciduous, fascicled in axils of branched spines (modified leaves). Branches angled. Petals entire. Introduced and naturalized.. **Subgenus *Berberis* . . . *B. vulgaris***

   Leaves pinnately compound, evergreen, alternate on the spineless, terete stem. Branches not angled or grooved. Petals apically notched. Native species .... **Subgenus *Mahonia*   2**

2. Leaflets palmately veined from base, dull-surfaced. Bud scales persistent, 2–4 cm long. Filaments not appendaged.. ***B. nervosa***

   Leaflets pinnately veined from base, often lustrous. Bud scales less than two cm long, deciduous or soon disintegrating. Filaments with a pair of small lateral appendages........... .... ***B. aquifolium***

Berberis aquifolium **Pursh**   **Tall Oregon-Grape**
   Mahonia aquifolium **(Pursh) Nuttall**
   Odostemon nutkanus **(DC.) Rydberg** = *B. aquifolium* **ssp.** *aquifolium*
   Odostemon aquifolium **(Pursh) Rydberg** = *B. aquifolium* **ssp.** *repens*

Rhizomatous shrub of variable habit: from low (10 cm) and creeping to tall (to 4 m) and erect or arching. Bud scales triangular, less than 2 cm long, commonly deciduous or disintegrating at the end of their first season.

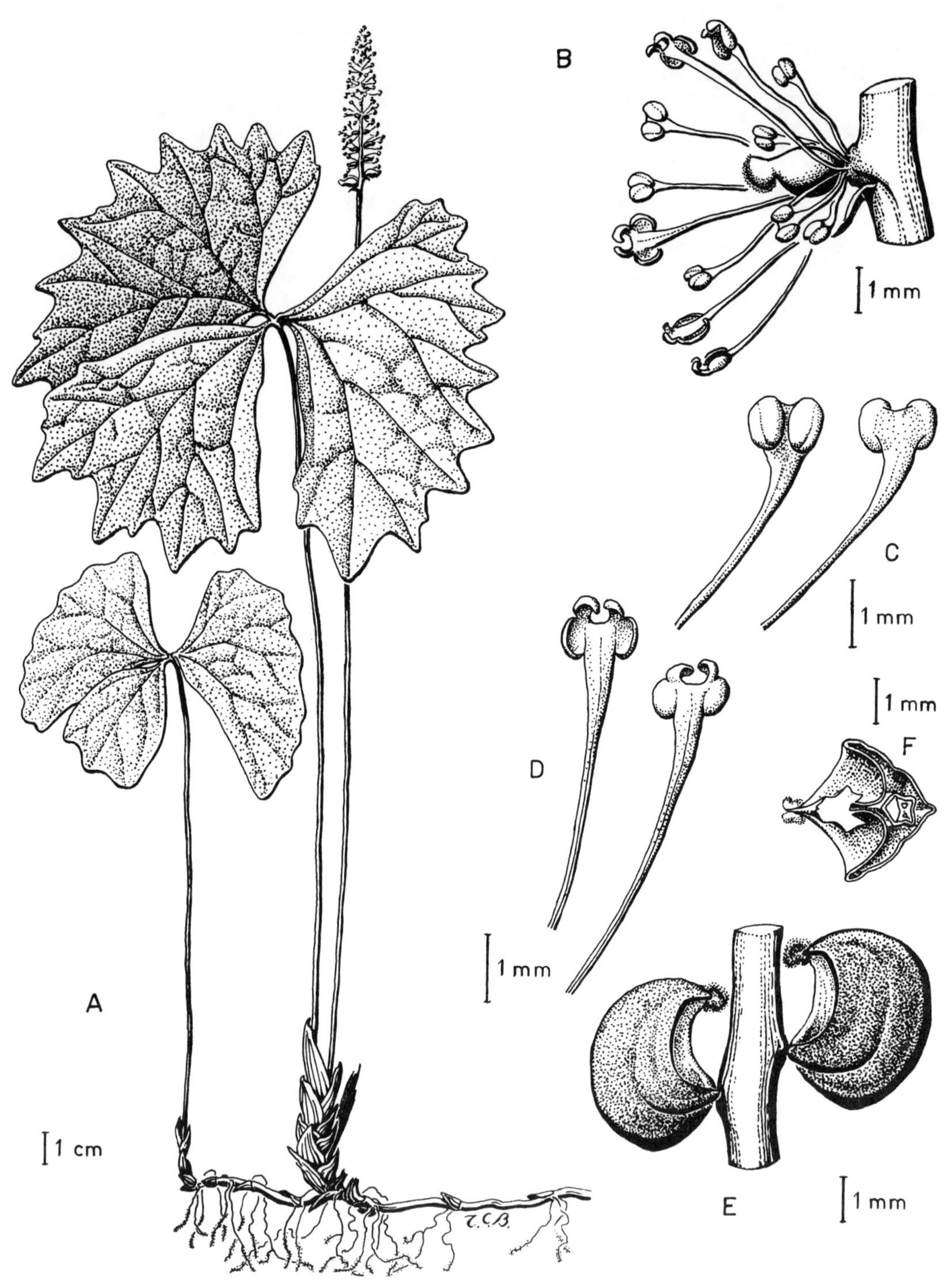

Figure 100. *Achlys triphylla*

A. flowering plant.
B. flower.
C. stamens with anthers closed.
D. stamens with anthers open.
E. 2 achenes.
F. cross-section of achene.

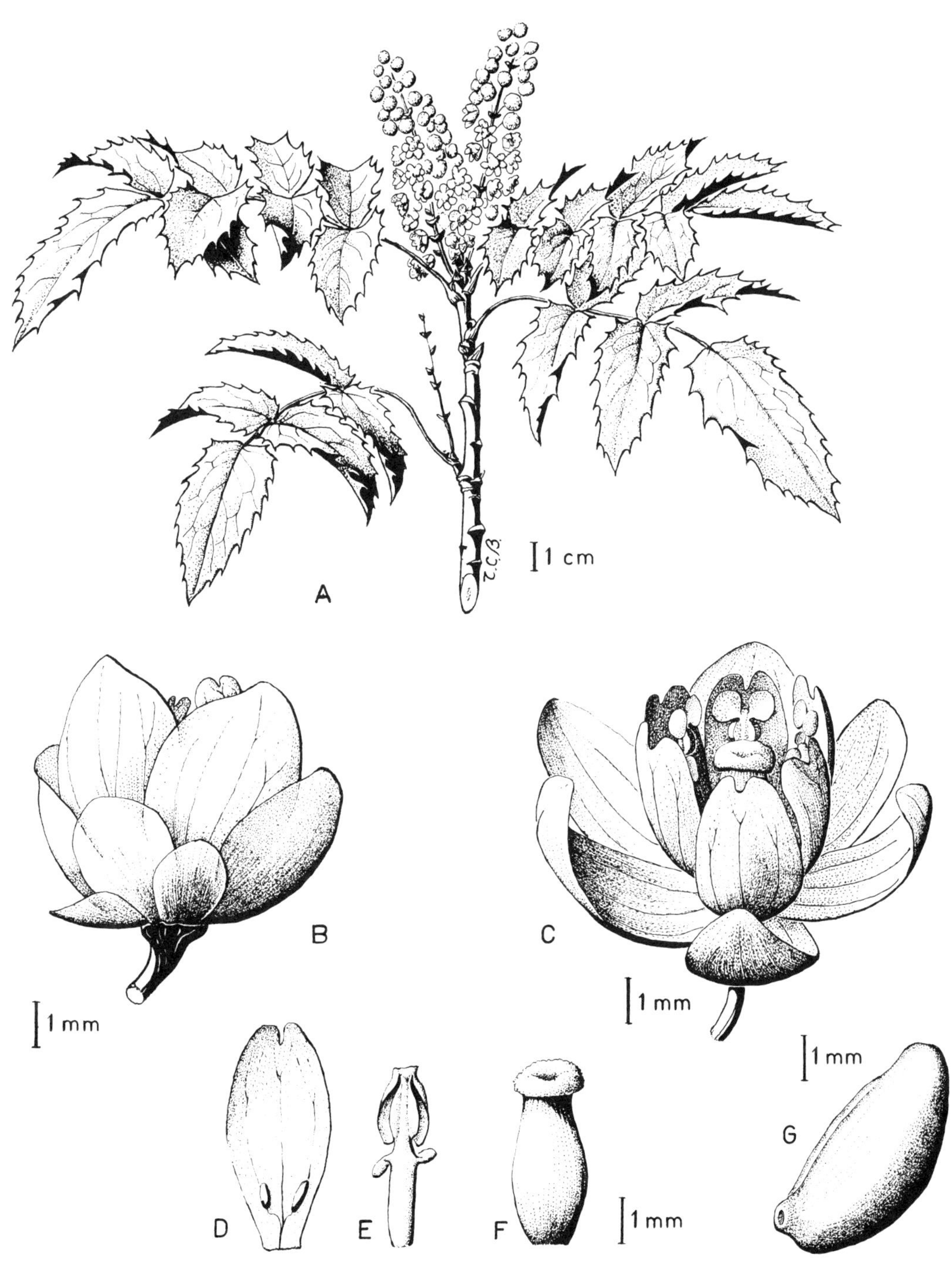

Figure 101. *Berberis aquifolium* ssp. *aquifolium*

A. top of flowering shoot.
B. opening flower bud.
C. flower.

D. petal.
E. stamen.

F. carpel from flower.
G. seed.

Leaves pinnately compound, with 5–11 leaflets. Leaflets oblong-ovate, pinnately veined from the base, sharply spinulose-toothed on the often undulate margin, commonly shiny above, and smooth and dull, and sometimes somewhat glaucous beneath.

Flowers in erect racemes usually shorter than those of *B. nervosa*, in the axils of bud scales, and sometimes of leaves. The racemes remaining erect or spreading as the fruit ripens. Perianth of 5 or 6 whorls; the outermost smaller than those within, and deciduous when the flower opens, leaving two whorls of persisting sepals and two whorls of petals. Petals apically notched between rounded lobes, and with pairs of near-basal nectaries. Stamens shorter than the petals, and about the same length (around 4 mm) as the ovary; the filaments bearing a pair of small lateral appendages just below the anthers. Carpel with a sessile, button-like stigma with a central depression, and containing 3–9 ovules.

Fruit a bluish purple berry 7–14 mm in diameter. 2n = 28.

A species of western North America, from central British Columbia and southwestern Alberta south to Texas and California, and from the Pacific coast eastward to South Dakota. An adaptable species, *B. aquifolium* is found in open, semi-arid grasslands and in woodland, on stony or gravelly, usually well-drained soils.

Commonly cultivated as an ornamental shrub, it has now established itself in eastern Canada and in Europe.

KEY TO SUBSPECIES AND VARIETIES

1. Leaflets (3–) 5–7 (–9), less than twice as long as wide, often dull above and papillose and slightly glaucous beneath; the margin with 10–40 fine spinulose teeth. Ovary with 5– 8 or 9 (average: 6) ovules. Plant low and creeping.. **ssp.** *repens*

   Leaflets 5–11, averaging more than twice as long as wide, usually shiny above, and dull and smooth, not glaucous beneath; the margins variously toothed with 8–20 teeth. Ovary with 3–6 (average: 5) ovules. Plant erect, commonly a metre tall, sometimes low and spreading.. **ssp.** *aquifolium*   2

2. Leaflets coarsely toothed, the spinules 2–4 mm long..................................... **var.** *aquifolium*

   Leaflets finely serrulate, the spinules 1 mm or less long................................ **var.** *lyallii*

Variety *aquifolium* [*B. aquifolium* Pursh var. *nutkana* DC., *B. nutkana* (DC.) Kearney, *Odostemon nutkanus* (DC.) Rydberg] ranges from southern British Columbia, including Vancouver Island, southward to Oregon, and eastward to southwestern Alberta (Scoggan, 1978) and Idaho.

Variety *lyallii* Ahrendt, recorded from the Cascade Range on the International Boundary, is in some way intermediate between subspecies *aquifolium* and *repens*.

Berberis aquifolium Pursh ssp. *repens*   **Creeping Oregon-Grape**
(Lindley) T.C. Brayshaw *stat. nov.*
    Berberis repens Lindley, Bot. Reg. 14:pl. 1176, 1828.
    Berberis aquifolium . . . α Torrey & Gray
    Mahonia repens var. *rotundifolia* Fedde
    Odostemon aquifolium (Pursh) Rydberg
    B. aquifolium forma *repens* (Lindley) Boivin
    B. aquifolium var. *repens* (Lindley) Scoggan

This subspecies, more widely ranging east of the Coast and Cascade Ranges than ssp. *aquifolium*, extends from central British Columbia and southwestern Alberta to South Dakota and Texas. It is associated with dry, exposed habitats to a greater degree than is ssp. *aquifolium*.

Subspecies *repens* is commonly treated in manuals as a species distinct from *B. aquifolium*. There has been much nomenclatural confusion over this complex: a problem that was well reviewed and sorted out by Piper (1922).

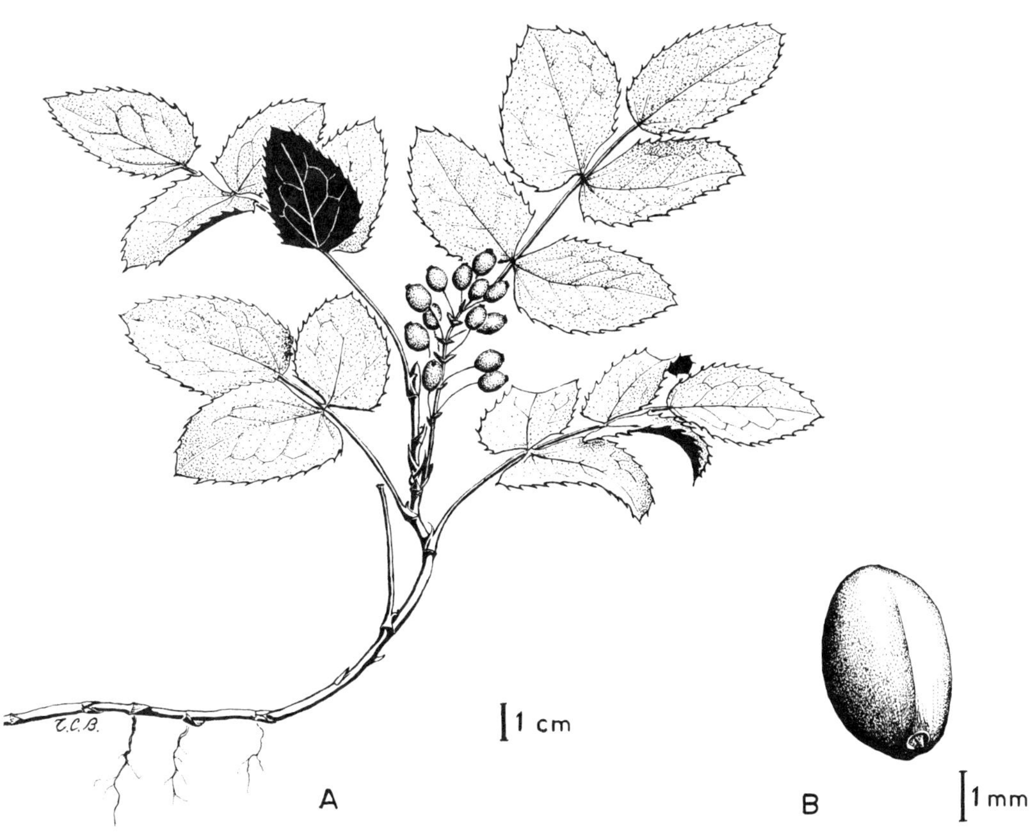

Figure 102. *Berberis aquifolium* ssp. *repens*

A. fruiting shoot.   B. seed.

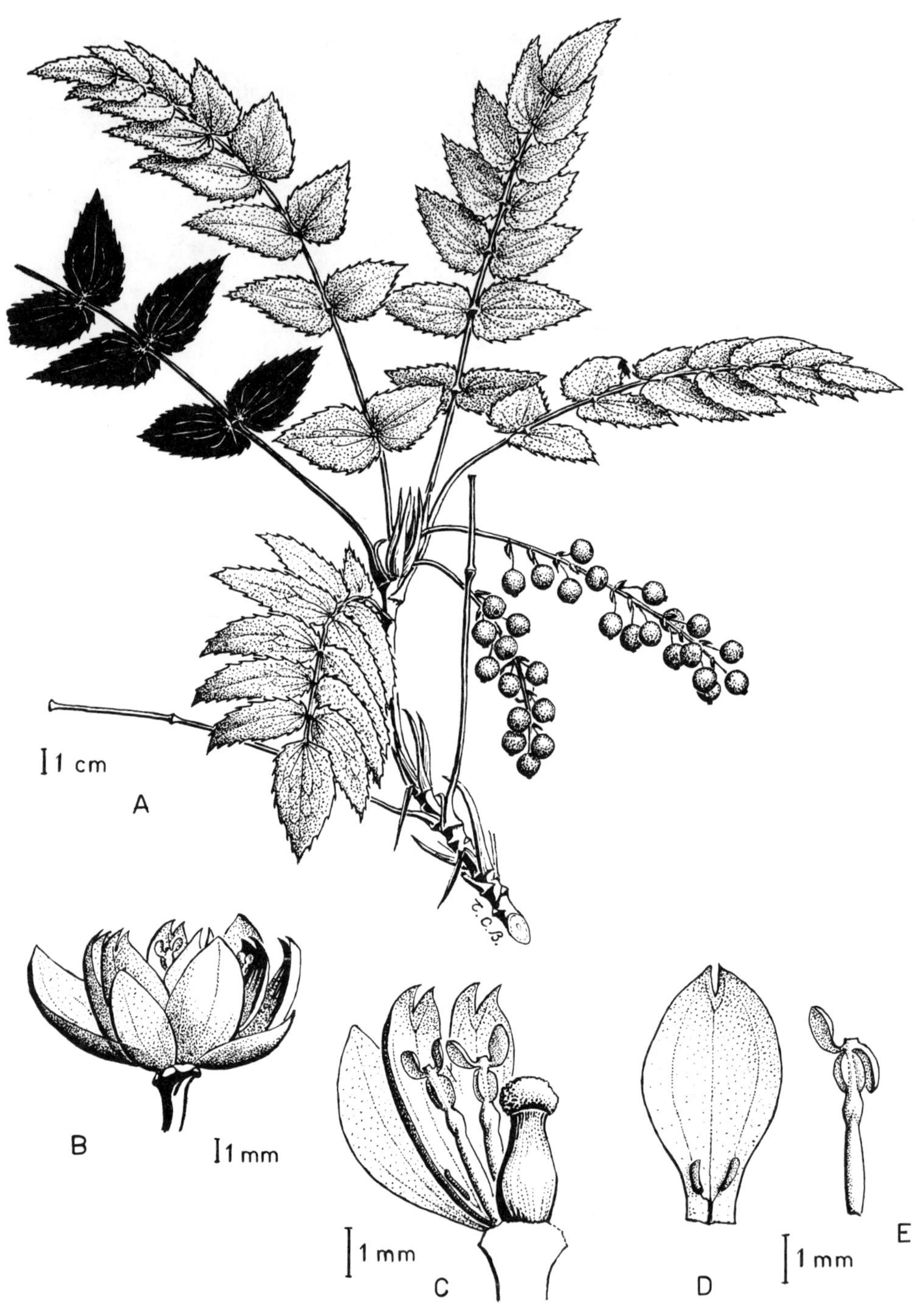

Figure 103. *Berberis nervosa*

A. fruiting shoot.
B. flower.
C. longitudinal section of part of flower, showing anthers opened.
D. petal, showing paired nectaries.
E. stamen, with anther open.

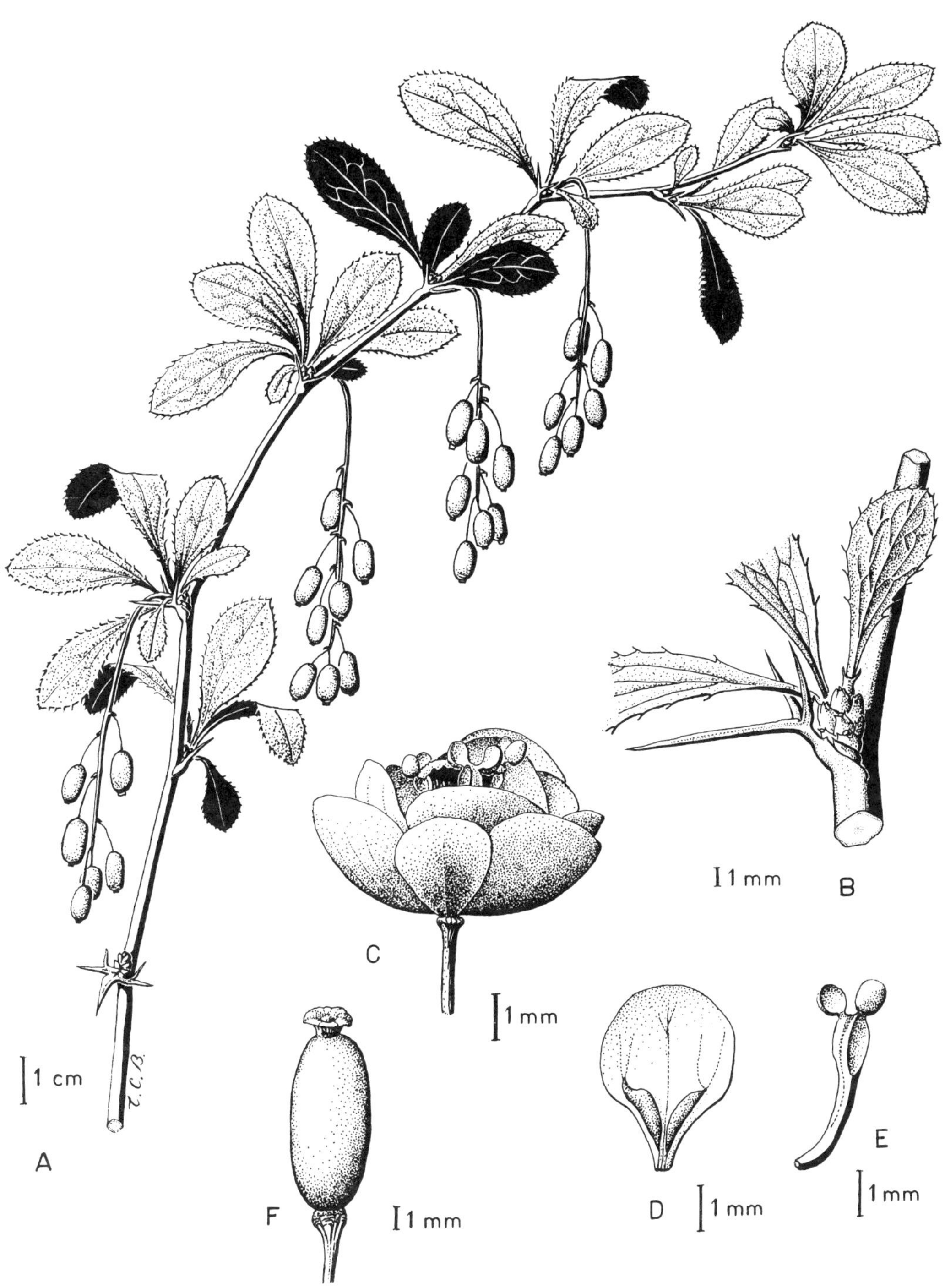

Figure 104. *Berberis vulgaris*

A. fruiting branch.
B. detail of node, spine, axillary bud and leaf fascicle.
C. flower.
D. petal, with nectaries.
E. stamen with anther open.
F. berry.

The distinctions between the two entities, though undoubtedly mainly genetically determined, are not absolutely clear-cut: the character-differences used to separate them not correlating perfectly with each other. The papillose leaf surface, visible at a magnification of X 25, may be absent from, or only spottily visible in, plants that by other criteria are *B. repens*; while it is sometimes visible on plants that otherwise are clearly *B. aquifolium*, even at the coast, where *B. repens* by any other criteria is clearly absent. The habit of the plants, though no doubt largely genetically based, is not free from environmental influence; and extreme exposure may force a low prostrate habit on a plant displaying all the other characteristics typical of ssp. *aquifolium*.

Because of the lack of a clear separation between these plants, it is felt that the appropriate treatment would be to follow the examples of Torrey and Gray (1838) and Scoggan (1978) in uniting them. It is probable that the correct name for *B. repens* when treated as a variety is var. *rotundifolia*, not var. *repens*, as pointed out by Hitchcock and Cronquist (1964).

Since varieties have already been described in both *aquifolium* and *repens* to contain entities differing from the norms of these taxa as species; and since the difference between ssp. *repens* and ssp. *aquifolium* is clearly greater than that between var. *aquifolium* and var. *lyallii*, it is felt that subspecific rank for *repens* would be more appropriate than varietal rank, to reflect the relative degrees of difference between these entities; and furthermore, would permit retention of the use of the varieties as such that have already been described within the subspecies.

Berberis nervosa Pursh   Oregon-Grape
Mahonia nervosa (Pursh) Nuttall

Rhizomatous, relatively low shrub, up to ½ m tall; the ascending shoots sparsely branched. Bud scales conspicuous, 2–4 cm long, lanceolate, persistent.

Leaves pinnately compound, with 9–19 leaflets (but sometimes with only one leaflet on seedlings). Leaflets 2–8 cm long, ovate to broadly lanceolate, palmately 5- to 7-veined at base, with regularly spinulose-toothed margins and a dull surface. Stipules inconspicuous, joined for most of their lengths to the petiole base.

Flowers in erect racemes in the axils of bud scales; 20 or more flowers to a raceme. The axes of the racemes elongate in fruit, and the racemes then often droop. Whorled outermost sepals (or bractlets) deciduous when the flower opens. Outer persisting sepals about 4–6 mm long, the inner ones 6–8 mm long. Petals apically notched between acute lobes, nearly as long as the inner sepals, and arching over the stamens. Stamens shorter than the petals, the filaments truncate apically and without lateral appendages. Carpel in flower about as long as the stamens (3–4 mm), with a sessile, button-like stigma.

Fruit a purplish blue, pruinose berry, 8–11 mm in diameter, tipped by the sessile stigma.

Berberis nervosa is found commonly on coarse, well-drained soils in Douglas-fir-dominated woods of the Coast Forest region from British Columbia to California; and locally in the Columbia Forest region of southeastern British Columbia.

In coastal areas where *B. nervosa* and *B. aquifolium* grow together, very occasional plants are found that appear to share characteristics of both species; e.g. palmately veined shiny leaflets and deciduous bud-scales. Such material has been found at Spectacle Lake, southeast of Duncan. This suggests that these species may very rarely interbreed and produce hybrid offspring. However, the author knows of no detailed work that could confirm or deny this impression.

The berries of *B. nervosa* are sometimes picked for home jelly or jam making.

Berberis vulgaris L.   Common Barberry

Tall (to 3 metres) shrub with arching, pale, longitudinally angled stems without terminal buds, with alternate nodal, 1–7 (usually 3) -pronged yellowish spines formed from modified leaves, and about 1–1½ cm long; and fascicles of apparently simple leaves and inflorescences in the axils of those spines.

Leaves deciduous, apparently simple, but actually compound and reduced to single leaflets; oblanceolate to obovate, glabrous, serrulate-margined with fine bristle-tipped teeth, acute or obtuse at apex, and tapering to a base that is separated from the very short petiole by an abscission joint that is just above the minute stipules, and is commonly concealed by the outer leaf bases and persistent bud scales. The short flat petiole remains attached after the leaflet is shed.

Flowers yellow, in pendulous racemes from the leaf fascicles. Sepals in three whorls, those of the outermost whorl very small and caducous. Petals entire, slightly longer than the inner sepals. Stamens without lateral teeth. Ovary with two or three basal ovules and a sessile, centrally depressed stigma.

Fruit an ellipsoid, scarlet (rarely purple), commonly one-seeded berry. $2n = 28$.

Purple-leafed plants are distinguishable as forma *atropurpurea* Regel.

Native of Europe; long cultivated both for ornament and for its edible berries; and now established as a weed across North America, including occasional occurrences in this province.

This species is an alternate host of the fungus, *Puccinia graminis*, that causes the Wheat Rust disease; and for many years its cultivation has been prohibited.

GLOSSARY OF TECHNICAL TERMS

Terms illustrated in Figure 105 are indicated by asterisks. Plant names following definitions indicate where those terms are illustrated in other figures.

***Achene:**   dry (i.e. non-fleshy) fruit, commonly composed of a single carpel, containing a single seed, and not opening to release the seed (*Ranunculus* spp.).

Acuminate: narrowly tapering to an acute apex.

Adnate: joined, in reference to unlike tissues or organs (e.g. petals to ovary in *Nymphaea odorata*).

Air canal: an extended passage among the cells of a stem or petiole, usually of an aquatic plant, to permit the diffusion of gases through submerged organs (*Nymphaea odorata*).

Alternate (leaves):   arising singly at nodes of stem, and usually spirally arranged.

Anastomosing (veins):   branching and rejoining, forming a net with closed loops.

Angiosperms: the flowering plants as a major division of the Plant Kingdom: characterized by flowers, bearing ovules and seeds enclosed in ovaries formed of closed carpels.

***Anther:**   pollen-containing case, often compound, terminating a stamen.

Anthesis: the time of opening of a flower, and its pollination.

Apetalous (flower):   lacking petals.

Apical: placed at the apex or tip of a supporting structure.

Apiculate: with a short, protruding apical point.

Aril: an outgrowth of the base of a seed, often forming a fleshy cup around the seed (*Nymphaea tetragona*).

***Axil:**   angle between the stem and the upper side of a leaf or petiole.

Axillary: placed in, or arising from, an axil.

Basal (leaf):   arising from the base of a stem, or from a rhizome.

Bipinnate (leaf):   compound leaf consisting of a pinnate arrangement of 'branches' which in turn bear pinnately arranged leaflets (*Actaea rubra*).

Biternate (leaf):   a leaf having three 'branches', each of which bears three leaflets (*Aquilegia formosa*).

***Blade:**   broad, flat, expanded part of leaf or petal.

Boreal (distribution):   northern, relating to the region of the northern coniferous forests.

Bract: a leaf, often reduced or otherwise modified, in the axil of which arises a flower or inflorescence (*Thalictrum dasycarpum*).

Bracteate (inflorescence or pedicel):   bearing bracts or bractlets.

Bractlet: small bract, commonly a secondary bract, i.e. arising on an inflorescence branch that has arisen in the axil of a bract (*Ranunculus uncinatus*).

Caducous: deciduous early in the growing season: shed before other leaves or floral appendages are normally shed.

Calyx: the aggregate of the sepals (i.e. the outermost whorl) in a flower.

Cambium: a longitudinal growing tissue in a stem, whose cells multiply radially to give rise to component cells of other tissues; usually xylem on its inner side and phloem on its outer side. In woody plants, the cambium forms a thin layer between the wood (xylem) and the inner bark (phloem).

Capsule: a multi-seeded fruit formed of two or more carpels, and maturing into a case that opens to release the seeds (*Nigella damascena*).

***Carpel:**   greatly modified leaf, enclosing one or more ovules (and ultimately seeds); the fundamental unit of the pistil or ovary.

Caudex: an ascending or erect, commonly buried, short, perennial stem, from which annual, aerial shoots and leaves arise (*Anemone patens*).

Cauline (leaf):   arising from a stem above the base.

Ciliate: bearing marginal hairs.

Circumboreal (distribution):   extending around the Northern Hemisphere in the latitudes principally of the northern coniferous (boreal) forest zone.

Circumpolar (distribution):   extending around the Northern Hemisphere in high arctic, and usually treeless, latitudes.

***Claw (of a petal):**   the slender basal segment of a petal.

***Compound: (leaf):**   leaf consisting of a number of distinct leaflets borne on a common petiole.

Connate: joined together, as referring to like tissues or organs, (e.g. carpels in *Nuphar polysepalum*).

Connective: the apex of the filament of a stamen: the part joining the halves of the anther (*Achlys triphylla*).

Cordate (leaf):   heart-shaped at base.

Cordilleran (region):   the western part of the North American continent; characterized by the array of mountain ranges.

Corolla: the aggregate of petals in a flower.

Corymb: an inflorescence in which the branch peduncles or pedicels are of unequal length.

Cotyledon: seed leaf: the first leaf (or one of the first pair of leaves) produced by a developing embryo. Commonly specialized for storage of food reserves for the developing seedling.

Crenate (leaf):   having a margin bearing rounded teeth (*Caltha biflora*).

Cyme: an inflorescence in which the terminal flower is the oldest, or is older than the nearby lateral flowers, and opens before them (*Aquilegia flavescens*).

Cymose: having the developmental characteristics of a cyme.

Deciduous: non-persistent; shed with the advance of the seasons.

Decumbent (stem):   lying prostrate below but with the apical part ascending (*Ranunculus ficaria*).

Dentate (leaf margin):   toothed with symmetric or irregular teeth.

Dicotyledons: a major class of flowering plants: characterized by embryos with two cotyledons, netted-veined leaves, sepals and petals in whorls of 4, 5, or more (with some exceptions).

Dioicous (Dioecious):   bearing functionally male and female organs on separate plants (*Thalictrum occidentale*).

Diploid: the number of chromosomes in a vegetative cell, comprising two basic complements: abbreviated as '2n'.

***Dorsal (leaf surface):**   the surface of a leaf facing downward and away from the axis of the stem.

Ellipsoid: three-dimensional form with an elliptic outline (widest at the middle) in the plane of its only long axis: football-shaped.

Endosperm: a tissue (actually a distinct plant) specialized for storage of food reserves for a developing embryo. Its absorption by the embryo may occur before the seed ripens or after germination.

Entire (leaf margin):   smoothly outlined; without teeth or lobes.

Erose (margin):   ragged-edged, irregularly toothed, or appearing worn.

Eutrophic (water):   having a high level of available nutrient chemicals; commonly as a result of pollution from waste or soluble fertilizer runoff from land.

Evergreen: retaining some green foliage at all seasons.

Extrorse (anthers):   on outer (dorsal) side of top of filament, and opening outward.

Fascicle: a lateral tuft of leaves, arising from a leaf axil (*Berberis vulgaris*).

***Filament:**   the part of a stamen supporting the anther.

***Follicle:**   a seed pod, formed from a carpel, and opening along one suture (usually the ventral one) to release the few to many seeds (*Cimicifuga elata*).

Fusiform: three-dimensional form, tapering to ends and circular in cross-section: carrot-shaped.

Glabrescent (surface):   initially hairy but becoming hairless or nearly so, with maturity.

Glabrous (surface):   without hairs or projecting glands.

Glandular: provided with glands.

Glaucous (surface):   whitened or greyish with waxy cuticle.

Gymnosperms: a division of seed-bearing plants without flowers. Ovules and seeds being attached to open, scale-like carpels aggregated in cones, or to stems.

Haploid: the number of chromosomes in a single complement: e.g. as in an egg or sperm or in the nuclei of pollen grains: abbreviated as 'x' for a genus or section, or as 'n' for species or individual.

Head: a compact globular aggregate of sessile flowers or fruits.

Herb: a plant with non-perennial shoots; dying back to the base seasonally.

Herbarium: a systematically filed collection of plant specimens.

Hexaploid: possessing, in the vegetative cell, six copies of the basic chromosome complement, i.e. of the haploid number of chromosomes for the genus or species group. Thus $2n = 6x$.

Hirsute: coarsely or roughly hairy.

Hispid: hairy with stiff, short, bristly hairs.

Hydathode: water gland, usually on a leaf tip or margin opposite the end of a vein, and secreting liquid water.

Hypocotyl: segment of stem between the root and the cotyledons.

Hypogynous (flower):   with outer parts arising below the carpels.

Hypsithermal (time interval):   an early post-Pleistocene period during which the climate was warmer and drier than at present. It is believed to have occurred between about 10,000 years and 6,500 years ago.

Inflorescence: a structured cluster of flowers, and subsequently, fruits.

***Internode:**   the interval, along a stem, between successive nodes.

Introrse (anthers):   position of anthers on the inner (ventral) side of the top of the filament, and opening inward.

Involucel: secondary involucre: a whorl of bracts or bractlets on an inflorescence branch arising from an involucre (*Anemone multifida*).

Involucre: a whorl of bracts at the base of an inflorescence (*Anemone multifida*).

Involucral leaves: leaves (i.e. bracts, but in form of unmodified leaves) making up an involucre (*Anemone* spp.).

***Keel (of an achene):**   protruding thin edge of achene (*Ranunculus repens*).

Lacunate (tissue):   containing air chambers and air canals.

Lanceolate (leaf blade):   lance-shaped; slender, tapering to tip, and widest at or below the middle.

***Leaflet:**   the individual blade of which a number comprise a compound leaf (*Aquilegia*).

Lignified (conductive tissue):   impregnated with lignin, and tough and woody in texture.

***Ligule:**   the freely projecting upper edge of a stipular sheath, formed of stipules joined across the ventral side of a petiole base.

Linear (leaf):   very slender and more or less parallel-sided (*Myosurus minimus*).

Locule: the chamber in an ovary or carpel, containing the ovules.

Monocolpate (pollen grain):   having a single germinal groove, where the pollen tube emerges on germination; characteristic of Gymnosperms, many Monocotyledons, and some primitive families of Dicotyledons.

Monocotyledons: a major class of flowering plants; characterized by seeds with a single cotyledon (seed leaf), parallel-veined leaves, scattered vascular strands in stems, no cambium, and sepals and petals in whorls of 3 (or 2).

Monoicous (Monoecious):   having functional male and female organs on the same plant.

Multifid (leaf blade):   divided into many divisions, but not a compound leaf, (*Ranunculus flabellaris*).

Multiovulate (carpel):   containing many ovules (*Aquilegia*).

Nectariferous: nectar-producing.

***Nectary:**   nectar-secreting gland.

***Nectary scale:**   scale-like outgrowth of a petal surface, overlying or surrounding, and protecting, a nectar gland.

***Node:**   point on a stem where a leaf is attached, and a lateral branch may depart.

Obovate: two-dimensional 'egg-shape', widest toward the apex.

Opposite (leaves):   departing from a stem opposite each other, in a pair at each node.

Orbicular: more or less circular in outline.

***Ovary:**   floral organ containing ovules, and consisting of the lower part of a carpel, or of an aggregate of connate carpels.

Ovate: two-dimensional 'egg-shape', widest toward the base.

Ovoid: three-dimensional 'egg-shape', thickest near base.

Ovule: egg-containing body in the ovary; becomes a seed after fertilization.

Palmately (veined or divided leaf blade):   having five or more veins or divisions radiating from a central or basal point.

Palmatifid: palmately divided or incised (leaf of *Aconitum delphiniifolium*).

Panicle: a much-branched and rebranched inflorescence (*Thalictrum occidentale*).

Paniculate (inflorescence):   having the character of a panicle.

Papilla (pl. papillae):   pimple-like projection of a surface (*Ranunculus sardous* achene).

Papillose (surface):   bearing papillae.

Paratype (specimen):   a specimen mentioned by the author of a species as typical, in addition to the designated type specimen.

Pedately divided (leaf):   resembling palmately divided but with the outer divisions arising from the next inner ones (*Ranunculus pedatifidus*).

***Pedicel:**   stem segment bearing an individual flower.

Peduncle: stem segment bearing an inflorescence.

Peltate (leaf blade):   having the petiole attached to the lower (dorsal) surface, not to the margin (*Brasenia schreberi*).

Penicillate: like a brush, or tuft of hairs (leaf of *Ranunculus trichophyllus*).

Perfect (flower):   bearing both functional stamens and ovary.

Perianth: the aggregate of the normally sterile floral appendages, ie. sepals and petals, considered together.

***Petal:**   a member of the inner whorl or series (the corolla) of sterile floral appendages.

Petaloid (sepal):   large, coloured and conspicuous, functioning visually as a petal (*Anemone*).

***Petiole:**   leaf stalk.

***Petiolule:**   leaflet stalk.

Phloem: food-conducting tissue in the vascular (conducting) system.

Pilose (surface):   with soft hairs more or less perpendicular to the surface.

Pinnately arranged (leaflets):   disposed in two rows along a central axis (*Berberis nervosa*).

Pistillate (flower):   having only functional ovaries and ovules, but not functional stamens; thus functionally female (*Thalictrum venulosum*).

Placenta: that part of the lining of the ovary to which the ovules and seeds are attached.

Plumose (achene):   feather-like, with a long style bearing abundant diverging short hairs (*Clematis* achenes).

Puberulent: finely and minutely pubescent.

Pubescent: hairy with short soft hairs.

Raceme: inflorescence having pedicels of equal length arising successively from a central axis (*Aconitum columbianum*).

Rachis: axis of a compound leaf above the petiole, i.e. beyond the lowest leaflets (*Berberis nervosa*).

***Receptacle:**   axis of a flower, that bears the carpels and other floral appendages (*Anemone occidentalis*).

Recurved: curved downward, or away from a floral axis or stem (achene beak of *Ranunculus uncinatus*).

Reflexed: strongly or sharply recurved.

Reniform (leaf blade):   kidney-shaped: rounded, cordate-based, commonly wider than long (*Ranunculus ficaria*).

Revolute (leaf margin):   rolled under (*Thalictrum venulosum*).

Rhizomatous (plant):   possessing a rhizome.

Rhizome: buried, creeping, usually horizontal, perennial stem, from which erect, aerial shoots arise (*Anemone parviflora*).

Scape: leafless stem, or peduncle, arising from ground level to support a flower or inflorescence (*Myosurus minimus*).

Scapose (plant):   characterized by bearing flowers on a scape.

Sclereid: a hard, thick-walled cell, commonly with radiating arms and a rough surface, scattered through other, softer, tissues (*Nymphaea tetragona*).

Semi-lunate: shaped like a half moon (*Ranunculus aquatilis* achene).

***Sepal:**   a member of the outer series or whorl (the calyx) of sterile floral appendages.

Sericeous (surface):   with oblique and parallel, overlapping, soft hairs, producing a silky texture.

Serrate (leaf margin):   having oblique teeth, like a saw.

Serrulate (leaf margin):   having very fine oblique teeth.

Sessile (leaf or flower):   stalkless: set against the main stem without a petiole or pedicel.

Sinuate (leaf margin):   having a wavy outline, with smoothly rounded lobes and sinuses.

***Sinus:**   re-entrant between adjacent lobes of a leaf.

Spatulate: spatula-shaped: slender, with a widened tip.

Spike: inflorescence in which the flowers are sessile on an elongated axis (*Achlys triphylla*); or a fruit in which achenes are sessile on an elongated receptacle (*Ranunculus pensylvanicus*).

Spinulose-toothed: (leaf margin): having stiff sharp teeth tapering into little spines (spinules), (*Berberis vulgaris*).

Spur (in a flower):   a drawn-out, hollow protuberance of a sepal or petal, associated with a nectary or with its protection (*Delphinium nuttallianum*).

*** Stamen:**   floral appendage specialized for bearing pollen in an anther.

Staminate (flower):   having functional stamens, but no functional carpels, and therefore functionally male (*Thalictrum occidentale*).

Staminode: sterile, non-functional stamen, usually lacking an anther (*Aquilegia formosa*).

***Stigma:**   pollen-receiving tip of carpel or ovary.

Stigmatic (beak):   formed from a stigma or stigmas (*Thalictrum venulosum*).

***Stipe:**   stalk-like base of a carpel or ovary.

Stipitate (carpel):   borne on a stipe.

***Stipule:**   one of a pair of small appendages associated with (and often joined to) the base of a petiole.

***Stipular sheath:** a broad, stem-clasping sheath involving a petiole base and combined stipules (*Ranunculus subrigidus*).

Stolon: a creeping, prostrate or arching stem on the ground surface, not buried (*Ranunculus cymbalaria*).

Stoloniferous: bearing stolons.

Striate: (surface): marked with parallel lines or grooves.

Strigose (surface): with stiff, hard, oblique, straight hairs; producing a rough feeling; harder to stroke in one direction than in others.

***Stylar beak:** beak of fruit composed mainly of the style (*Ranunculus orthorhynchus*).

***Style:** slender apical projection of a carpel or ovary, empty of ovules and bearing the stigma.

Subsessile (flower or leaf): almost sessile; having only a very short pedicel or petiole.

Subspecies: the largest subdivision of a species.

Subulate (beak): stiff, sharp pointed, tapering (*Ranunculus testiculatus* achene).

***Suture:** line where the two sides of a carpel meet.

Sympatric: growing in the same, or largely overlapping, geographic ranges.

Taproot: single, vertical root system (*Aconitum columbianum*).

Terete (stem): cylindrical, and circular in cross-section.

Ternate (leaf): divided into three parts (*Ranunculus repens*).

Tetraploid: possessing, in the vegetative cell, four copies of the basic chromosome complement; i.e. of the haploid number of chromosomes for the genus or species group. Thus $2n = 4x$.

Tomentose (surface): with matted short soft hairs; closely woolly (*Ranunculus testiculatus*).

Trachea: see Vessel.

Tricolpate (pollen grain): having three germinal grooves, whence pollen tube may emerge.

Trifoliolate (leaf): consisting of three distinct leaflets (*Coptis trifolia*).

Tundra: treeless vegetation of cold regions, either arctic or alpine.

Type species: species formally acknowledged as typical of a genus. When a genus is divided, the part containing the type species must retain the original generic name.

Type specimen: specimen on which the original description of a species, or of a subspecific entity, was based, and so designated by the author of the species. If a species is divided, the original name is retained by that part that is still typified by the original type specimen.

Umbel: inflorescence in which all primary branches originate from one node at the apex of the peduncle.

Undulate (leaf margin): wavy; undulating above and below the plane of the leaf blade.

Unicellular: consisting of a single cell.

Unisexual (flower): flower in which either only stamens or only carpels function, but not both (*Thalictrum occidentale*).

Unlignified (xylem): xylem tissue in which the cell walls are not impregnated with lignin.

Valvate (sepals in bud): meeting edge to edge; their margins not overlapping (*Clematis vitalba*).

Vascular strand: a strand of elongated cells, specialized for water conduction, food conduction, and often for support. The veins in leaves are vascular strands.

Velutinous (surface): velvety; with dense, short, erect, moderately stiff hairs.

***Ventral (surface):** surface facing toward the apex of a stem or axis of a flower.

***Ventral suture (of carpel or derived fruit):** suture on the upper or inner side of a carpel; representing the joined or appressed margins of the ancestral open carpellary leaf.

Vessel (Trachea, pl. Tracheae): multicellular, linear series of wide-diameter xylem cells, joined end-to-end with the intervening partitions perforated, forming a long tube. Functions to expedite water transport. Found in vascular strands of most flowering plants, but absent from most Gymnosperms.

Vestigial (organ):   reduced to a minute vestige of its normal size and functions as found in other species.

Villous (surface):   with long soft hairs, often curly or shaggy, but not matted.

Whorl: a ring of three or more leaves, bracts, sepals, etc. around a stem at one node (*Ceratophyllum demersum*).

Whorled: occurring in whorls.

Xylem: vascular strand tissue specialized for water conduction, and sometimes for support; often lignified and forming wood.

Figure 105 (opposite)
ILLUSTRATIONS OF STRUCTURES

A. Simple leaf of *Caltha leptosepala*.
B. Compound leaf and stem of *Ranunculus macounii*.
C. Flower of *Ranunculus hyperboreus*.
D. Petal of *Berberis vulgaris*.
E. Petal of *Ranunculus pensylvanicus*.
F. Carpel from flower of *Thalictrum venulosum*.
G. Receptacle with achene of *Ranunculus orthorhynchus*.
H. Follicle of *Coptis aspleniifolia*.

KEY TO STRUCTURES

1. Main stem
2. Axil
3. Axillary branch
4. Node
5. Internode
6. Stipule
7. Stipular sheath
8. Ligule
9. Petiole
10. Petiolule
11. Leaf blade
12. Leaflet
13. Ventral surface
14. Dorsal surface
15. Mid-vein
16. Lateral vein
17. Lobe
18. Sinus
19. Basal sinus
20. Sepal
21. Petal
22. Anther
23. Filament
24. Stamen
25. Carpel
26. Nectary (bare)
27. Nectary scale
   (covering nectary)
28. Claw
29. Blade
30. Ovary
31. Style
32. Stigma
33. Receptacle
34. Achene body
35. Ventral keel
36. Dorsal keel
37. Stylar beak
38. Stipe
39. Ventral suture
40. Pedicel

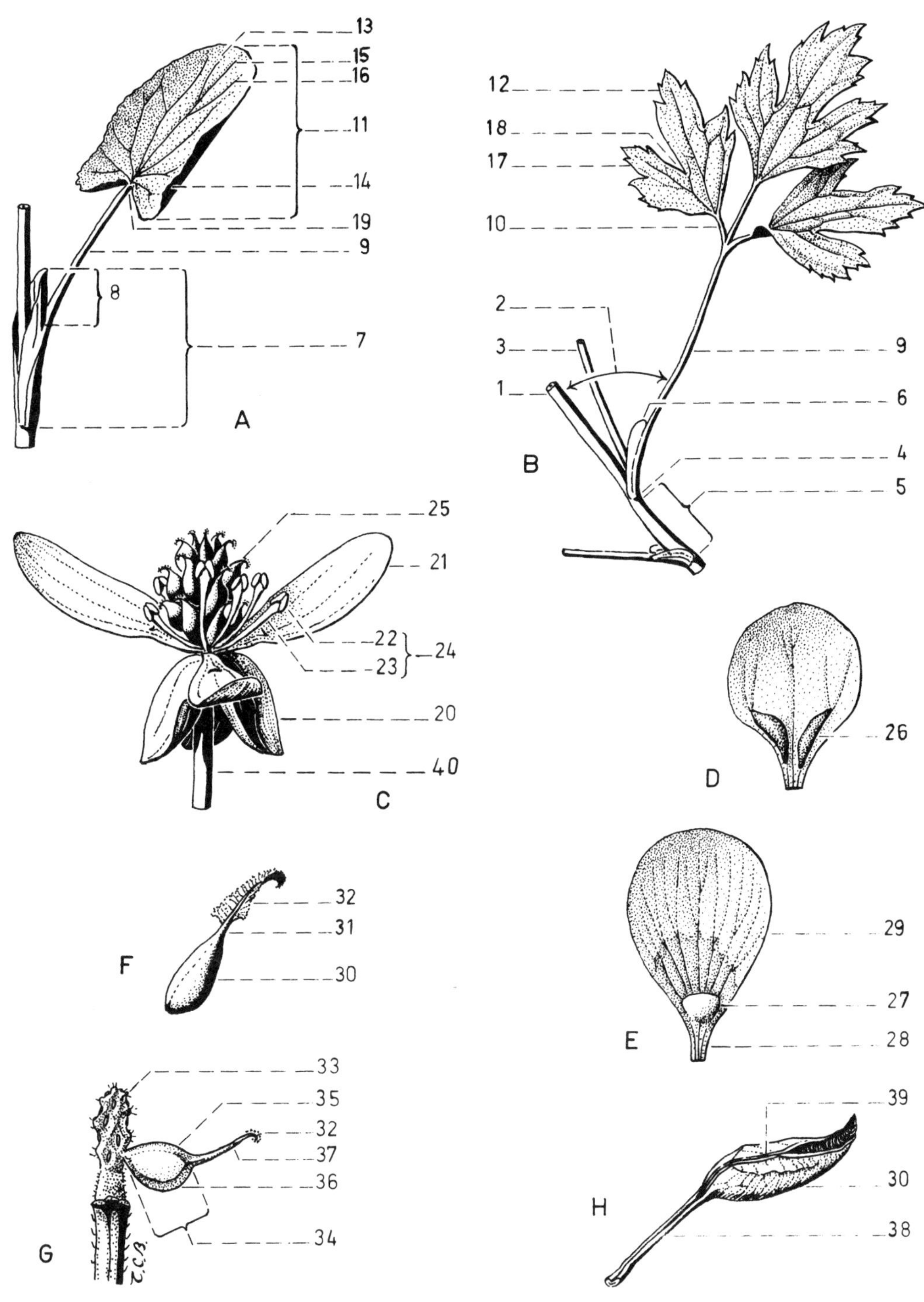

195

ABBREVIATIONS AND SYMBOLS

cm—centimetre(s).

DC.—de Candolle, Augustin Pyramus.

dm—decimetre(s).

f.—forma (a subdivision of a variety).

L.—Linnaeus, Carolus (the elder).

L.f.—Linnaeus, Carolus (the younger).

m—metre(s).

mm—millimetre(s).

nom. cons.—'*nomen conservandum*': a name retained in use by general agreement in spite of the existence of older names; usually on the grounds of extensive use and familiarity. Applied to genera and higher categories.

nom. nov.—'*nomen novum*': a new name, published here for the first time.

spp.—species (plural).

ssp.—subspecies.

stat. nov.—'*status novus*': a new rank for an existing name.

subgen.—subgenus.

var.—variety (Latin: "*varietas*").

X—(1) "crossed with" in a formula for a hybrid pedigree when used between the specific names of the parents.
(2) when used as a prefix, indicating the hybrid origin of a plant with the following name.

x—the basic haploid (reduced) chromosome number characteristic of a genus or a series of species: e.g. $x = 8$ for most species of *Ranunculus*.

2n—the diploid (doubled) number of chromosomes found in each cell nucleus of a flowering plant, or animal, body. Eggs and sperms in animals, and spores and pollen nuclei as well in flowering plants, possess the haploid (single, or "n") number of chromosomes, which is doubled at fertilization.

±—more or less.

BRITISH COLUMBIA
Brasenia schreberi
130° W
60° N
120° W
55° N
55° N
50° N
50°
N
50°
N
120° W
130° W
MILES
0   100   200
0   100   200   300
KILOMETRES

1.

BRITISH COLUMBIA
Nuphar polysepalum
130° W
60° N
120° W
55° N
55° N
50° N
50°
N
50°
N
120° W
130° W
MILES
0   100   200
0   100   200   300
KILOMETRES

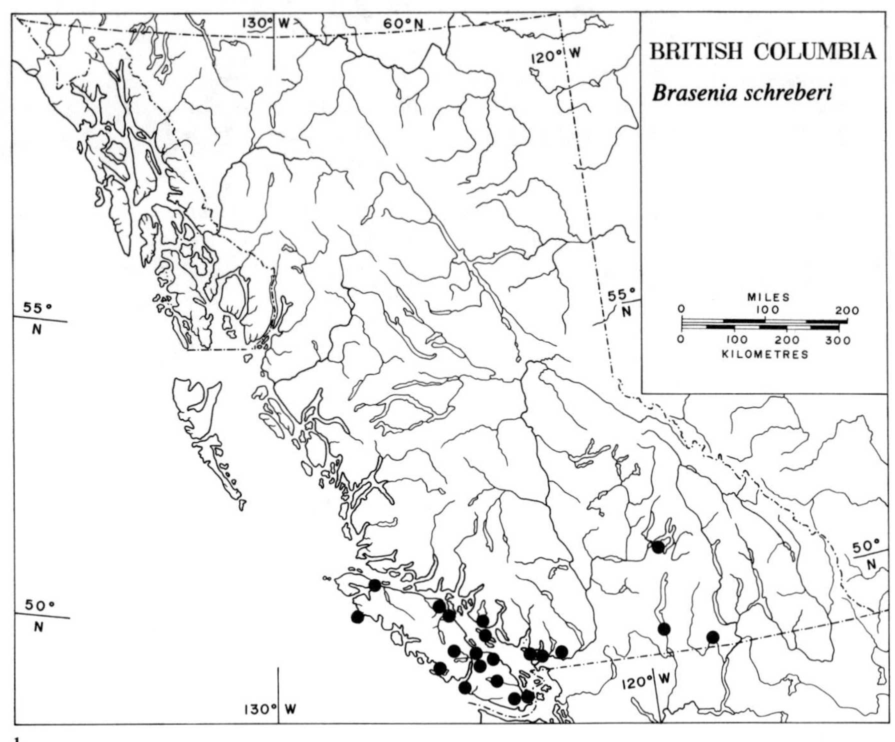

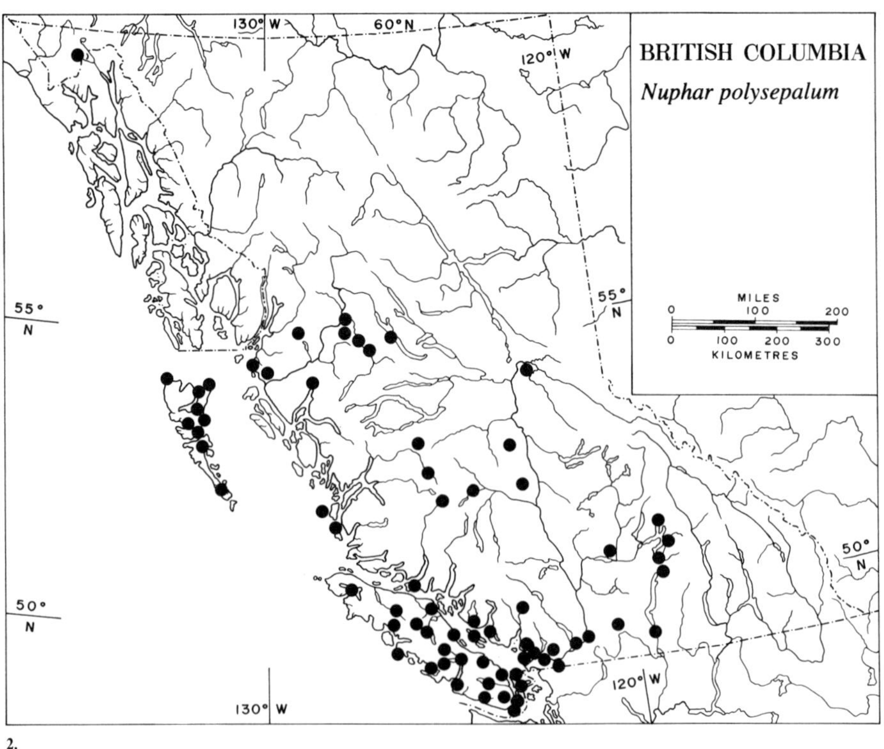

2.

BRITISH COLUMBIA
Nuphar variegatum
130° W
60° N
120° W
55°
N
50°
N
130° W
120° W
MILES
0   100   200
0   100   200   300
KILOMETRES

3.

BRITISH COLUMBIA
Nuphar: intermediates between
N. polysepalum
& N. variegatum
130° W
60° N
120° W
55°
N
50°
N
130° W
120° W
MILES
0   100   200
0   100   200   300
KILOMETRES

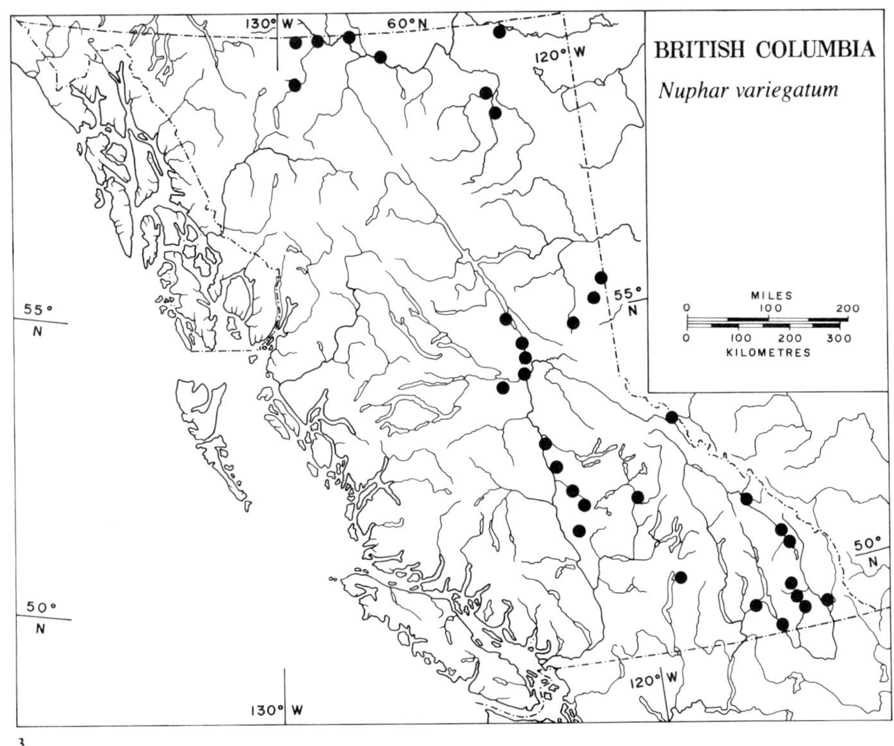

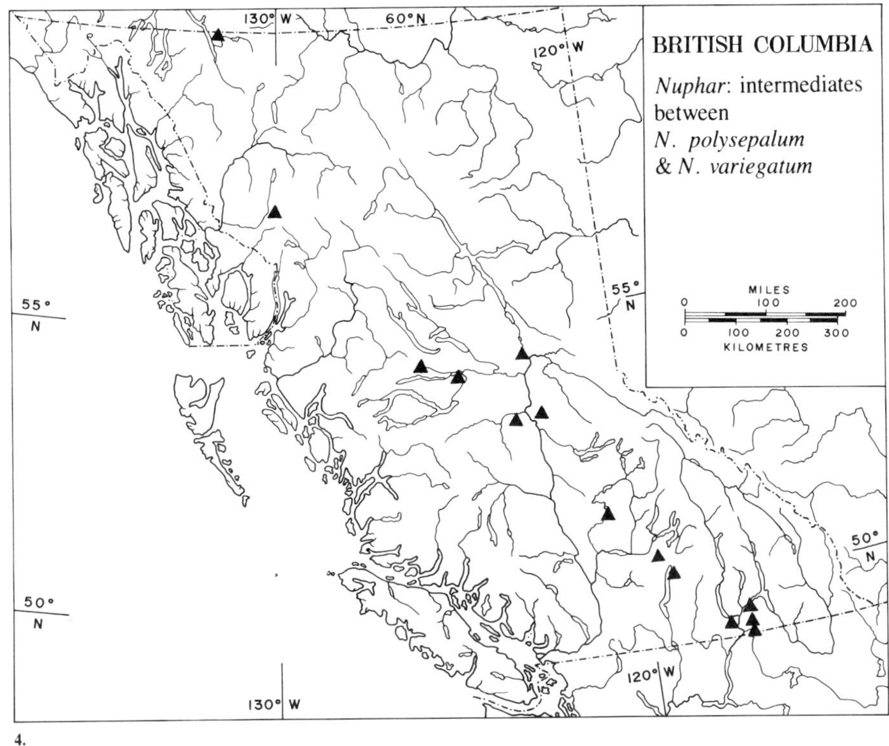

4.

130° W
60° N
120° W
BRITISH COLUMBIA
Nymphaea tetragona
55°
N
55°
N
MILES
100
200
100
200
300
KILOMETRES
50°
N
50°
N
130° W
120° W

5.

130° W
60° N
120° W
BRITISH COLUMBIA
Nymphaea odorata
55°
N
55°
N
MILES
100
200
100
200
300
KILOMETRES
50°
N
50°
N
130° W
120° W

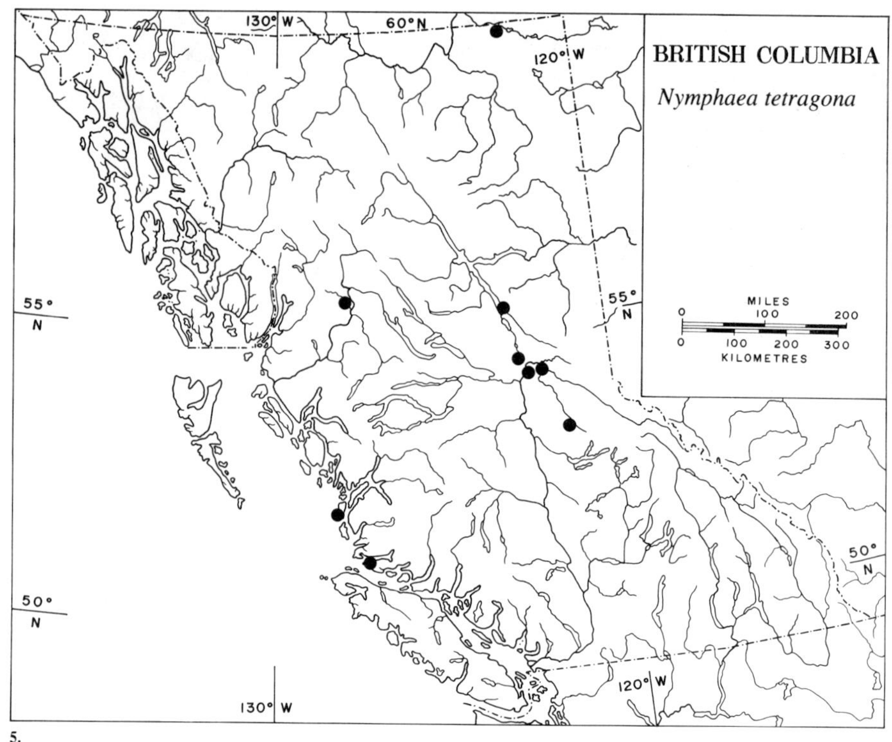

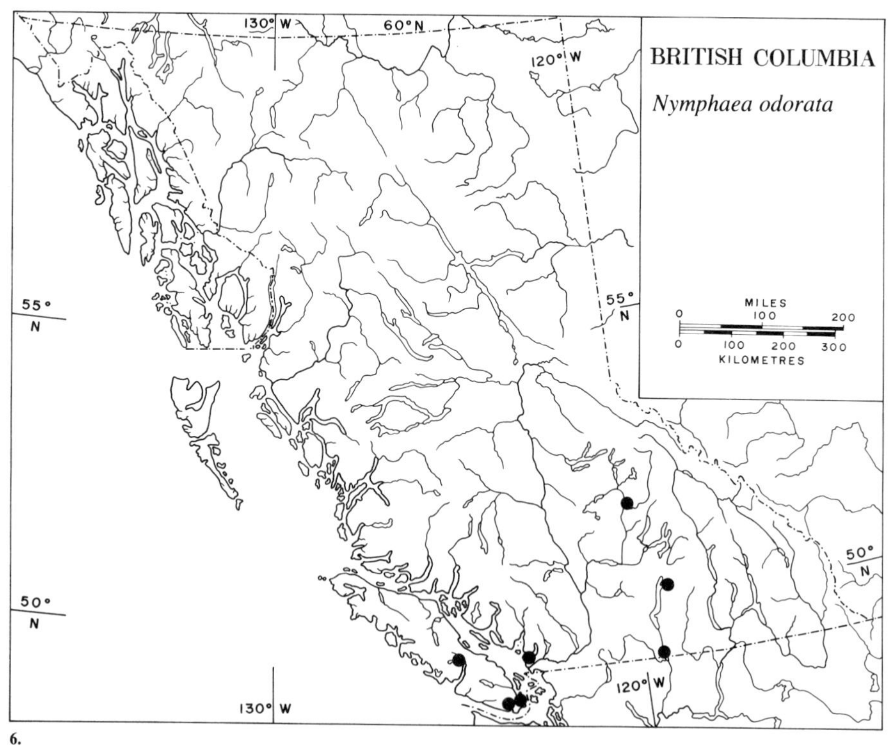

6.

BRITISH COLUMBIA
Ceratophyllum
demersum
MILES
0   100   200
0   100   200   300
KILOMETRES
60°N
120° W
55° N
50° N
50° N
55° N
130° W
120° W

7.

BRITISH COLUMBIA
Ceratophyllum
echinatum
MILES
0   100   200
0   100   200   300
KILOMETRES
130° W
60°N
120° W
55° N
50° N
50° N
55° N
130° W
120° W

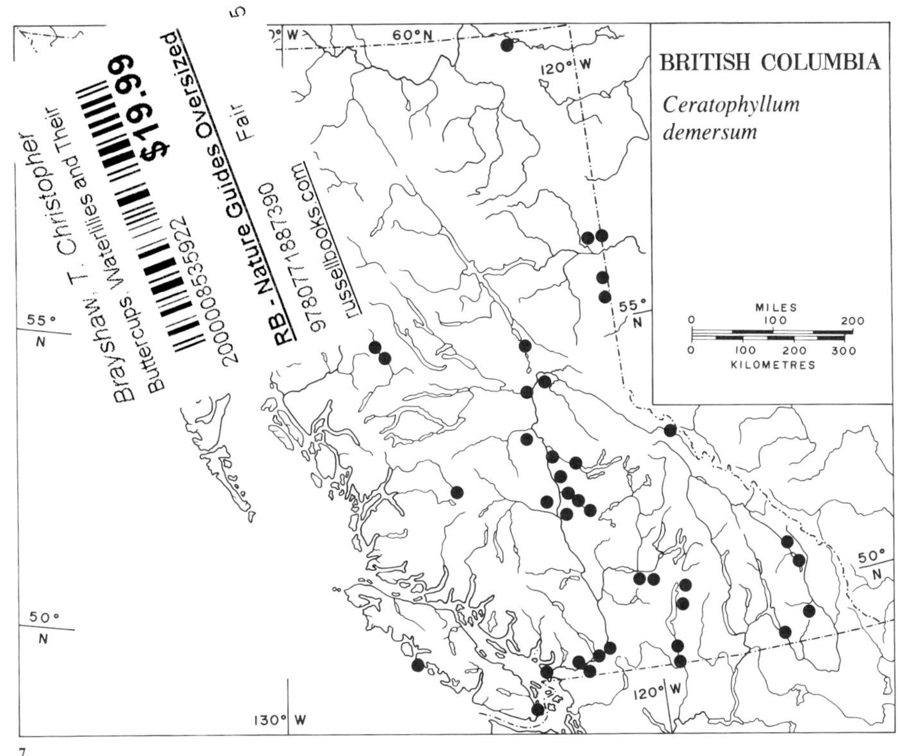

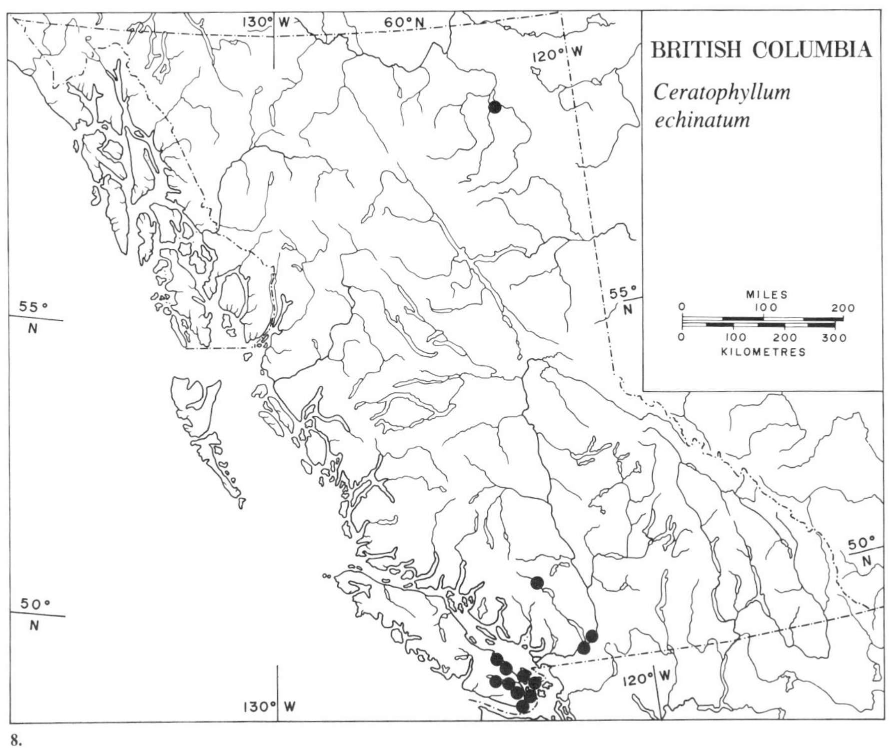

8.

130° W
60° N
120° W
BRITISH COLUMBIA
Aconitum columbianum
55° N
55° N
MILES
100
0
200
0
100
200
300
KILOMETRES
50° N
50° N
50° N
120° W
130° W

9.

130° W
60° N
120° W
BRITISH COLUMBIA
Aconitum
delphiniifolium
55° N
55° N
MILES
100
0
200
0
100
200
300
KILOMETRES
55° N
50° N
50° N
50° N
120° W
130° W

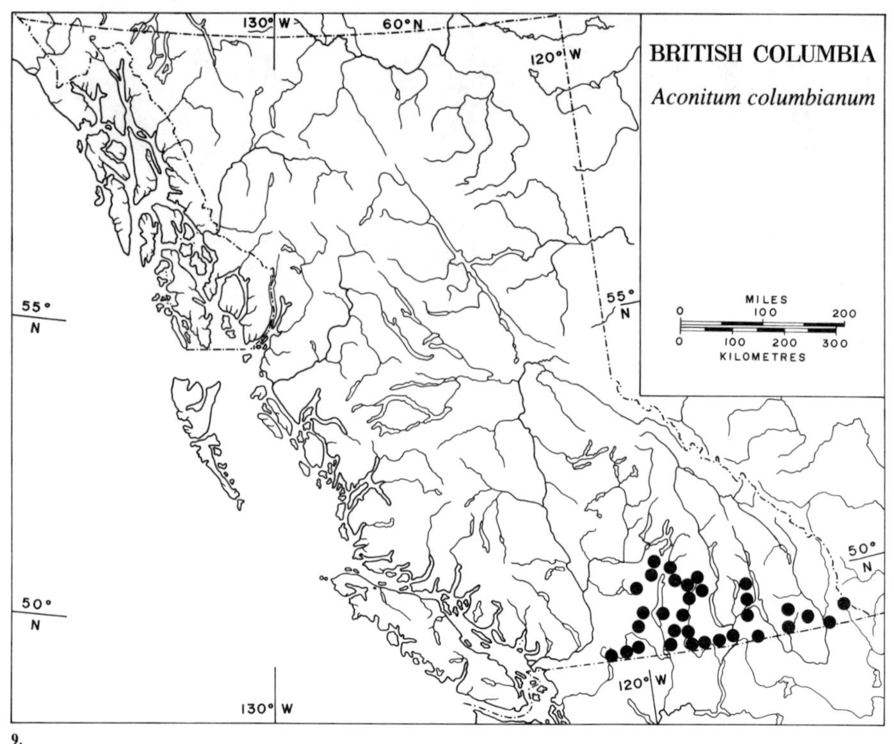

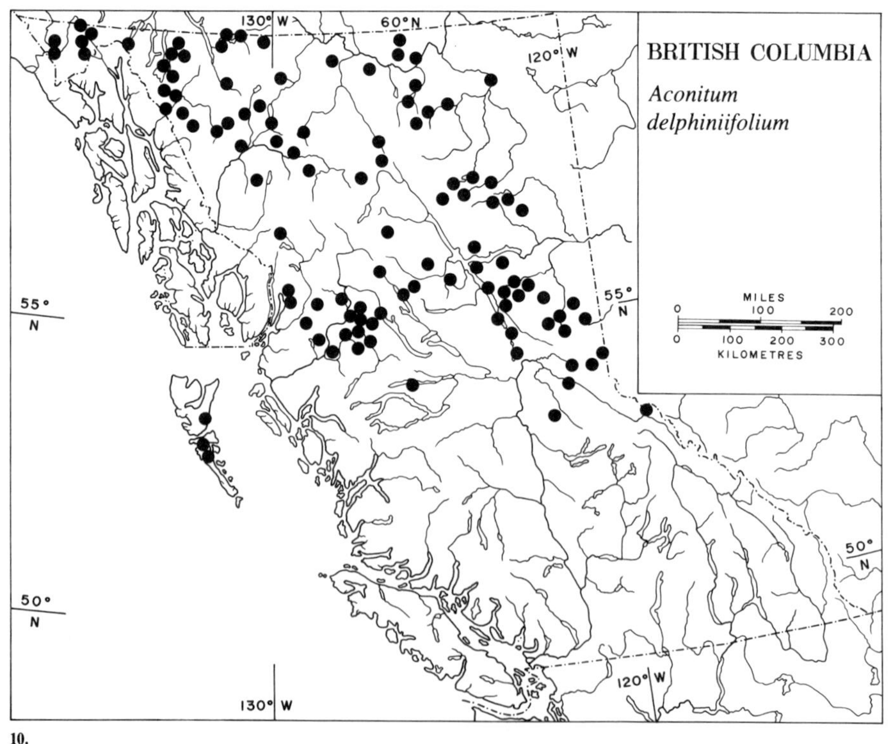

10.

202

130° W
60° N
120° W
BRITISH COLUMBIA
Actaea rubra
55° N
MILES
100
200
0
100
200
300
KILOMETRES
55° N
50° N
50° N
130° W
120° W

11.

130° W
60° N
120° W
BRITISH COLUMBIA
Anemone cylindrica
55° N
MILES
100
200
0
100
200
300
KILOMETRES
55° N
50° N
50° N
130° W
120° W

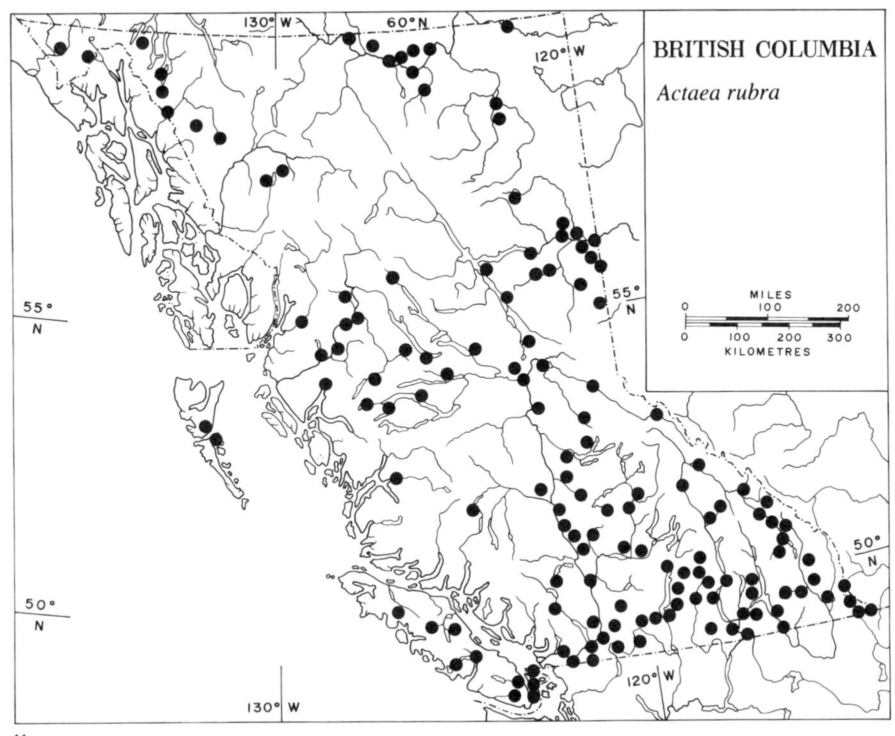

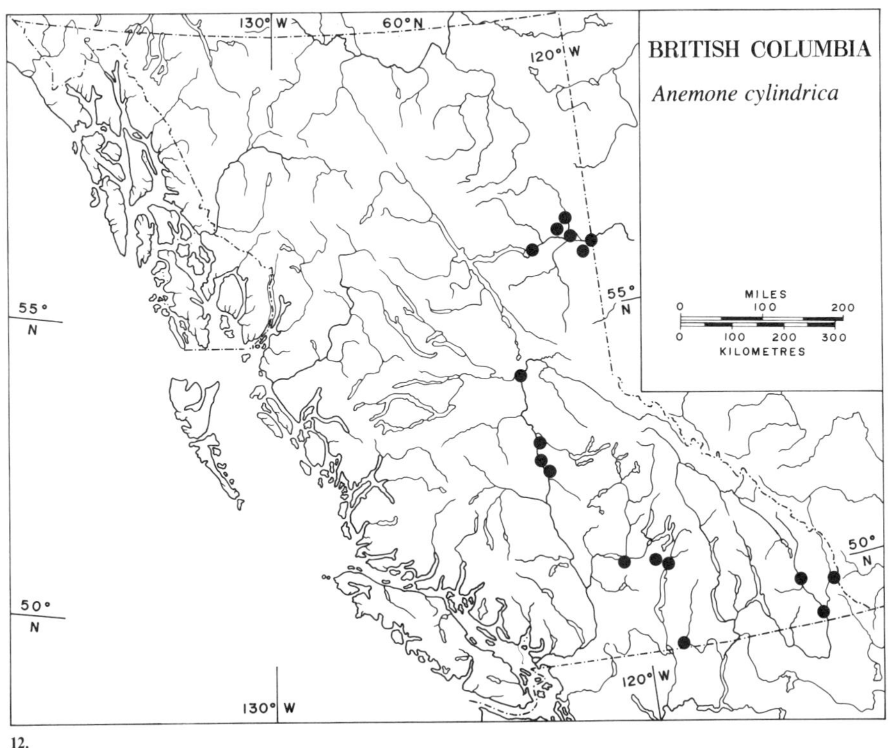

12.

BRITISH COLUMBIA
Anemone riparia
130° W
60° N
120° W
55° N
50° N
55° N
50° N
130° W
120° W
MILES
100
200
KILOMETRES
100
200
300

13.

BRITISH COLUMBIA
Anemone drummondii
var. drummondii
var. lithophila
130° W
60° N
120° W
55° N
50° N
55° N
50° N
130° W
120° W
MILES
100
200
KILOMETRES
100
200
300

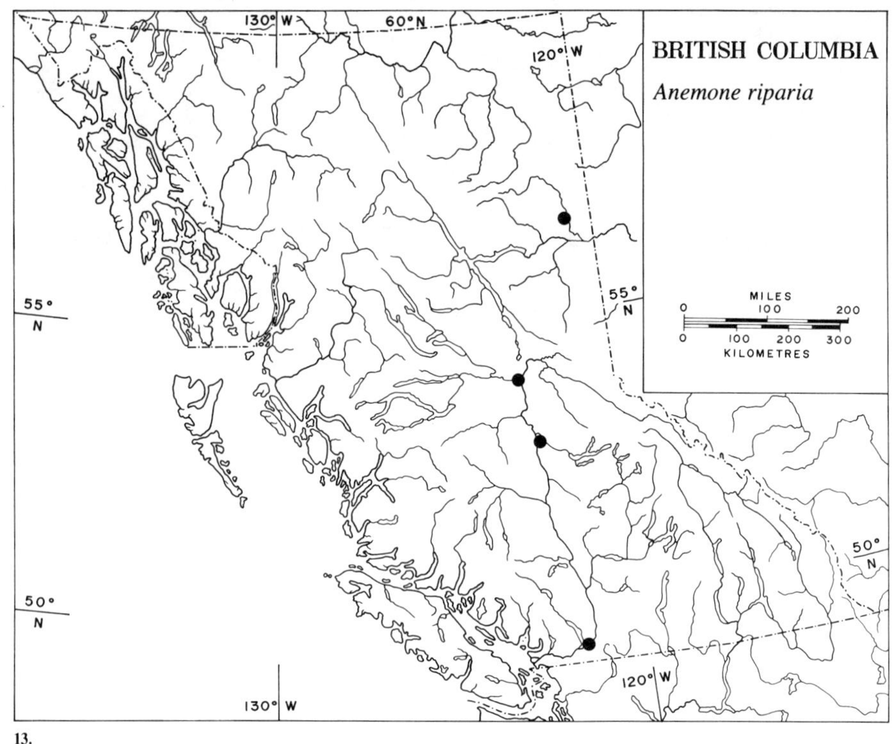

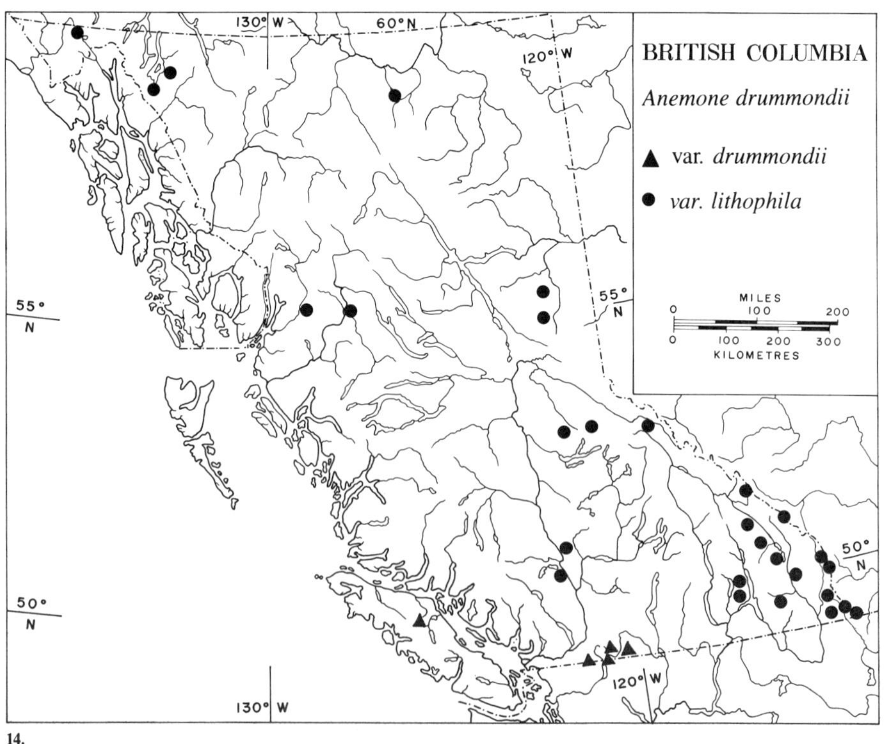

14.

BRITISH COLUMBIA
Anemone multifida
130° W
60° N
120° W
55° N
55° N
50° N
50° N
120° W
130° W
MILES
0   100   200
0   100   200   300
KILOMETRES

15.

BRITISH COLUMBIA
Anemone parviflora
130° W
60° N
120° W
55° N
55° N
50° N
50° N
120° W
130° W
MILES
0   100   200
0   100   200   300
KILOMETRES

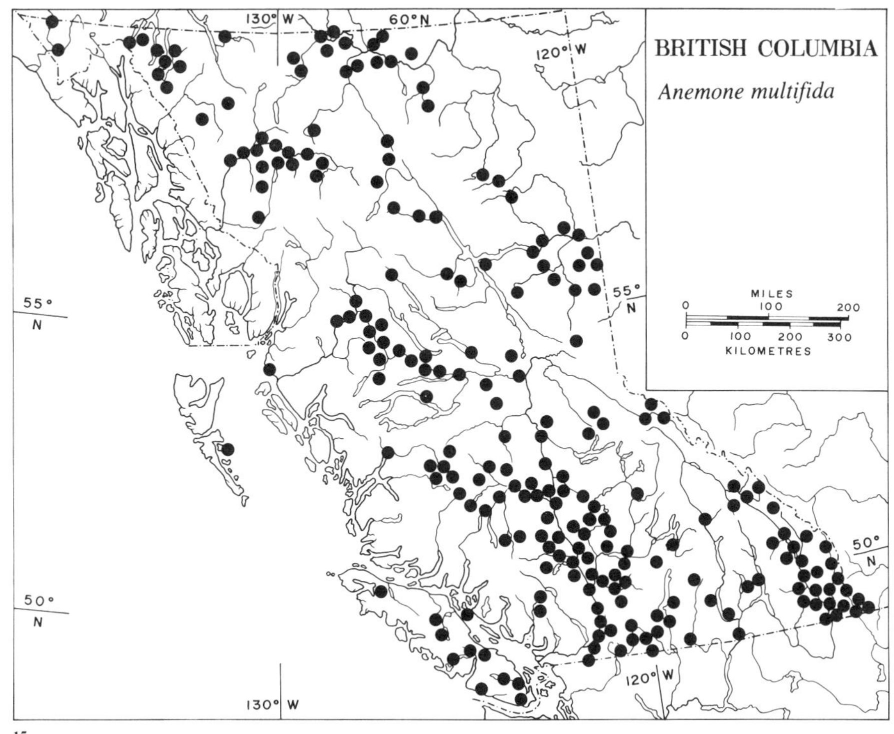

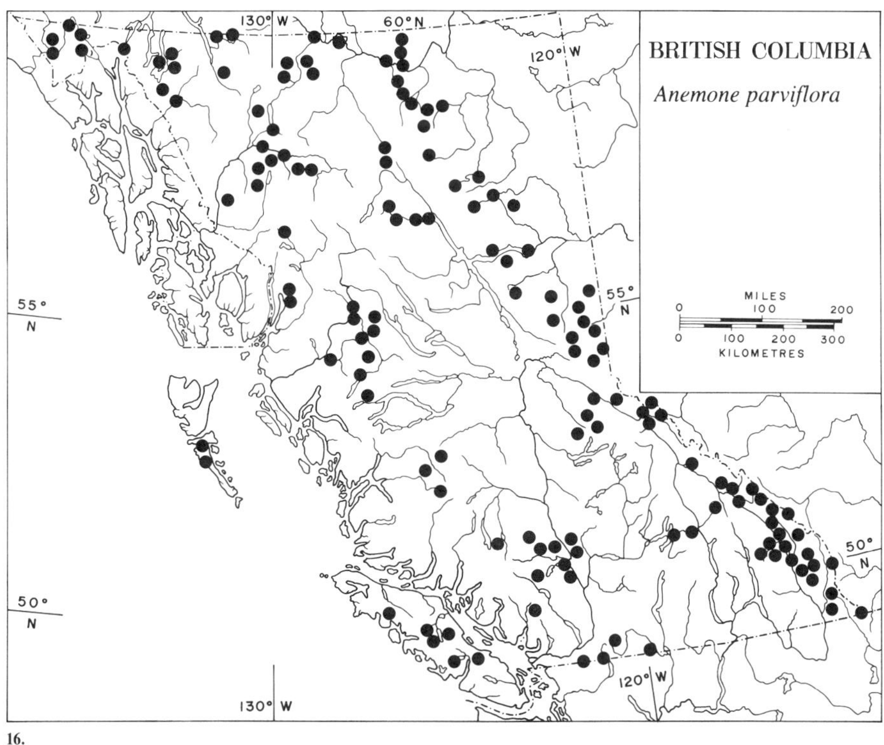

16.

BRITISH COLUMBIA
Anemone lyallii.
130° W
60° N
120° W
55° N
50° N
55° N
50° N
130° W
120° W
MILES
0   100   200
0   100   200   300
KILOMETRES

17.

BRITISH COLUMBIA
Anemone piperi
130° W
60° N
120° W
55° N
50° N
55° N
50° N
130° W
120° W
MILES
0   100   200
0   100   200   300
KILOMETRES

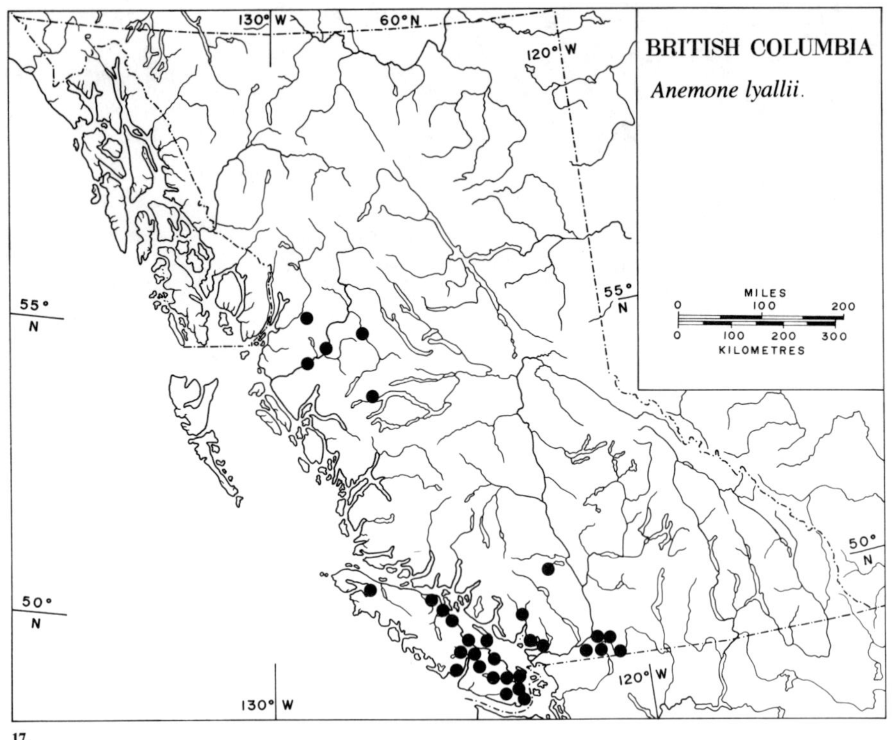

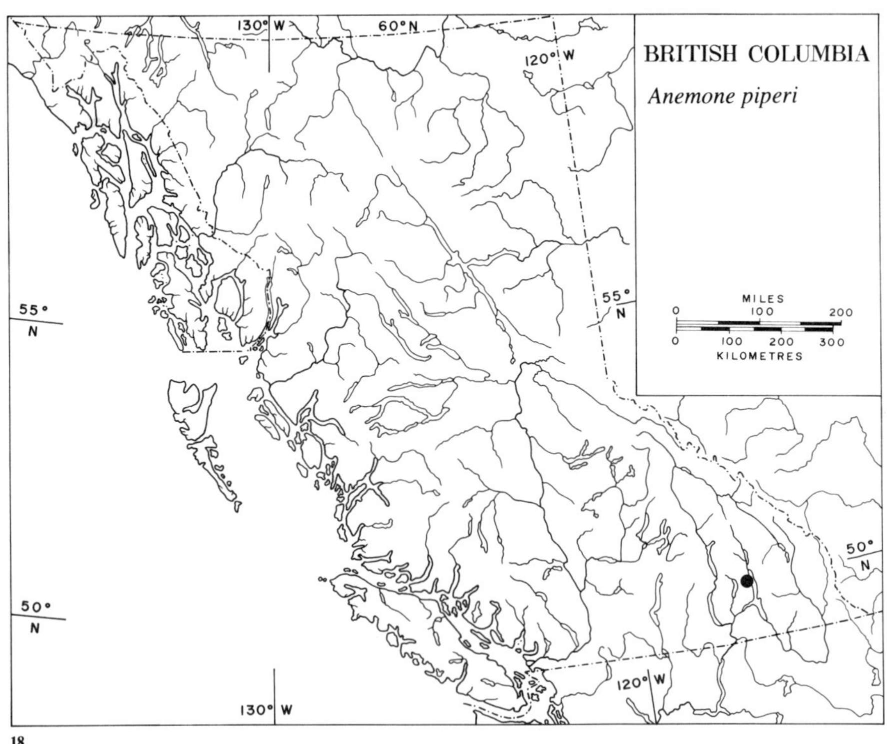

18.

BRITISH COLUMBIA
Anemone canadensis
130° W
60° N
120° W
55° N
50° N
55°
N
50°
N
130° W
120° W
MILES
0   100   200
0   100   200   300
KILOMETRES

19.

BRITISH COLUMBIA
Anemone narcissiflora
130° W
60° N
120° W
55° N
50° N
55°
N
50°
N
130° W
120° W
MILES
0   100   200
0   100   200   300
KILOMETRES

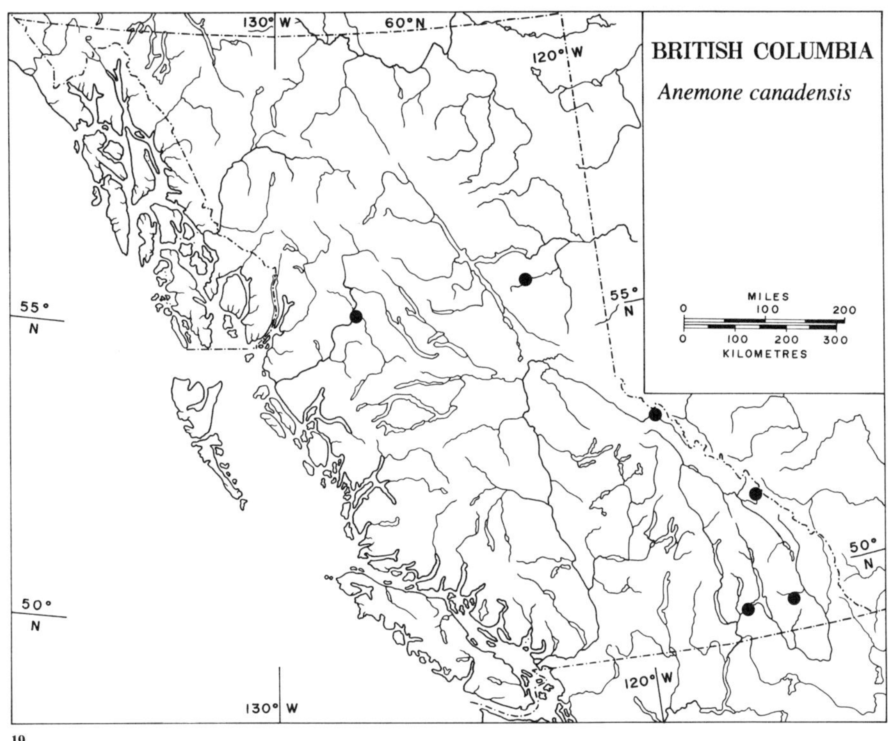

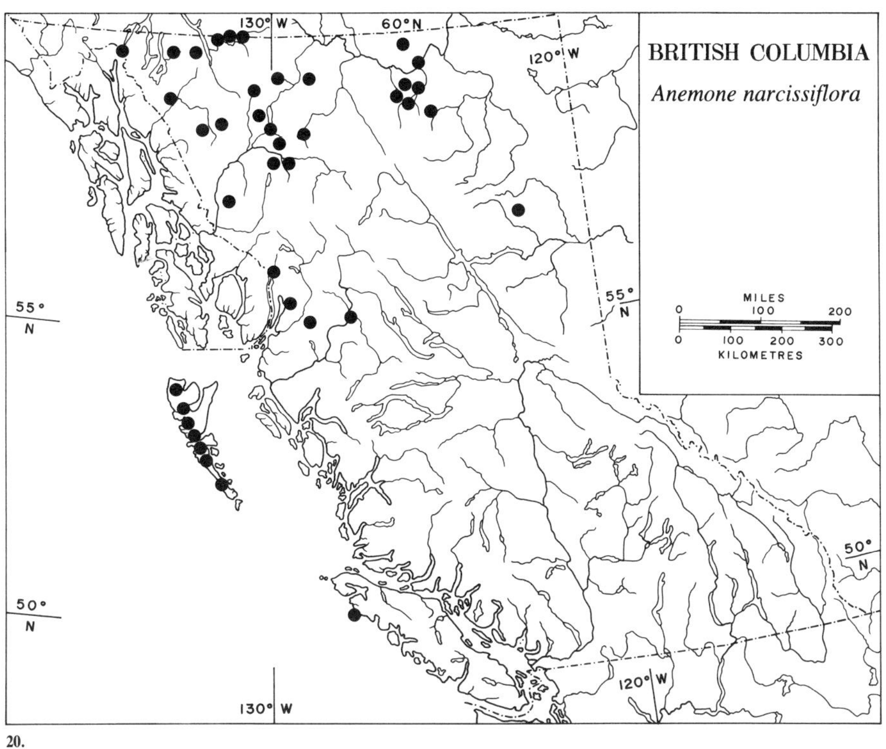

20.

BRITISH COLUMBIA
Anemone richardsonii
130° W
60° N
120° W
55° N
50° N
130° W
120° W
MILES
100
200
0
100
200
300
KILOMETRES

21.

130° W
60° N
120° W
BRITISH COLUMBIA
Anemone occidentalis
55° N
50° N
130° W
120° W
MILES
100
200
0
100
200
300
KILOMETRES

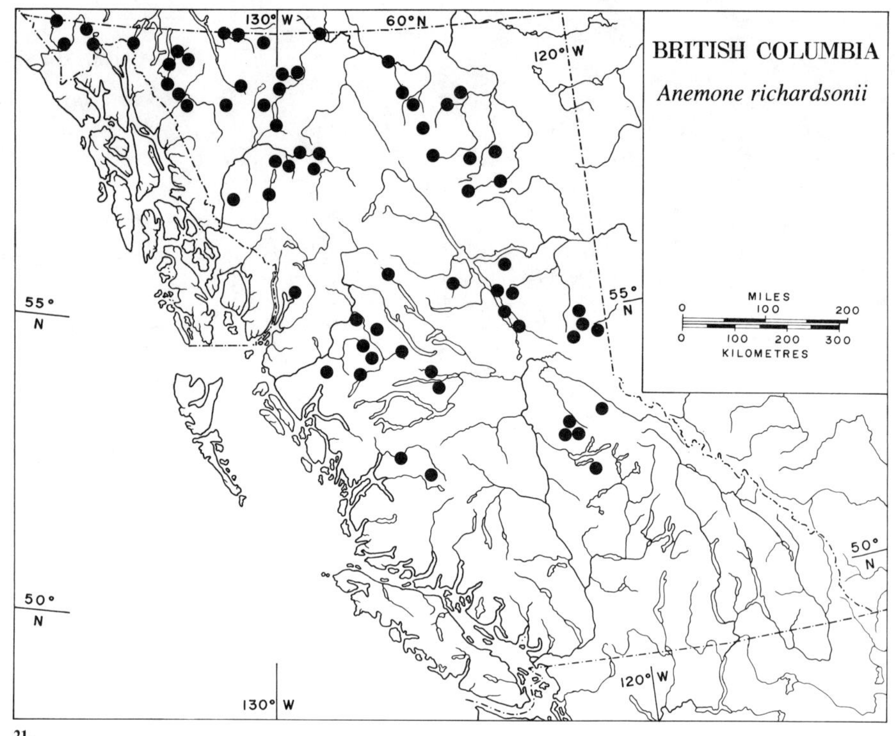

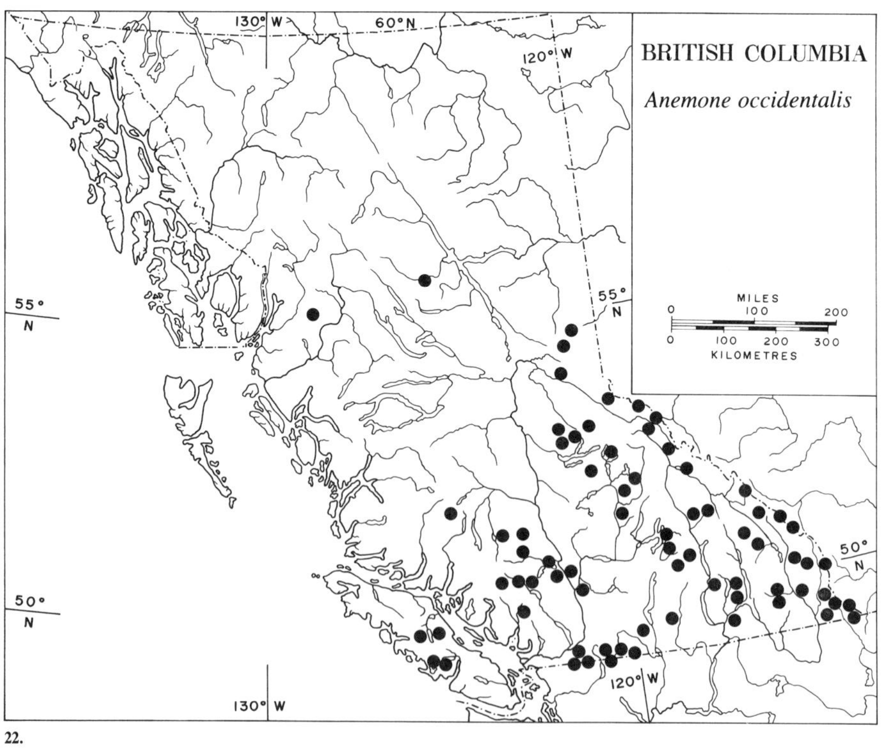

22.

BRITISH COLUMBIA
Anemone patens
130° W
60°N
120° W
55°
N
55°
N
50°
N
50°
N
130° W
120° W
130° W
MILES
0   100   200
0   100   200   300
KILOMETRES

23.

BRITISH COLUMBIA
Aquilegia brevistyla
130° W
60°N
120° W
55°
N
55°
N
50°
N
50°
N
130° W
120° W
130° W
MILES
0   100   200
0   100   200   300
KILOMETRES

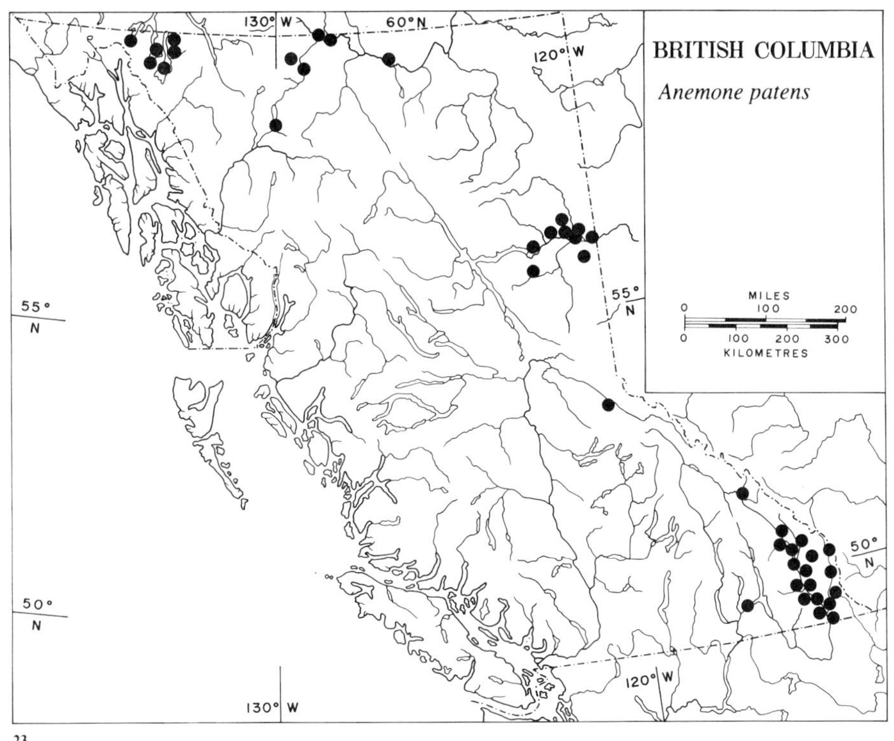

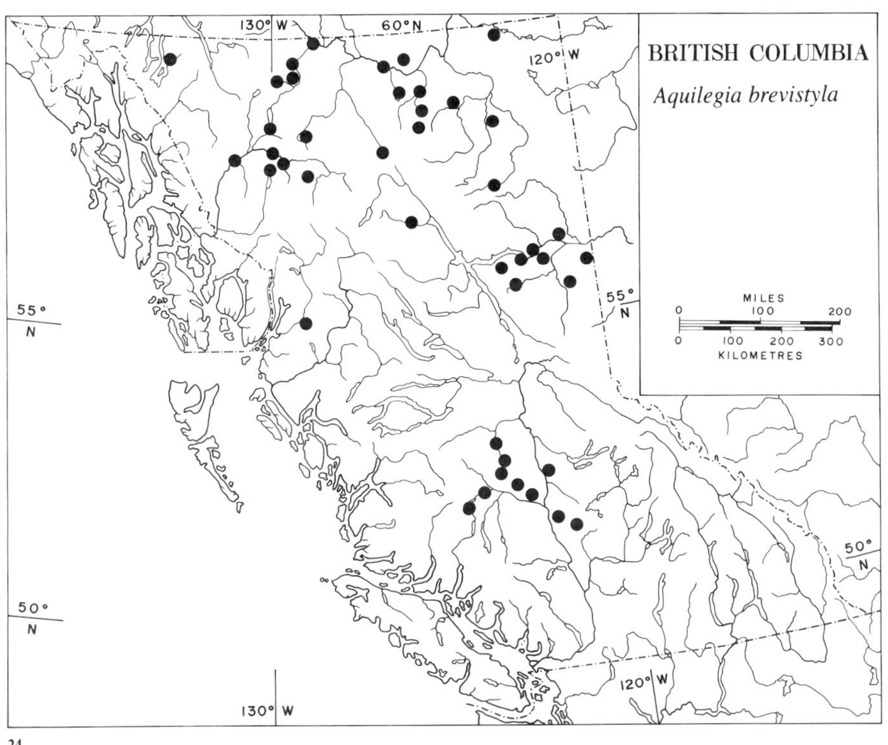

24.

209

BRITISH COLUMBIA
Aquilegia flavescens
var. flavescens
var. miniana
MILES
100
200
KILOMETRES
100 200 300
130° W
60° N
120° W
55° N
50° N
130° W
120° W

25.

BRITISH COLUMBIA
Aquilegia formosa
MILES
100
200
KILOMETRES
100 200 300
130° W
60° N
120° W
55° N
50° N
130° W
120° W

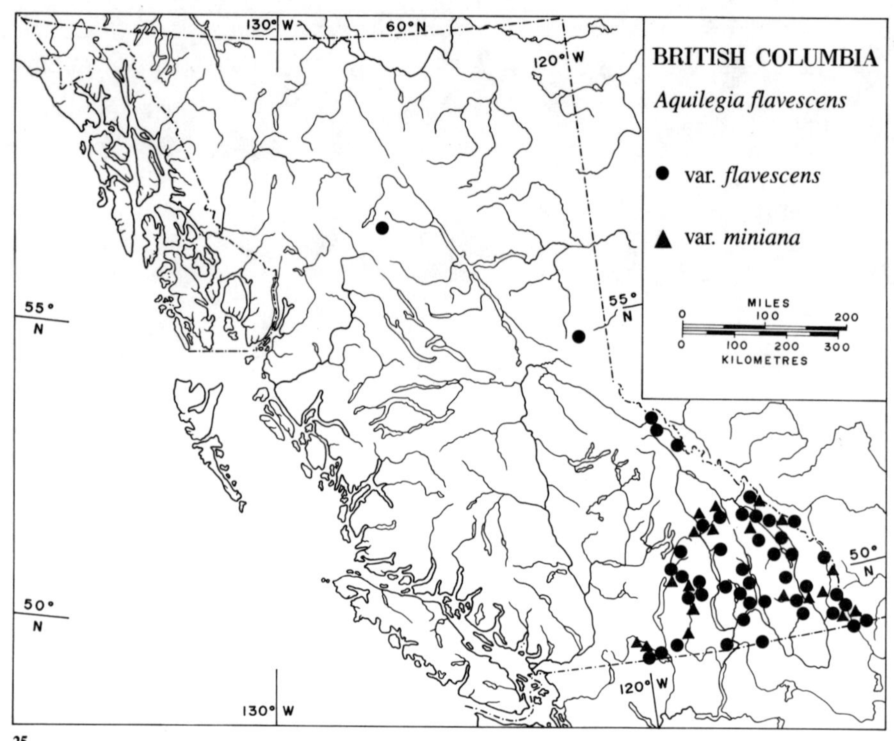

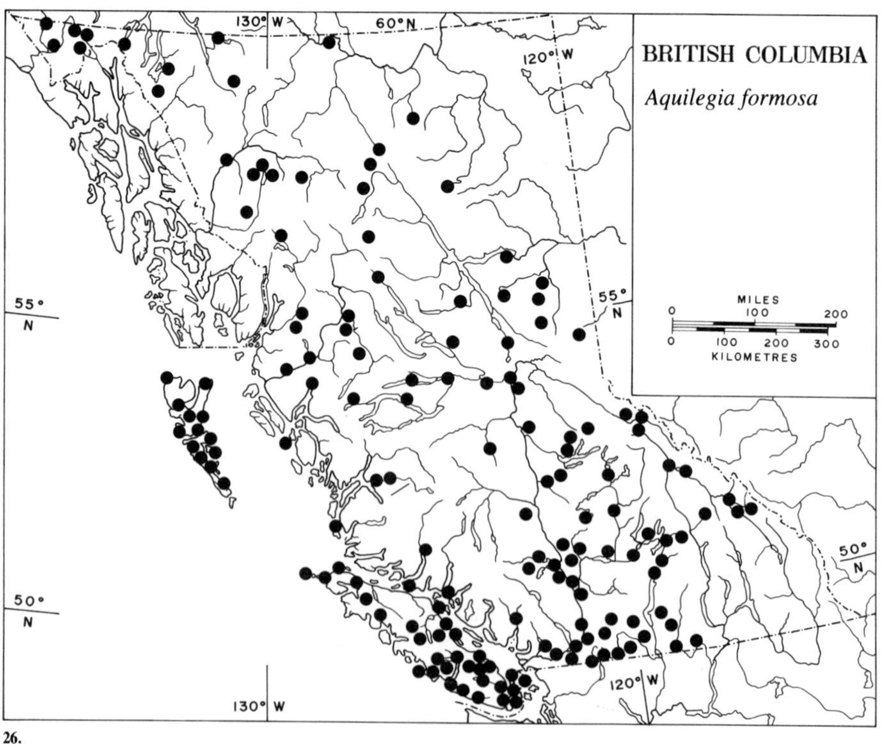

26.

130° W
60°N
120° W
55° N
55° N
50° N
50° N
130° W
120° W
BRITISH COLUMBIA
Caltha biflora
MILES
0   100   200
0   100   200   300
KILOMETRES

27.

130° W
60°N
120° W
55° N
55° N
50° N
50° N
130° W
120° W
BRITISH COLUMBIA
Caltha leptosepala
MILES
0   100   200
0   100   200   300
KILOMETRES

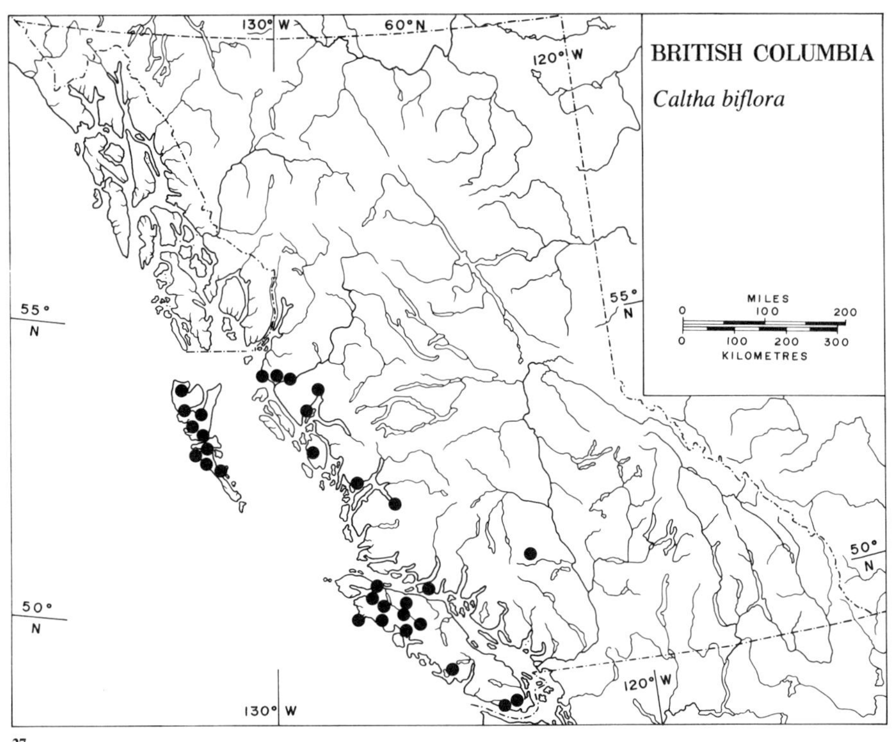

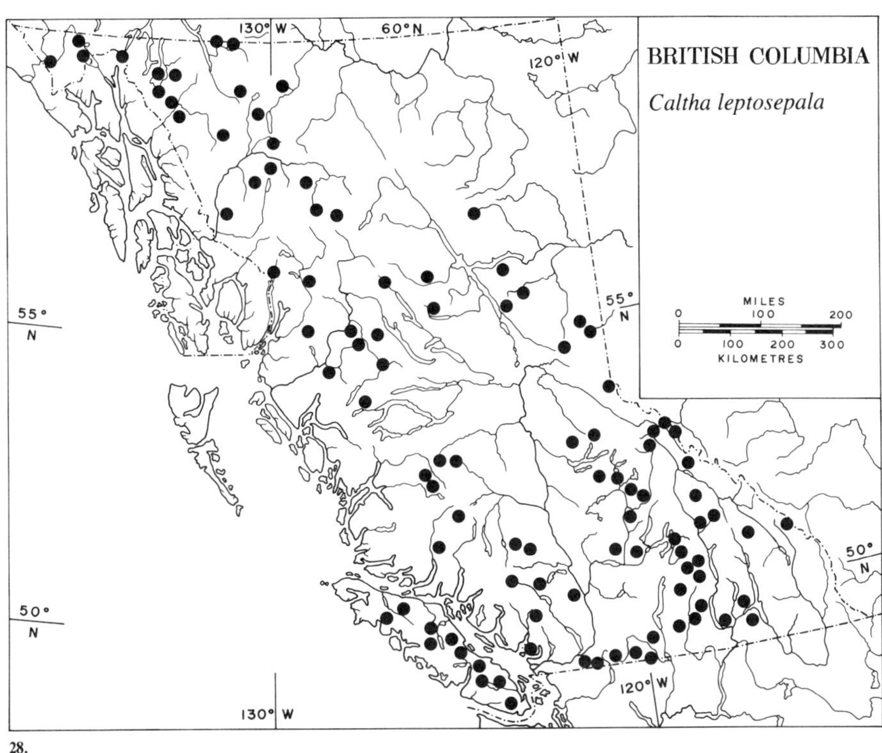

28.

BRITISH COLUMBIA
Caltha natans
130° W
60° N
120° W
55° N
50° N
MILES
0   100   200
0   100   200   300
KILOMETRES

29.

BRITISH COLUMBIA
Caltha palustris
130° W
60° N
120° W
55° N
50° N
MILES
0   100   200
0   100   200   300
KILOMETRES

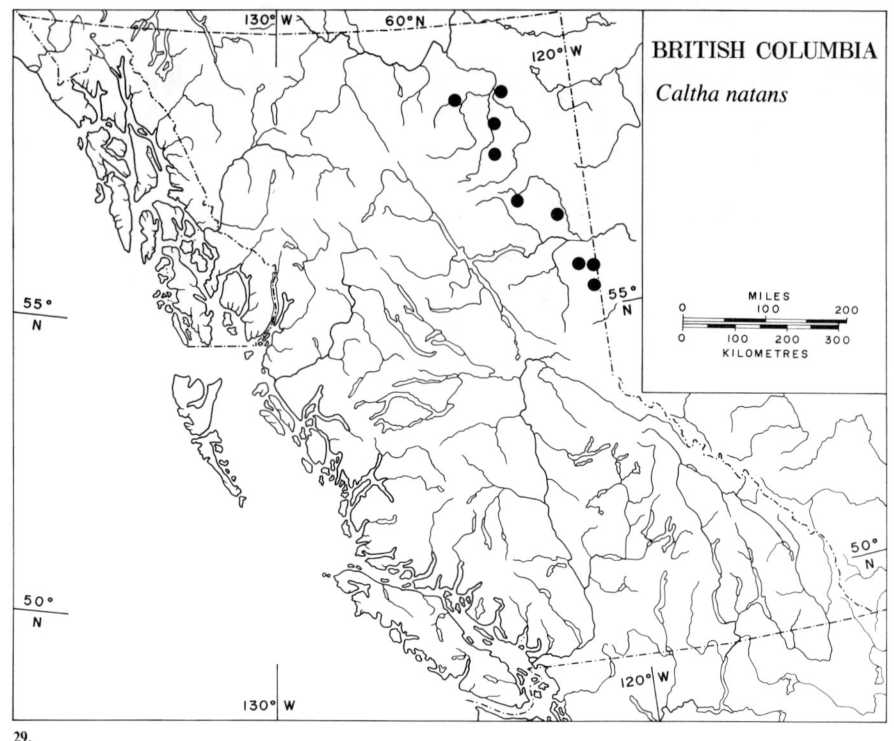

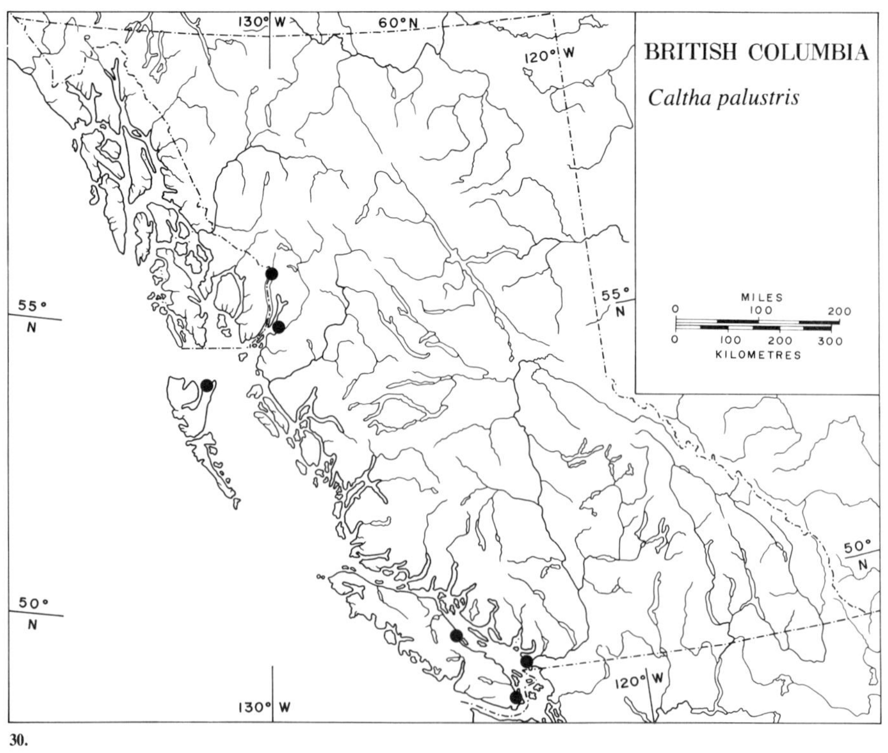

30.

212

BRITISH COLUMBIA
Cimicifuga elata
130° W
60° N
120° W
55°
N
55°
N
50°
N
50°
N
120° W
130° W
MILES
0   100   200
0   100   200   300
KILOMETRES

31.

BRITISH COLUMBIA
Clematis ligusticifolia
130° W
60° N
120° W
55°
N
55°
N
50°
N
50°
N
120° W
130° W
MILES
0   100   200
0   100   200   300
KILOMETRES

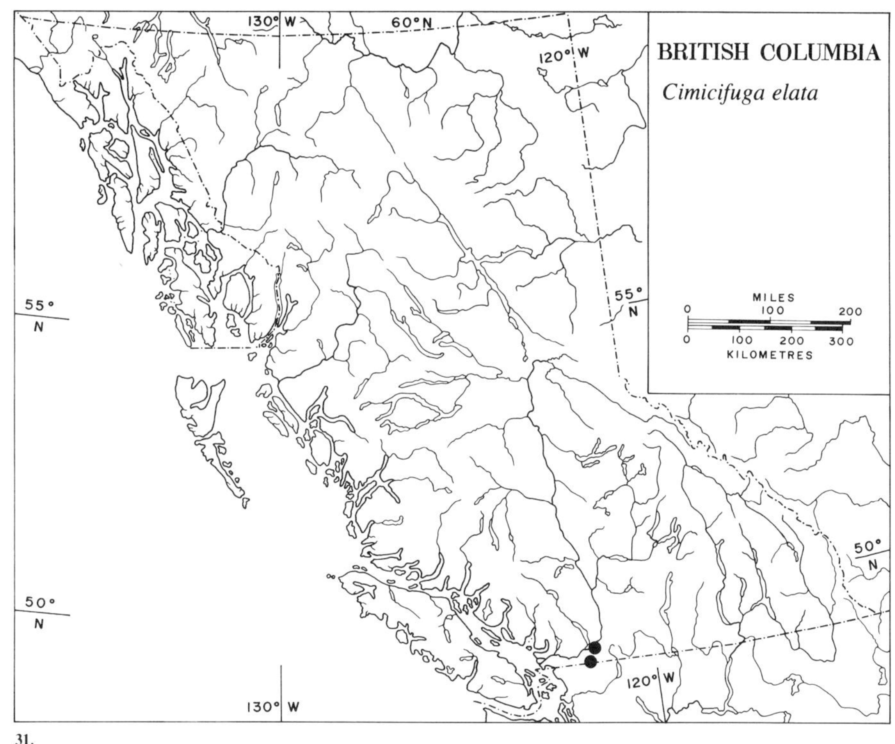

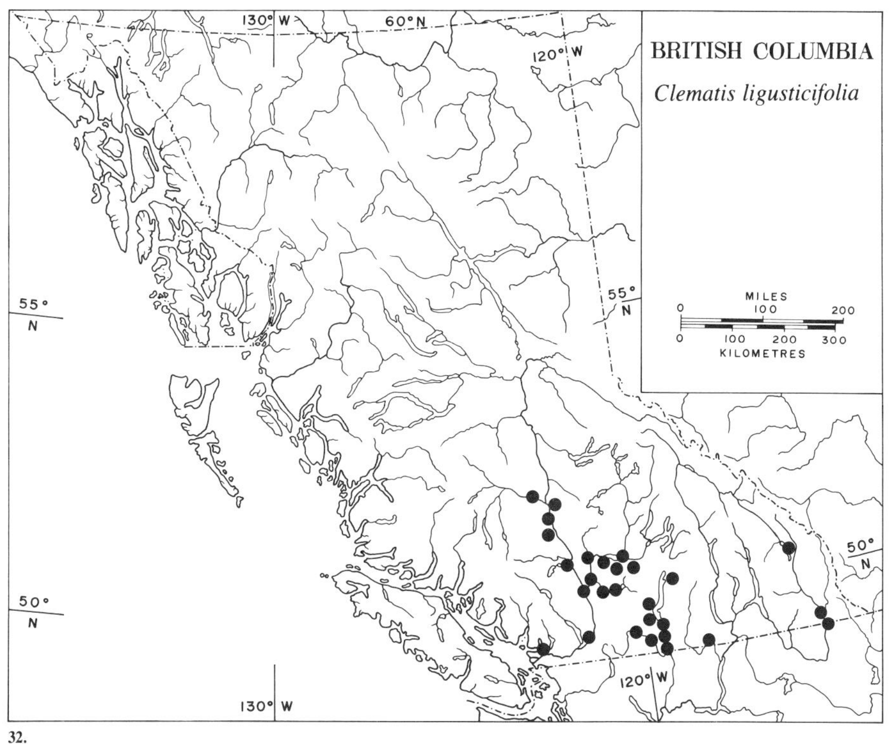

32.

213

BRITISH COLUMBIA
Clematis occidentalis
130° W
60° N
120° W
55° N
50° N
130° W
120° W
MILES
0   100   200
0   100   200   300
KILOMETRES

33.

BRITISH COLUMBIA
Clematis tangutica
130° W
60° N
120° W
55° N
50° N
130° W
120° W
MILES
0   100   200
0   100   200   300
KILOMETRES

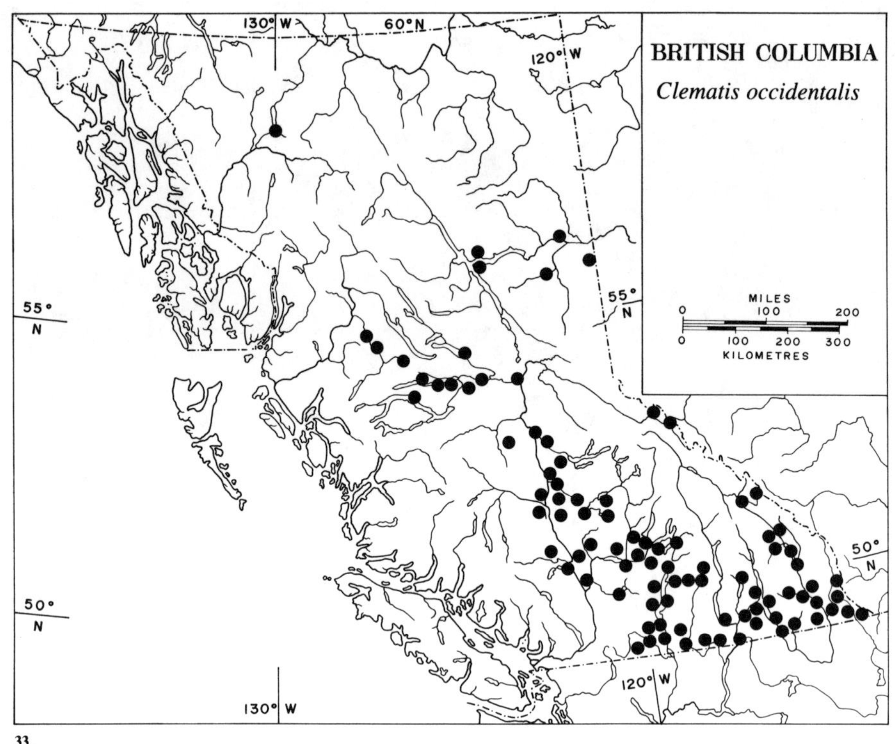

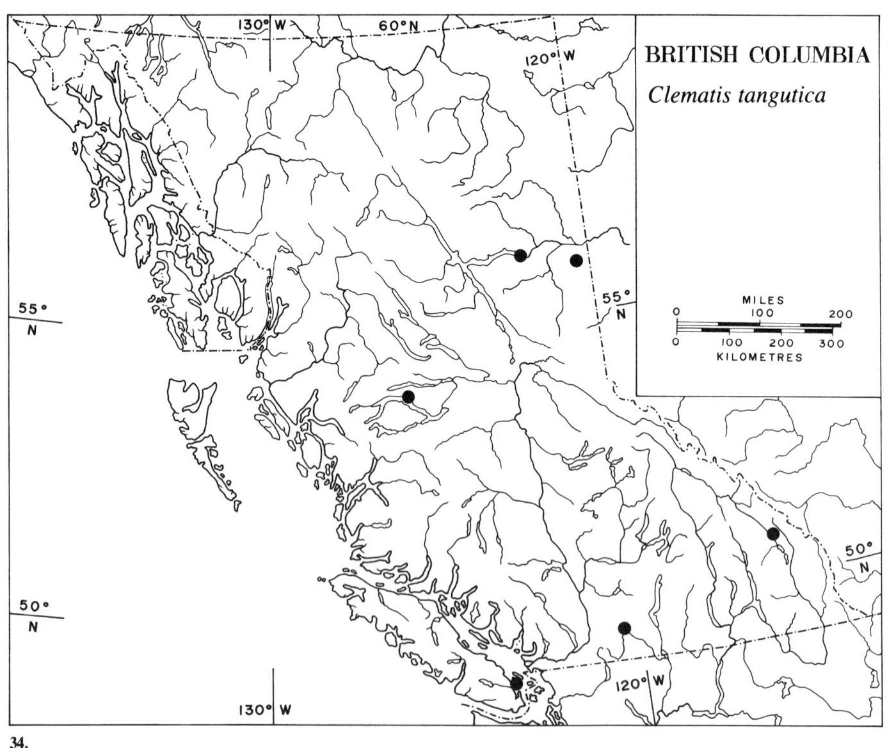

34.

BRITISH COLUMBIA
Clematis vitalba
130° W
60° N
120° W
55° N
50° N
55° N
50° N
120° W
130° W
MILES
0   100   200
0   100   200   300
KILOMETRES

35.

BRITISH COLUMBIA
Coptis aspleniifolia
130° W
60° N
120° W
55° N
50° N
55° N
50° N
120° W
130° W
MILES
0   100   200
0   100   200   300
KILOMETRES

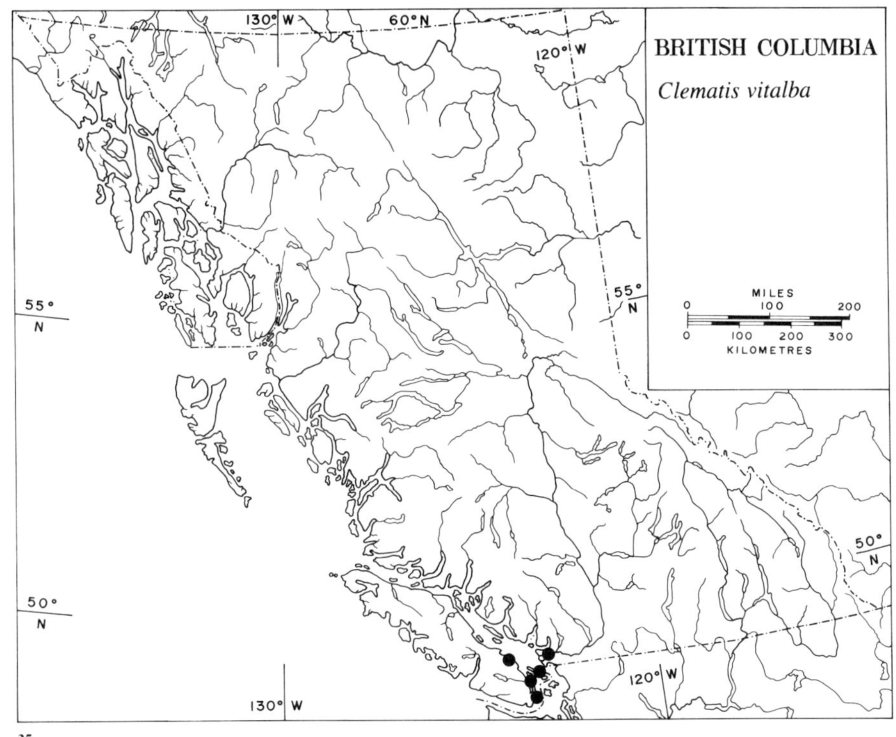

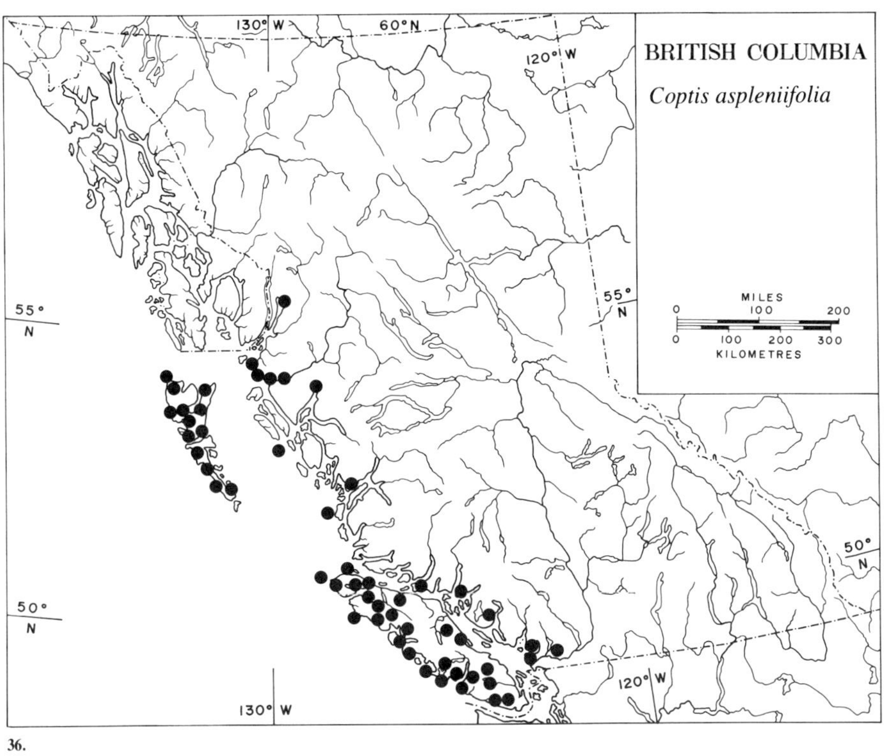

36.

130° W
60°N
120° W
BRITISH COLUMBIA
Coptis trifolia
55°
N
55°
N
MILES
100
200
0
100
200
300
KILOMETRES
50°
N
50°
N
120° W
130° W

37.

130° W
60°N
120° W
BRITISH COLUMBIA
Delphinium ambiguum
55°
N
55°
N
MILES
100
200
0
100
200
300
KILOMETRES
50°
N
50°
N
120° W
130° W

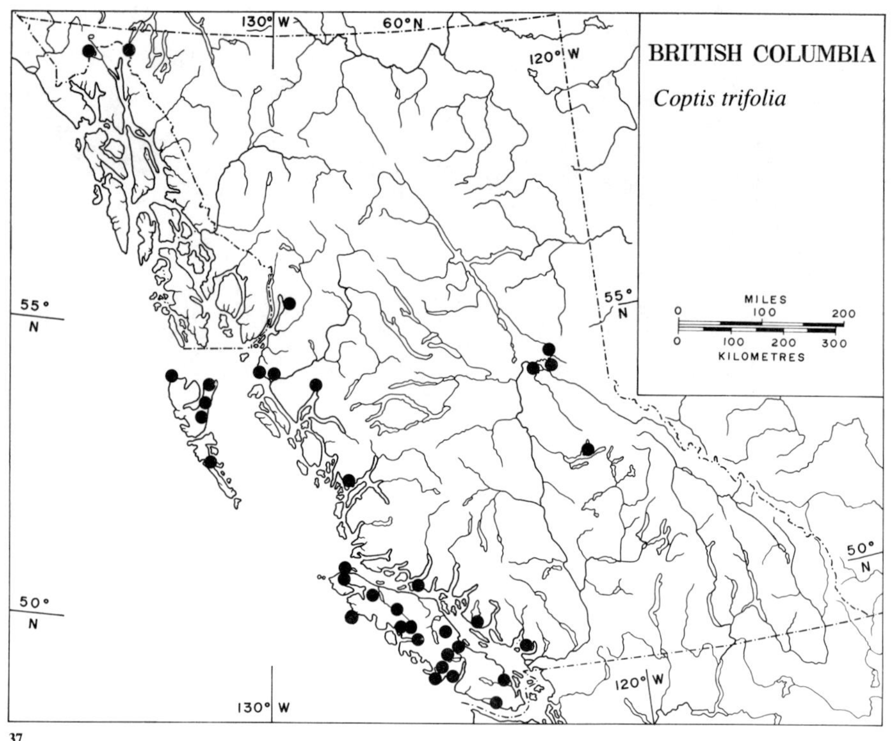

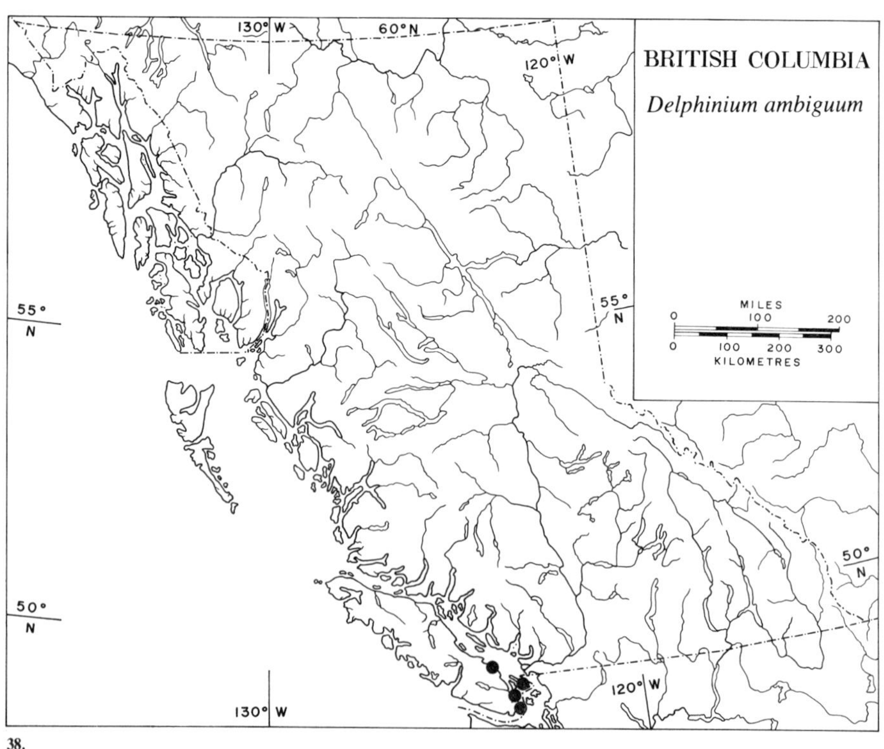

38.

BRITISH COLUMBIA
Delphinium bicolor
130° W
60° N
120° W
55° N
50° N
50° N
55° N
50° N
120° W
130° W
MILES
0 100 200
0 100 200 300
KILOMETRES

39.

BRITISH COLUMBIA
Delphinium burkei
130° W
60° N
120° W
55° N
50° N
55° N
50° N
120° W
130° W
MILES
0 100 200
0 100 200 300
KILOMETRES

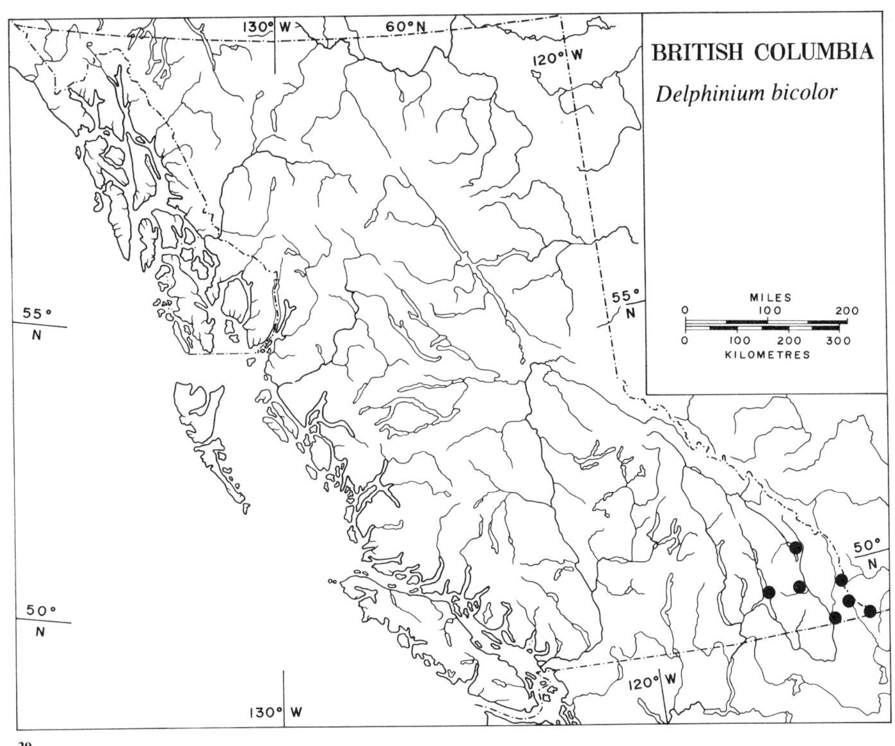

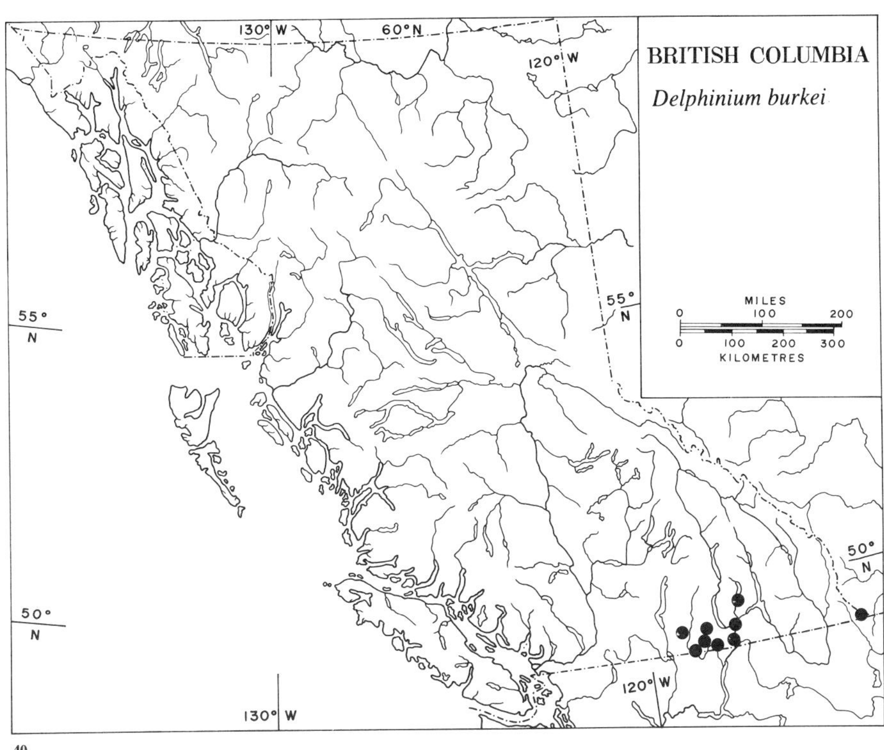

40.

130° W
60° N
120° W
55° N
50° N
BRITISH COLUMBIA
Delphinium glaucum
MILES
0   100   200
0   100   200   300
KILOMETRES
55° N
50° N
130° W
120° W

41.

130° W
60° N
120° W
55° N
50° N
BRITISH COLUMBIA
Delphinium menziesii
MILES
0   100   200
0   100   200   300
KILOMETRES
55° N
50° N
130° W
120° W

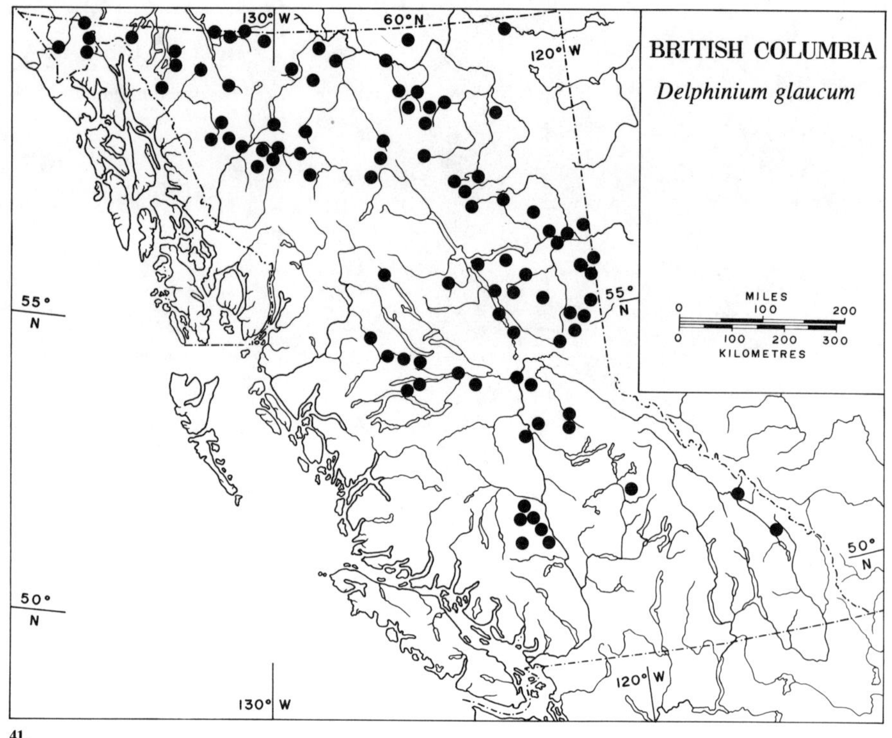

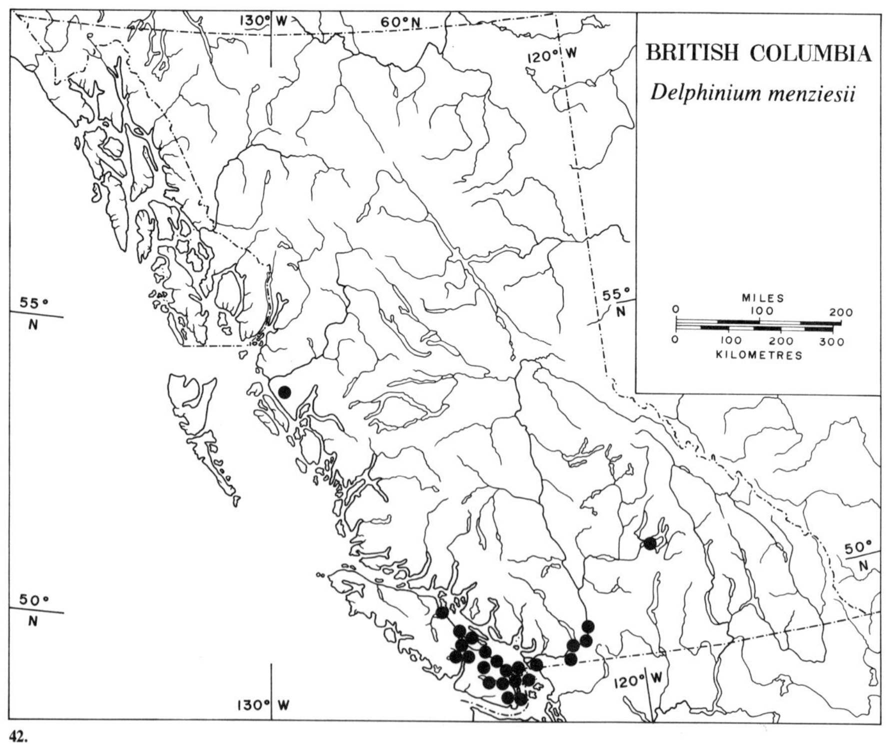

42.

218

BRITISH COLUMBIA
Delphinium
nuttallianum
MILES
KILOMETRES

43.

BRITISH COLUMBIA
Isopyrum savilei
MILES
KILOMETRES

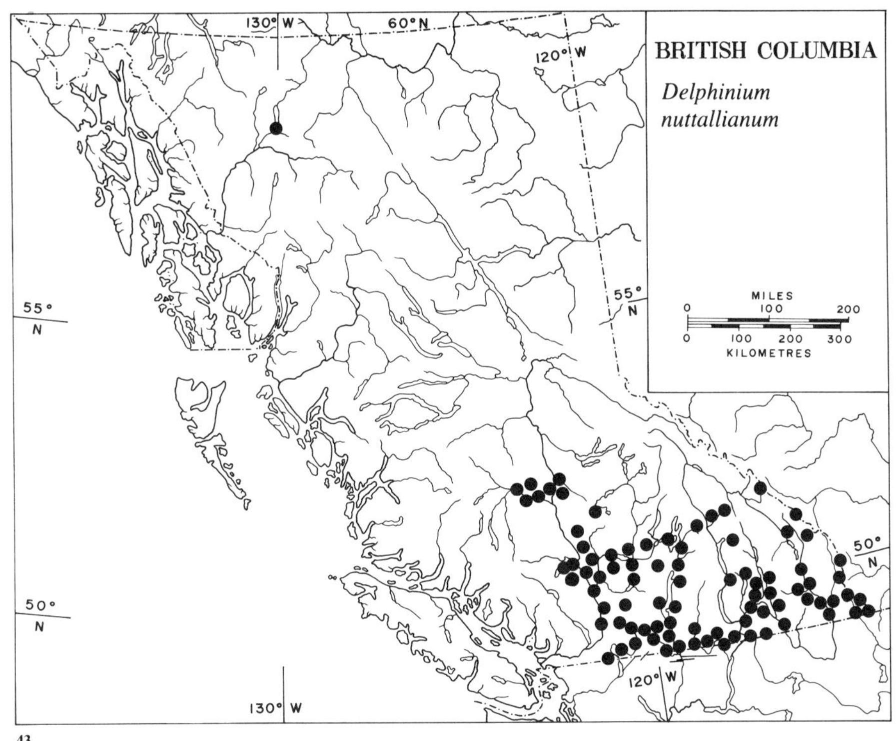

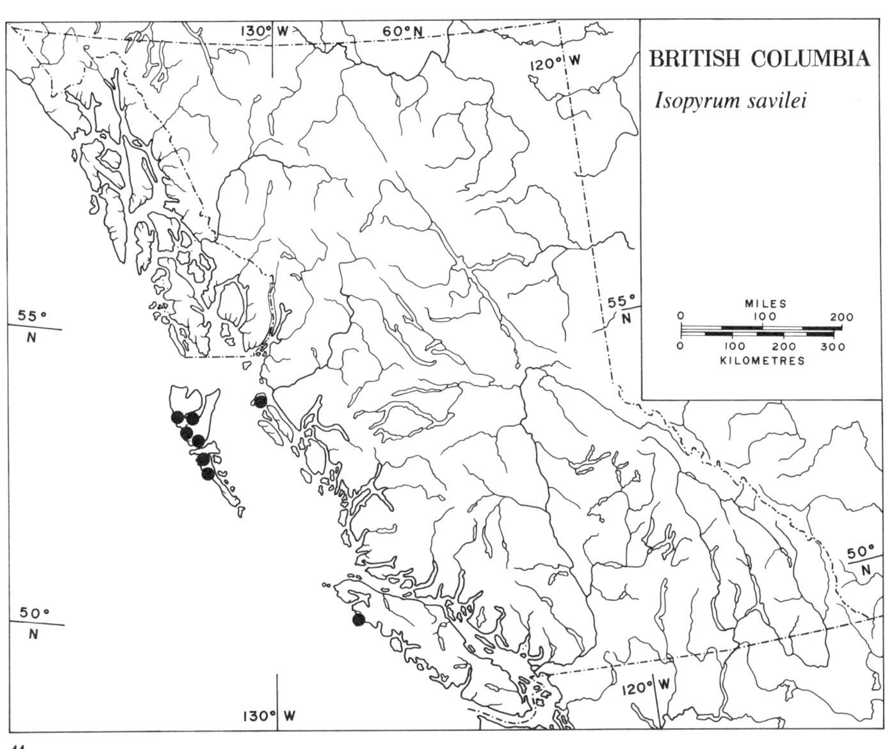

44.

BRITISH COLUMBIA
Myosurus aristatus
MILES
100
200
100 200 300
KILOMETRES

45.

BRITISH COLUMBIA
Myosurus minimus
MILES
100
200
100 200 300
KILOMETRES

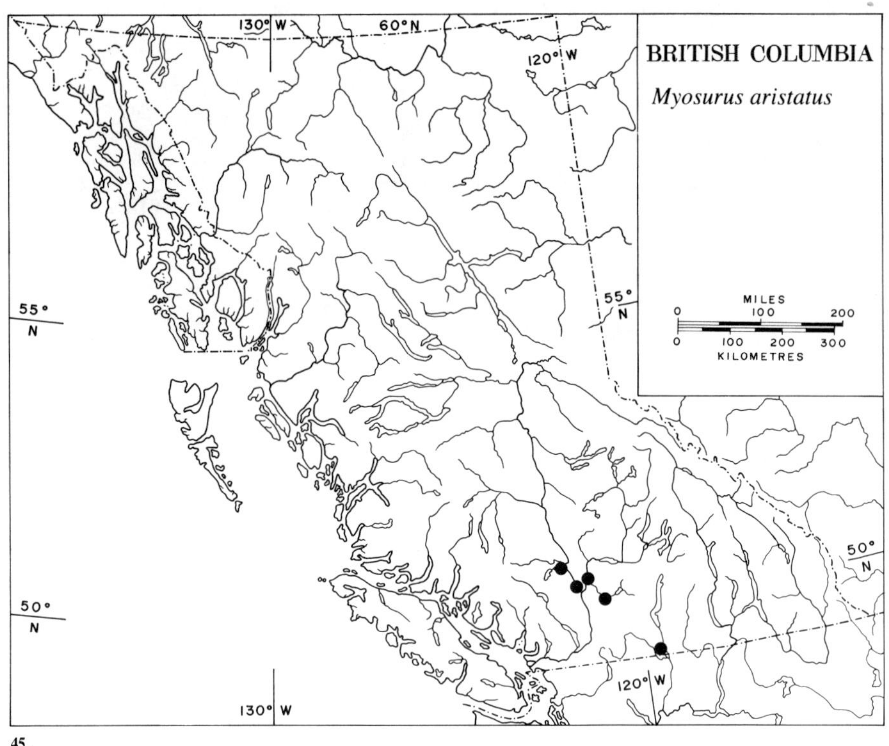

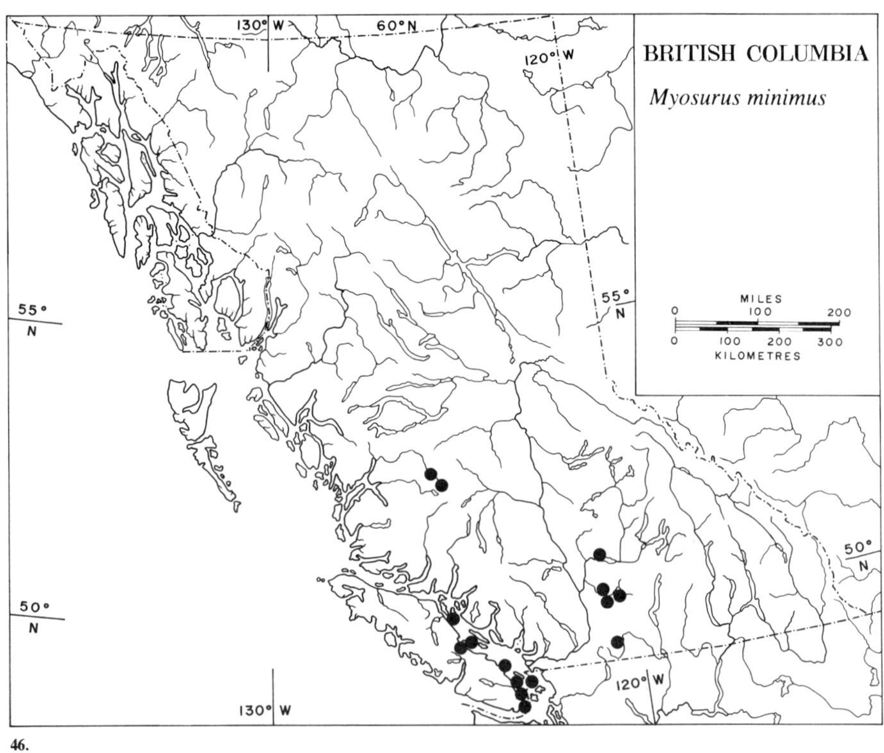

46.

220

130° W
60° N
120° W
BRITISH COLUMBIA
Nigella damascena
55° N
55° N
50° N
MILES
0 100 200
0 100 200 300
KILOMETRES
50° N
50° N
120° W
130° W

47.

130° W
60° N
120° W
BRITISH COLUMBIA
Ranunculus
pensylvanicus
55° N
55° N
MILES
0 100 200
0 100 200 300
KILOMETRES
50° N
50° N
120° W
130° W

48.

221

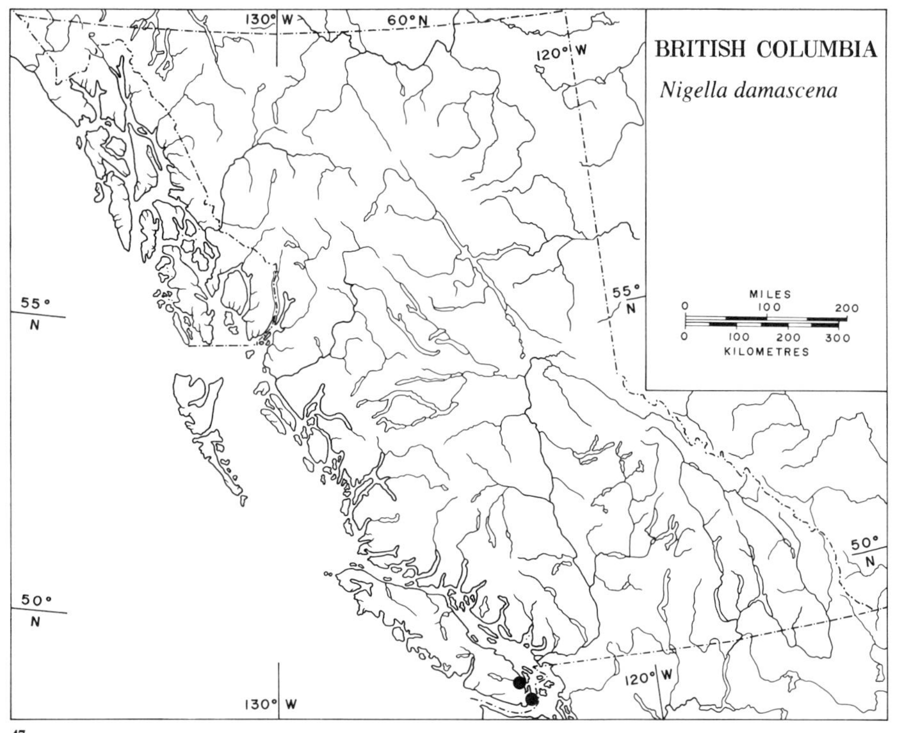

BRITISH COLUMBIA
Ranunculus macounii
130° W
60° N
120° W
55° N
50° N
50° N
55° N
50° N
130° W
120° W
MILES
100
0
200
0
100
200
300
KILOMETRES

49.

BRITISH COLUMBIA
Ranunculus repens
130° W
60° N
120° W
55° N
50° N
50° N
55° N
50° N
130° W
120° W
MILES
100
0
200
0
100
200
300
KILOMETRES

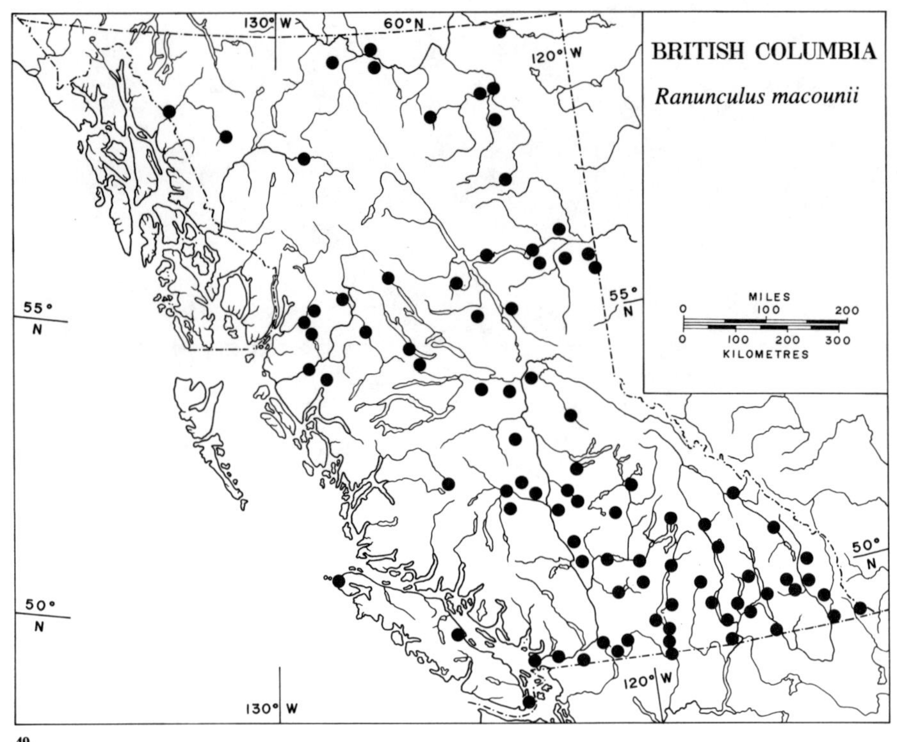

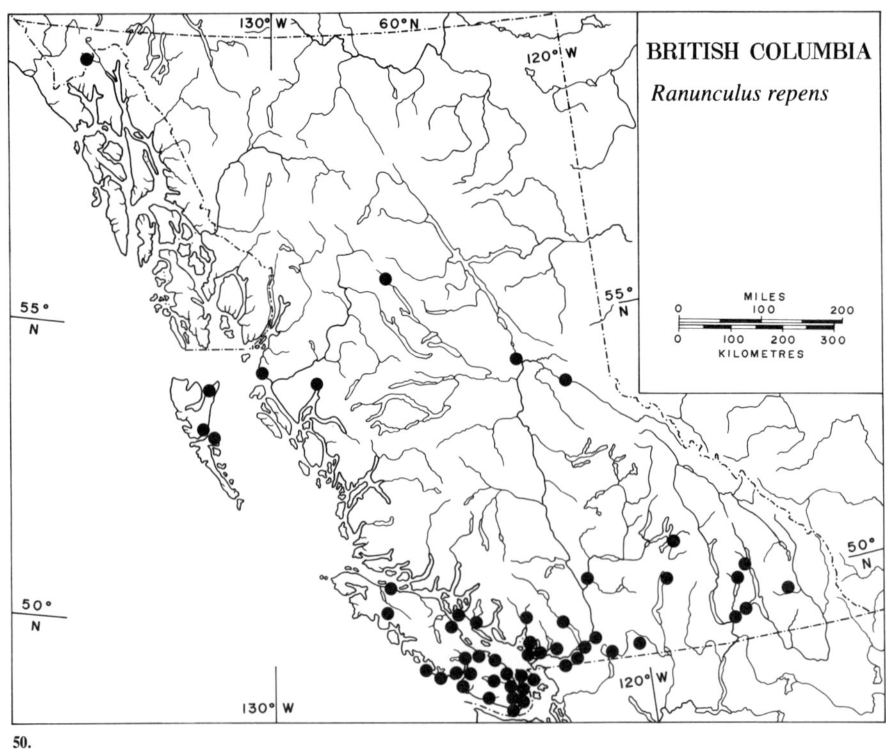

50.

BRITISH COLUMBIA
Ranunculus
orthorhynchus
130° W
60° N
120° W
55°
N
55°
N
50°
N
50°
N
120° W
130° W
MILES
0   100   200
0   100   200   300
KILOMETRES

51.

BRITISH COLUMBIA
Ranunculus bulbosus
130° W
60° N
120° W
55°
N
55°
N
50°
N
50°
N
120° W
130° W
MILES
0   100   200
0   100   200   300
KILOMETRES

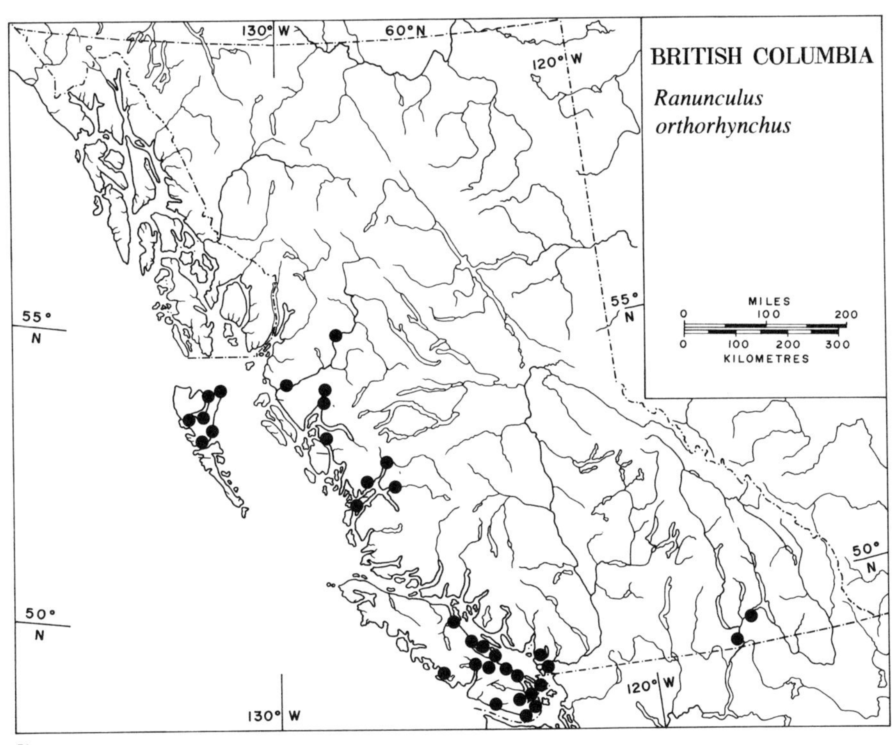

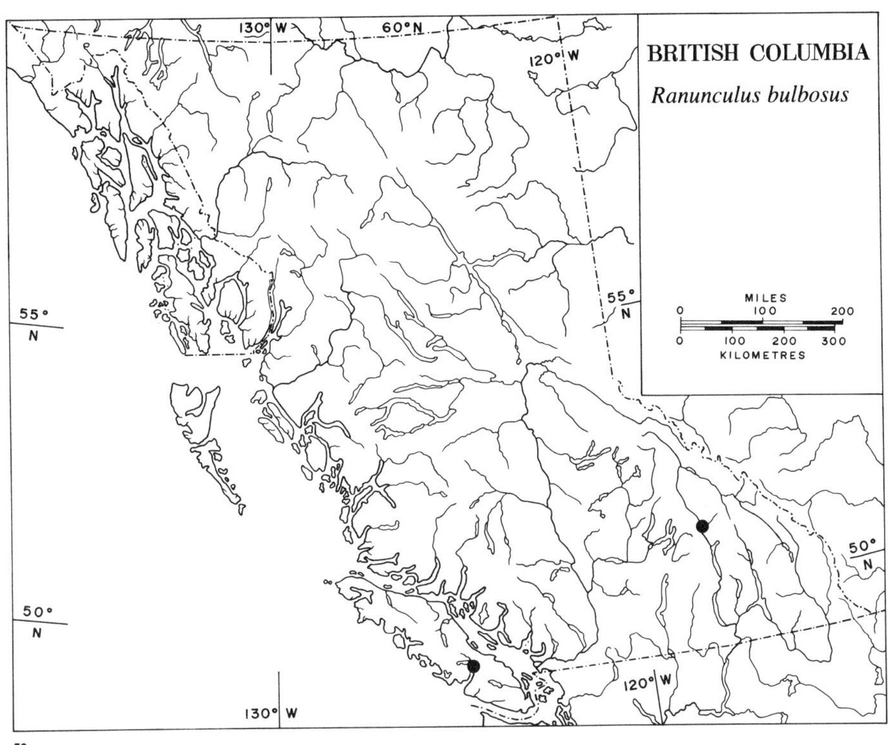

52.

223

BRITISH COLUMBIA
Ranunculus sardous
MILES
0   100   200
0   100   200   300
KILOMETRES

53.

BRITISH COLUMBIA
Ranunculus acris
MILES
0   100   200
0   100   200   300
KILOMETRES

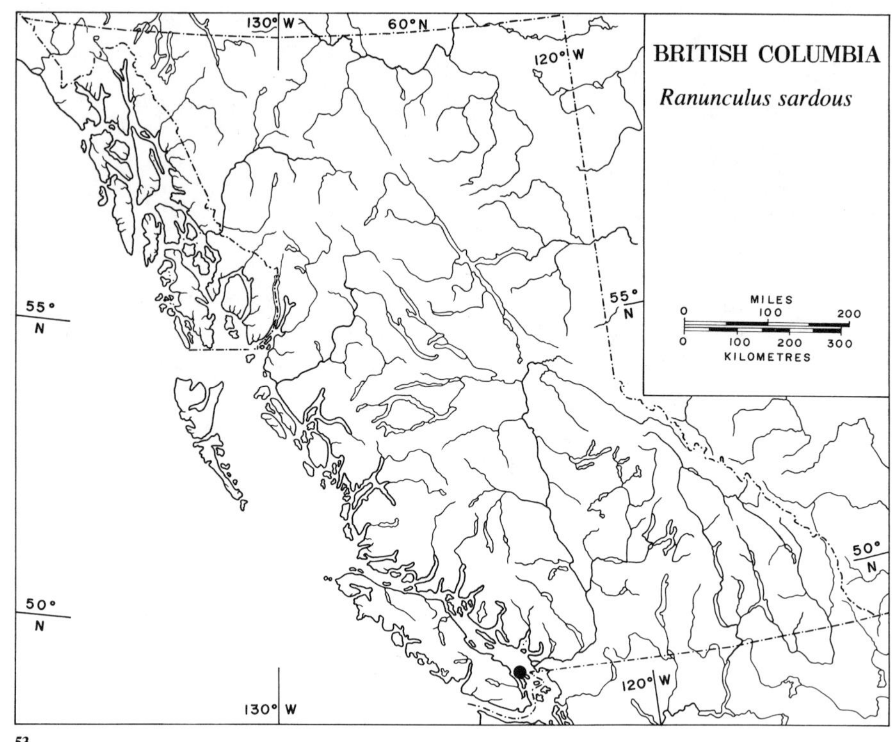

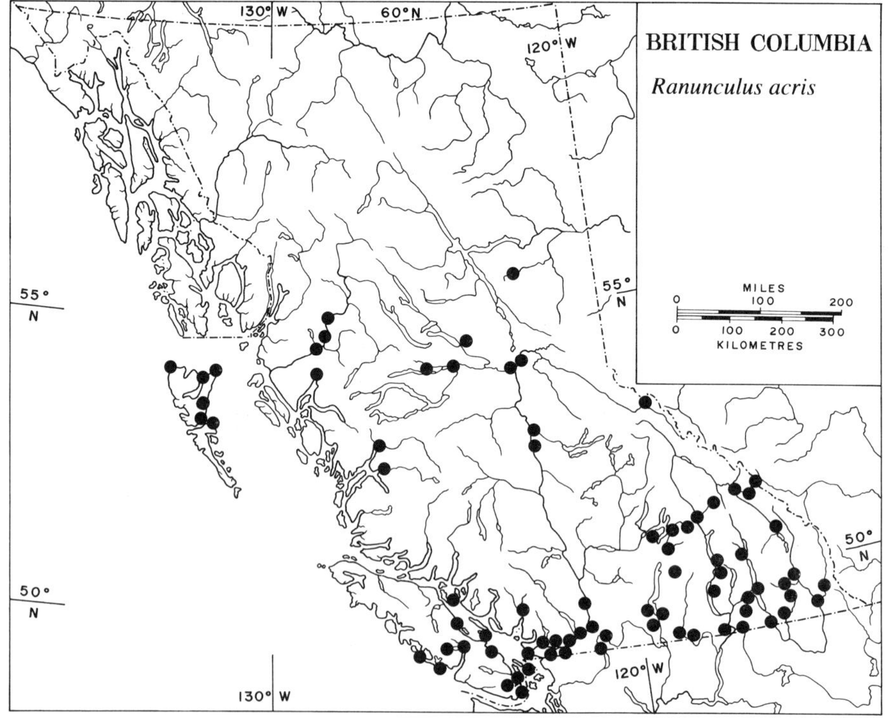

54.

130° W
60° N
120° W
BRITISH COLUMBIA
Ranunculus uncinatus
55°
N
55°
N
MILES
0 100 200
0 100 200 300
KILOMETRES
50°
N
50°
N
120° W
50°
N
130° W

55.

130° W
60° N
120° W
BRITISH COLUMBIA
Ranunculus
occidentalis
55°
N
55°
N
MILES
0 100 200
0 100 200 300
KILOMETRES
50°
N
50°
N
120° W
130° W

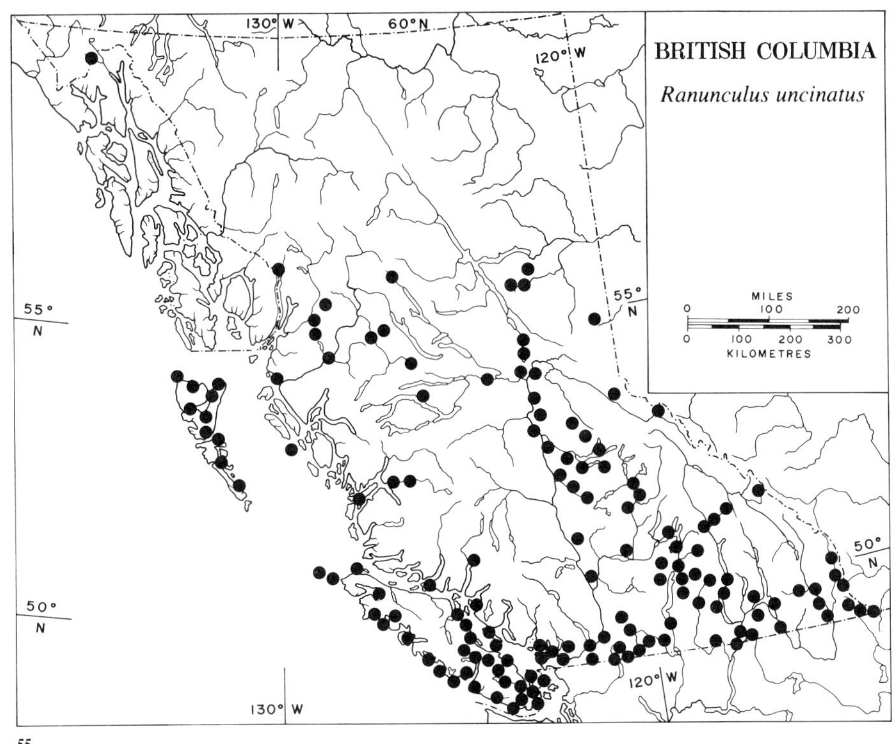

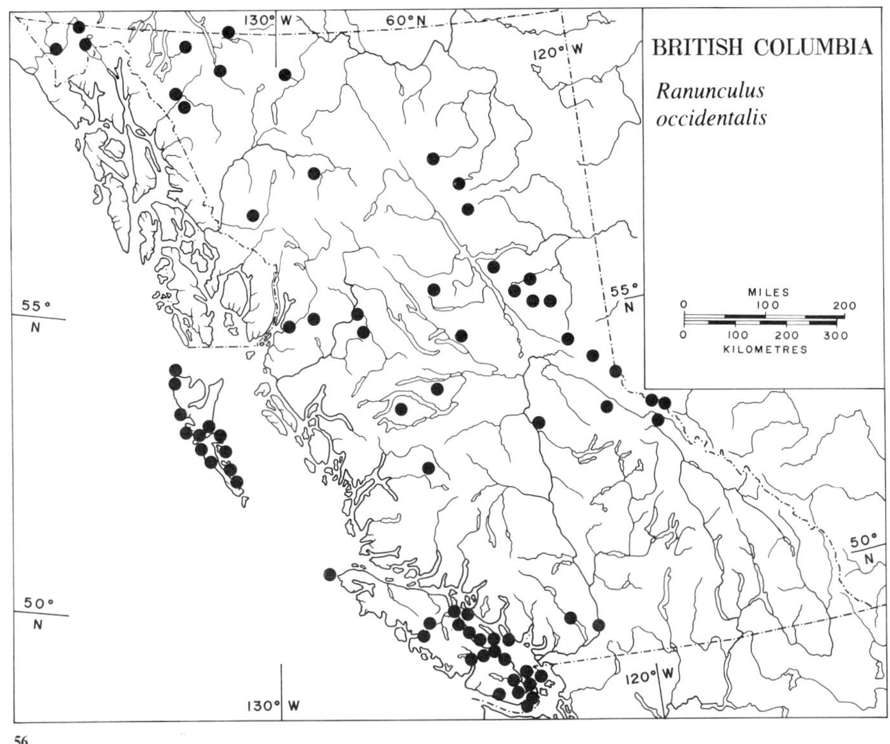

56.

BRITISH COLUMBIA
Ranunculus
californicus
130° W
60° N
120° W
55° N
50° N
130° W
120° W
MILES
100
200
0
100 200 300
KILOMETRES

57.

130° W
60° N
120° W
BRITISH COLUMBIA
Ranunculus abortivus
55° N
50° N
130° W
120° W
MILES
100
200
0
100 200 300
KILOMETRES

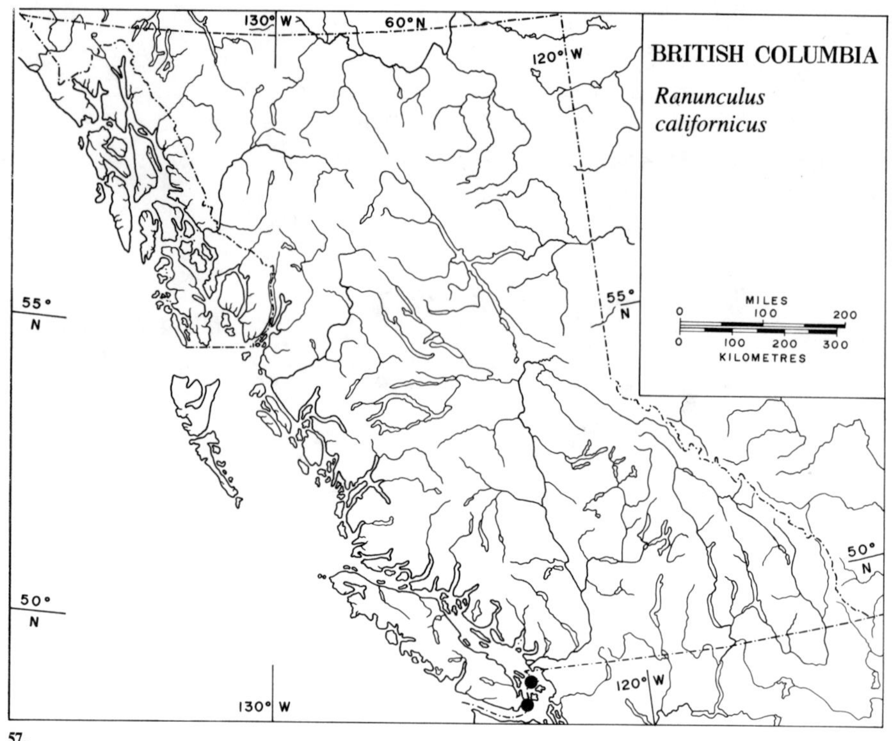

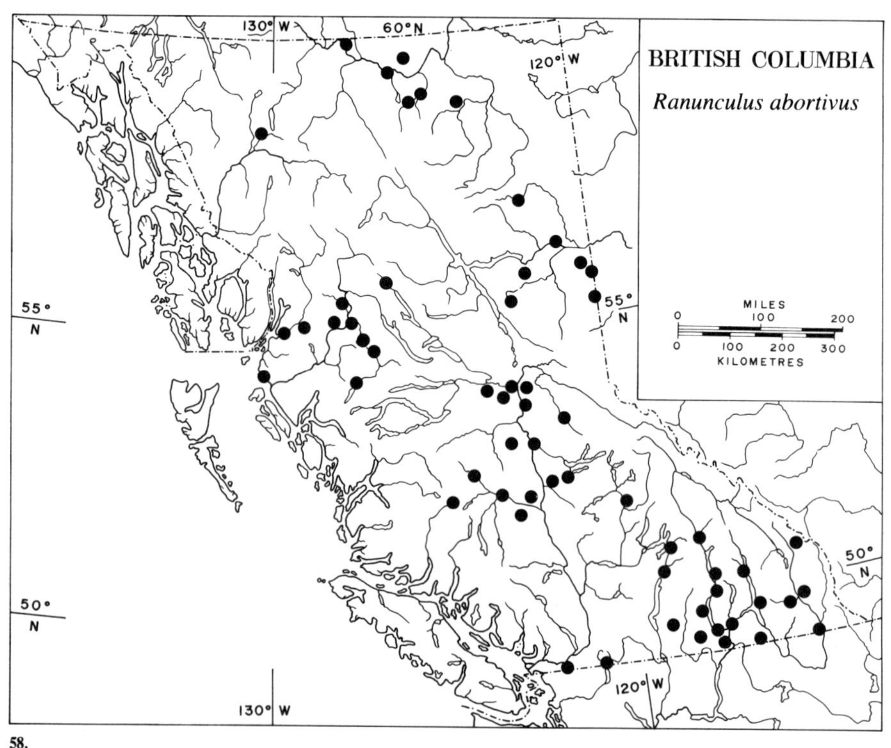

58.

130° W
60° N
120° W
BRITISH COLUMBIA
Ranunculus inamoenus
55°
N
55°
N
MILES
0   100   200
0   100   200   300
KILOMETRES
50°
N
50°
N
130° W
120° W

59.

130° W
60° N
120° W
BRITISH COLUMBIA
Ranunculus pedatifidus
55°
N
55°
N
MILES
0   100   200
0   100   200   300
KILOMETRES
50°
N
50°
N
130° W
120° W

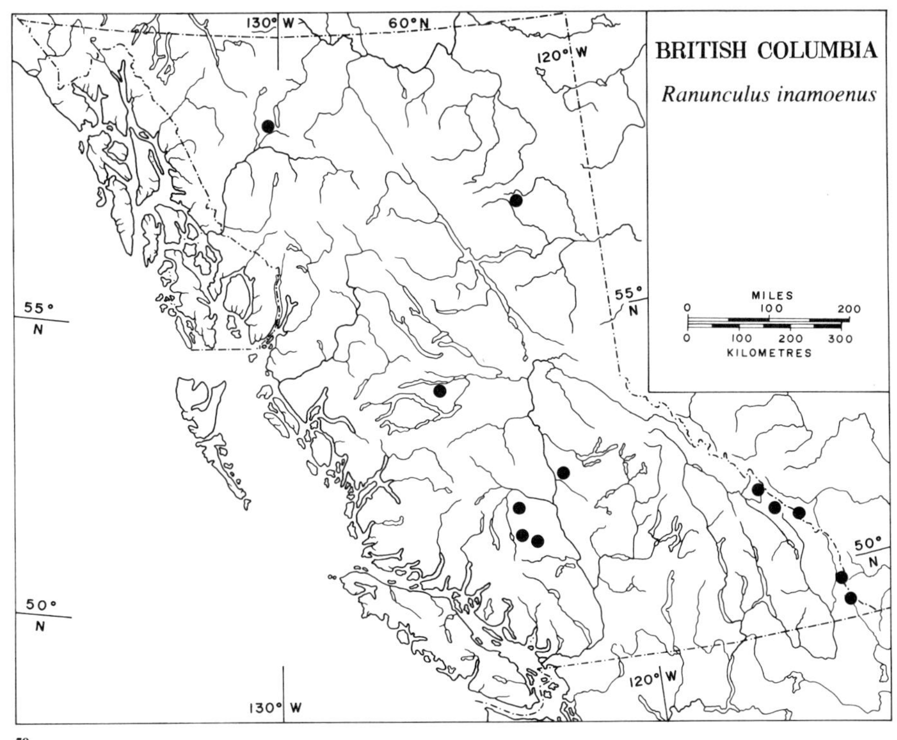

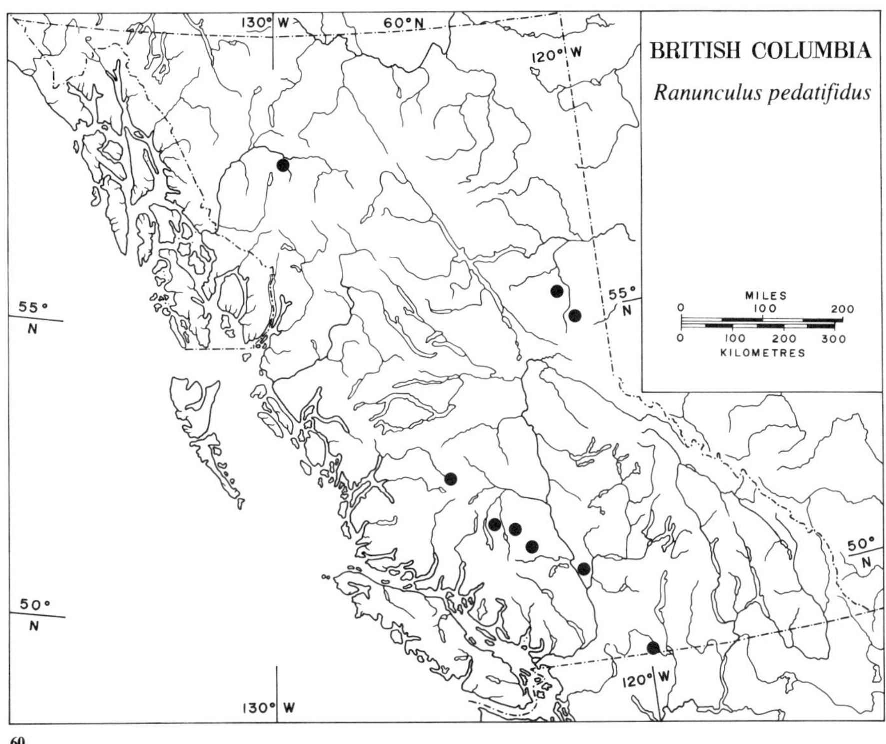

60.

BRITISH COLUMBIA
Ranunculus
cardiophyllus
130° W
60° N
120° W
55° N
55° N
50° N
50° N
130° W
120° W
MILES
0 100 200
0 100 200 300
KILOMETRES

61.

BRITISH COLUMBIA
Ranunculus
glaberrimus
130° W
60° N
120° W
55° N
55° N
50° N
50° N
130° W
120° W
MILES
0 100 200
0 100 200 300
KILOMETRES

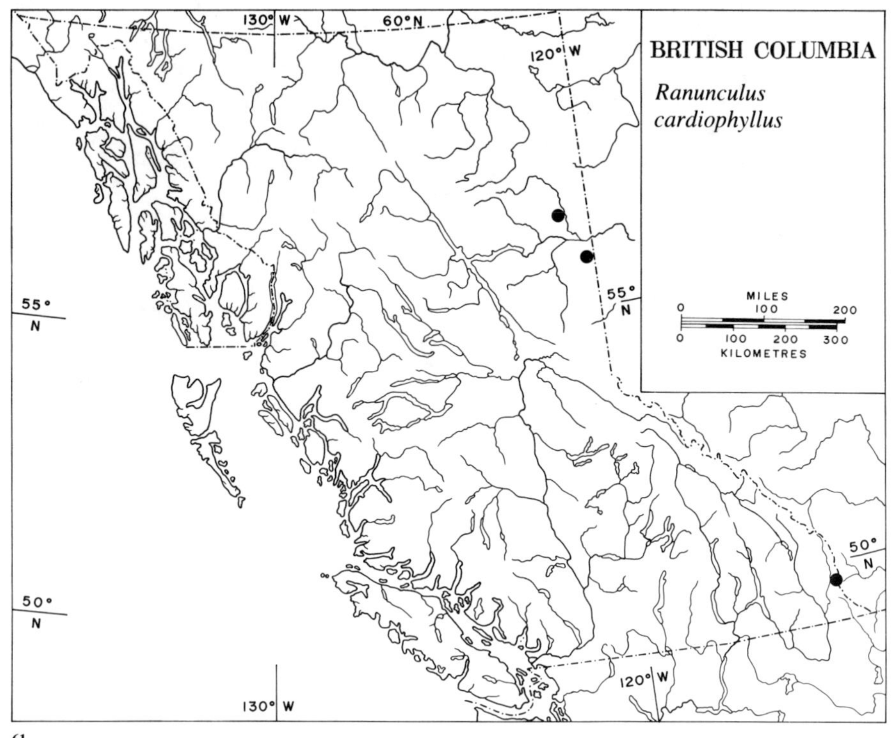

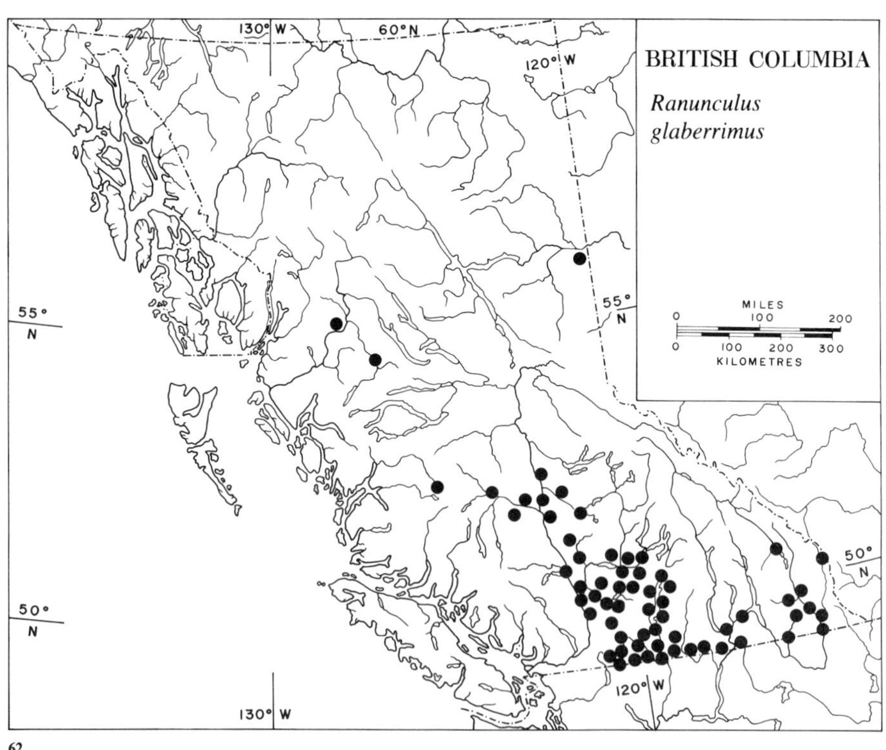

62.

228

BRITISH COLUMBIA
Ranunculus
rhomboideus
MILES
0 100 200
0 100 200 300
KILOMETRES
130° W
60° N
120° W
55° N
55° N
50° N
50° N
50° N
130° W
120° W

63.

BRITISH COLUMBIA
Ranunculus
eschscholtzii
ssp. eschscholtzii
ssp. suksdorfii
MILES
0 100 200
0 100 200 300
KILOMETRES
130° W
60° N
120° W
55° N
55° N
50° N
50° N
50° N
130° W
120° W

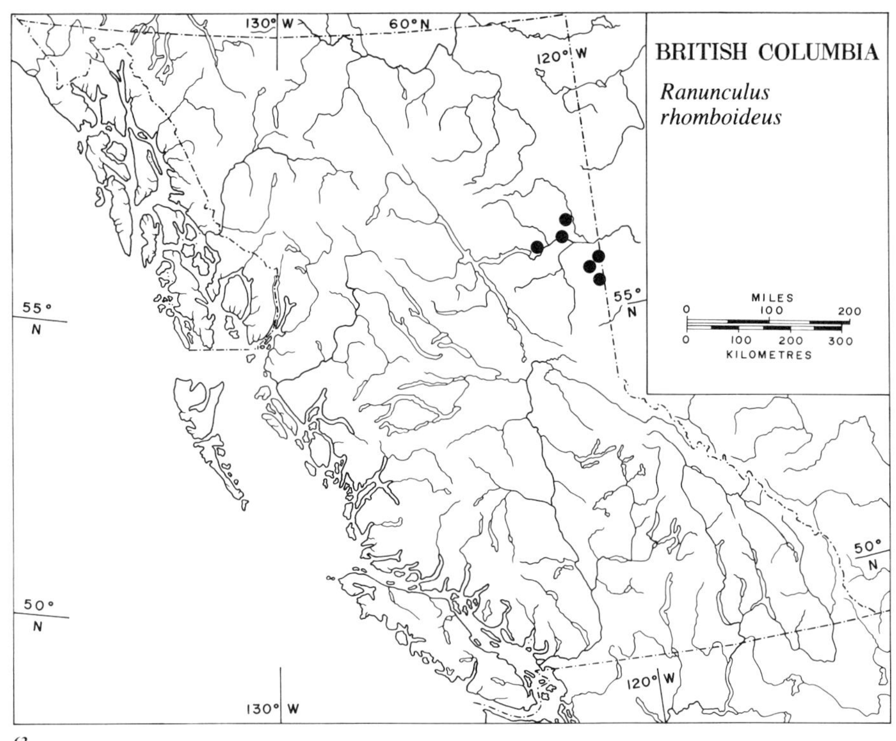

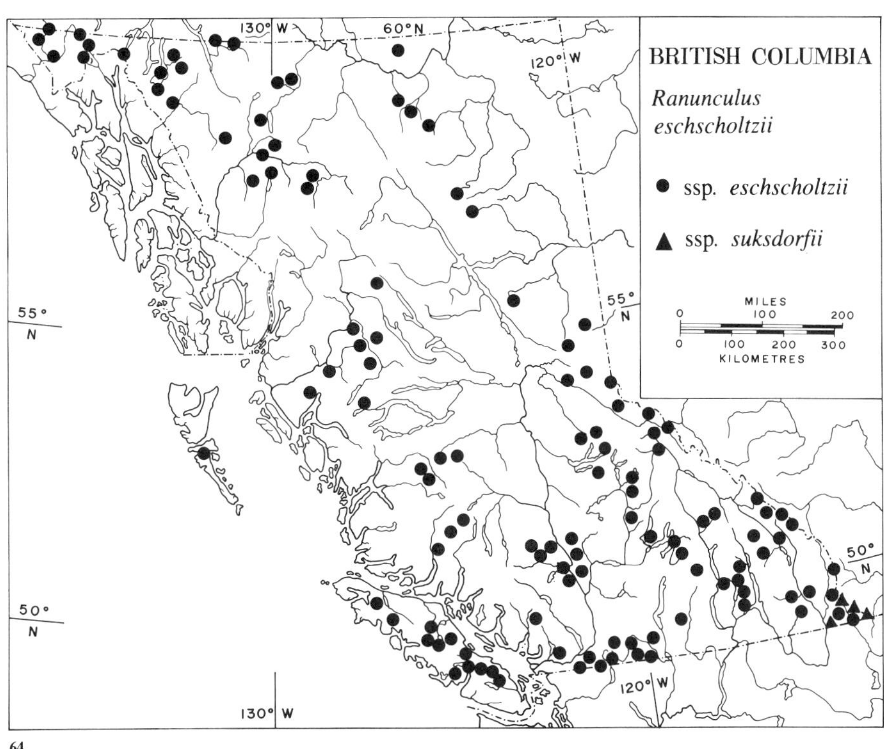

64.

BRITISH COLUMBIA
Ranunculus eximius
130° W
60° N
120° W
55° N
50° N
50° N
55° N
130° W
120° W
MILES
0   100   200
0   100   200   300
KILOMETRES

65.

BRITISH COLUMBIA
Ranunculus gelidus
130° W
60° N
120° W
55° N
50° N
50° N
55° N
130° W
120° W
MILES
0   100   200
0   100   200   300
KILOMETRES

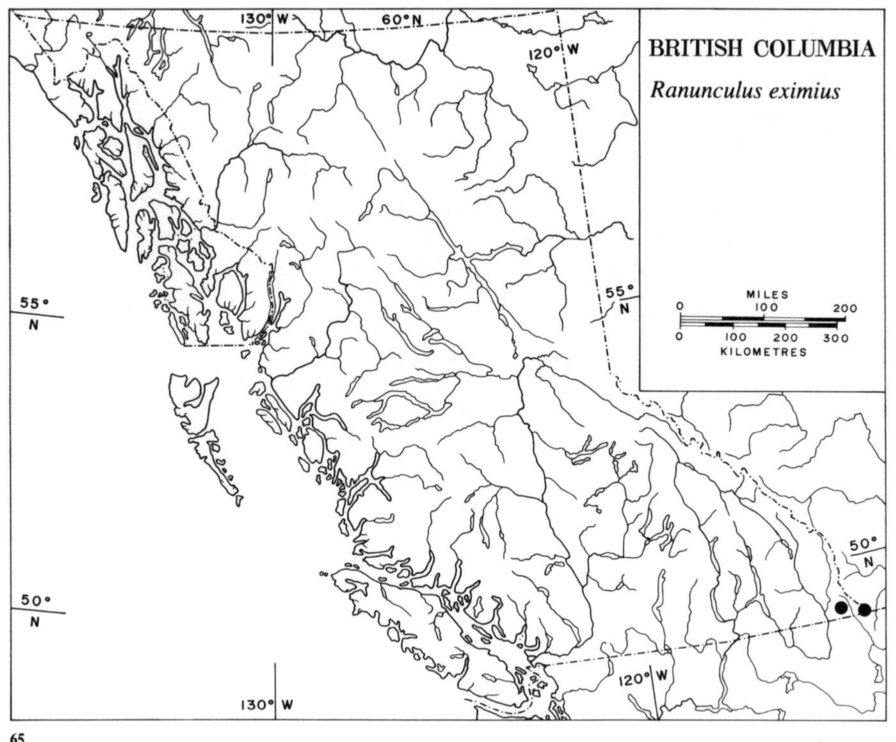

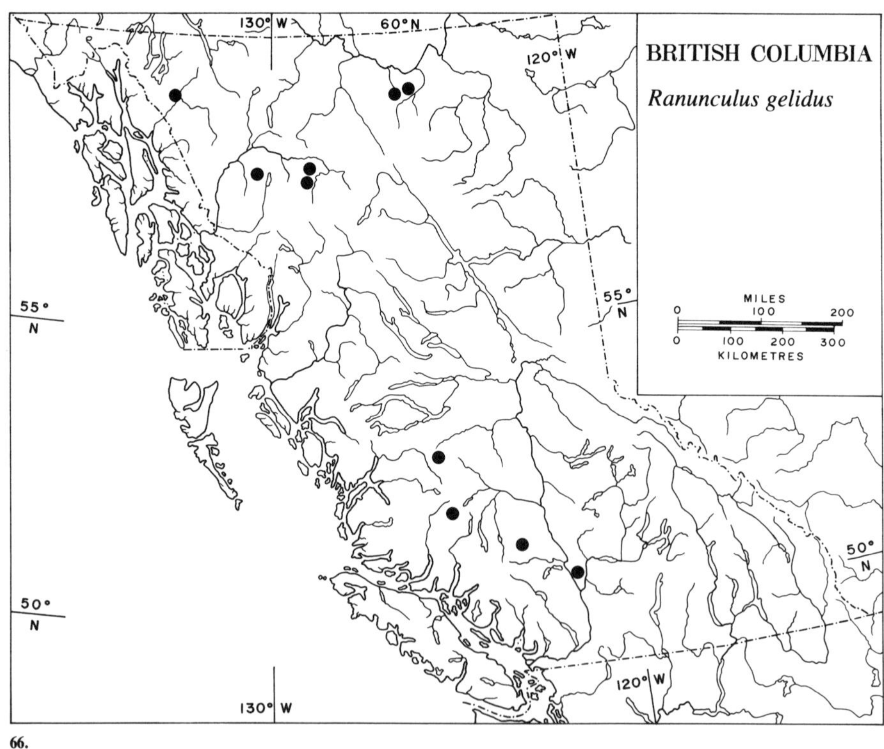

66.

BRITISH COLUMBIA
Ranunculus verecundus
MILES
100
200
KILOMETRES
100
200
300

67.

BRITISH COLUMBIA
Ranunculus pygmaeus
MILES
100
200
KILOMETRES
100
200
300

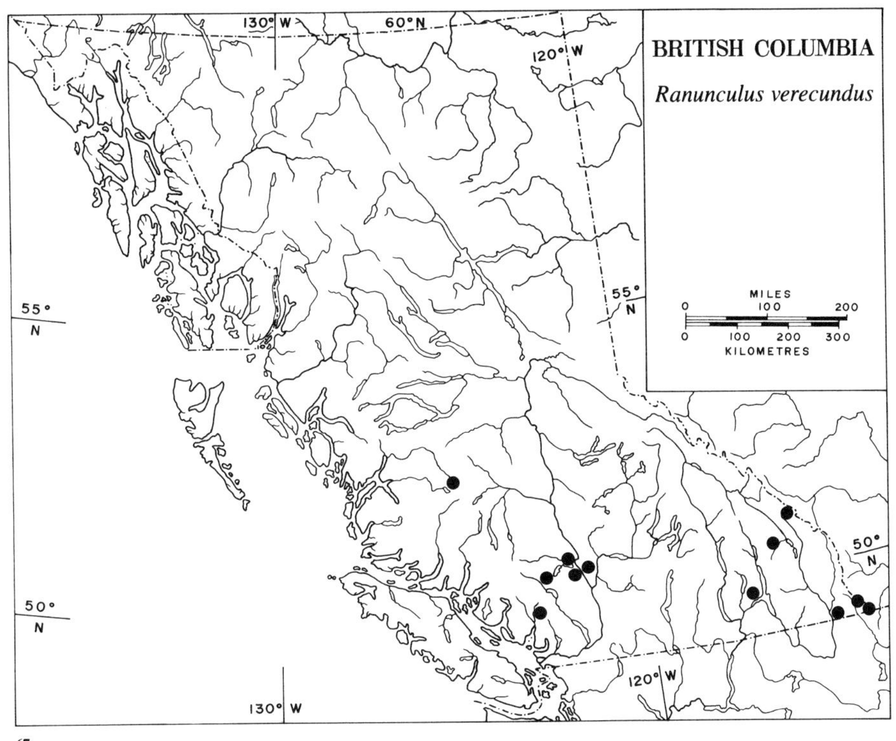

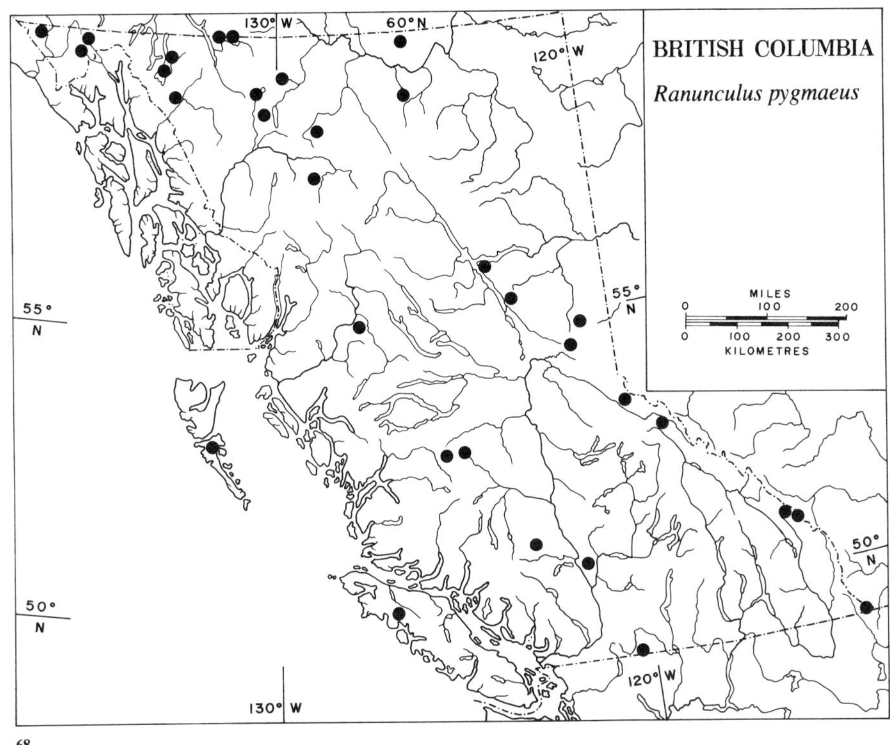

68.

130° W
60° N
120° W
55°
N
50°
N
130° W
120° W
BRITISH COLUMBIA
Ranunculus nivalis
MILES
0   100   200
0   100   200   300
KILOMETRES

69.

130° W
60° N
120° W
55°
N
50°
N
130° W
120° W
BRITISH COLUMBIA
Ranunculus sulphureus
MILES
0   100   200
0   100   200   300
KILOMETRES

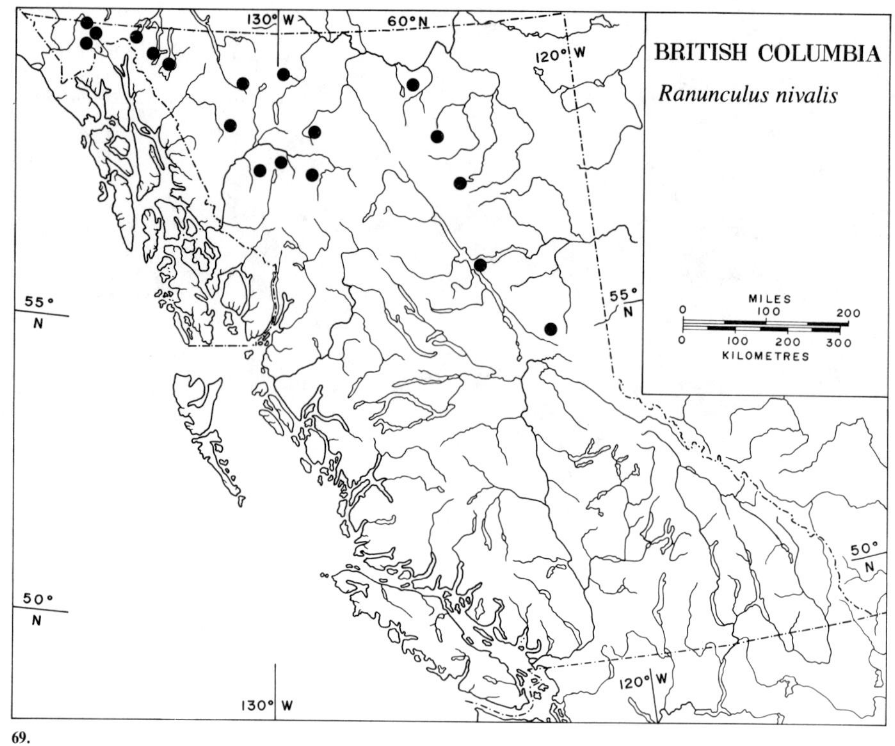

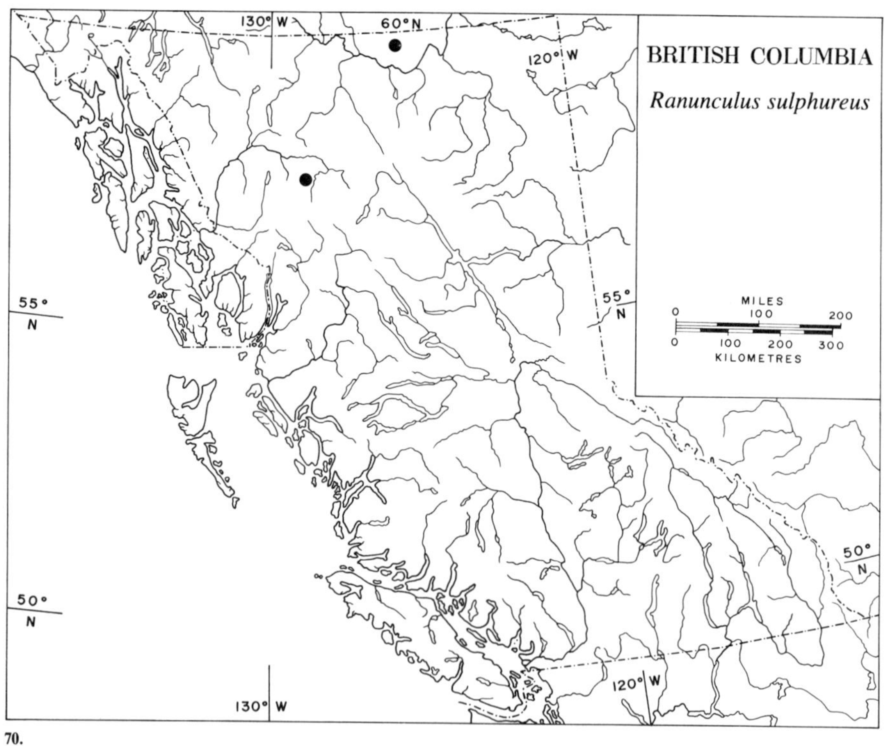

70.

BRITISH COLUMBIA
Ranunculus flammula
MILES
100
200
KILOMETRES
100 200 300
130° W
60° N
120° W
55° N
50° N

71.

BRITISH COLUMBIA
Ranunculus
alismifolius
MILES
100
200
KILOMETRES
100 200 300
130° W
60° N
120° W
55° N
50° N

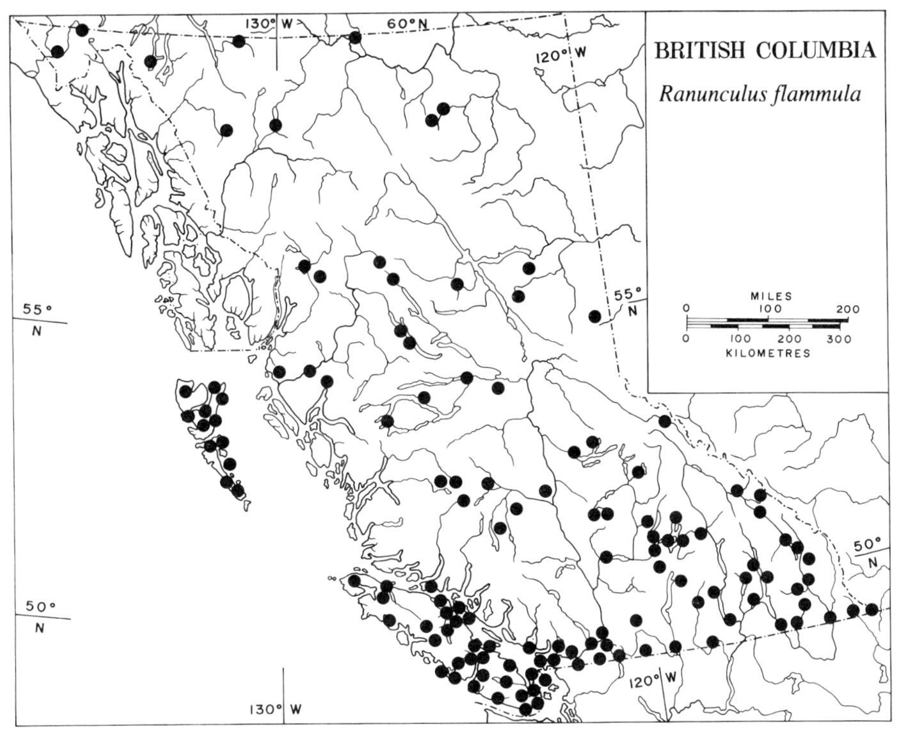

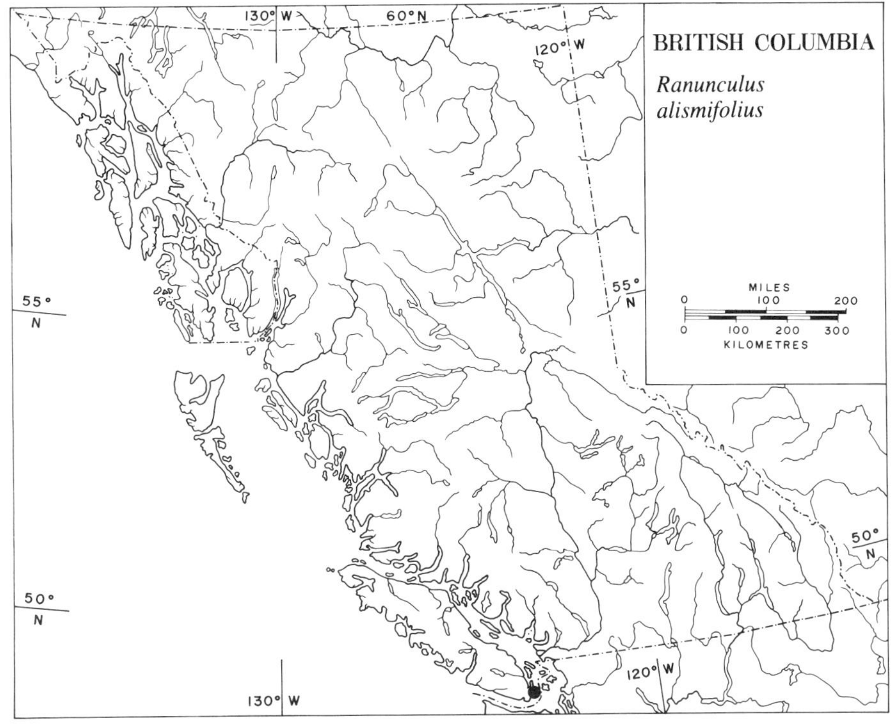

72.

BRITISH COLUMBIA
Ranunculus ficaria
MILES
KILOMETRES

73.

BRITISH COLUMBIA
Ranunculus lapponicus
MILES
KILOMETRES

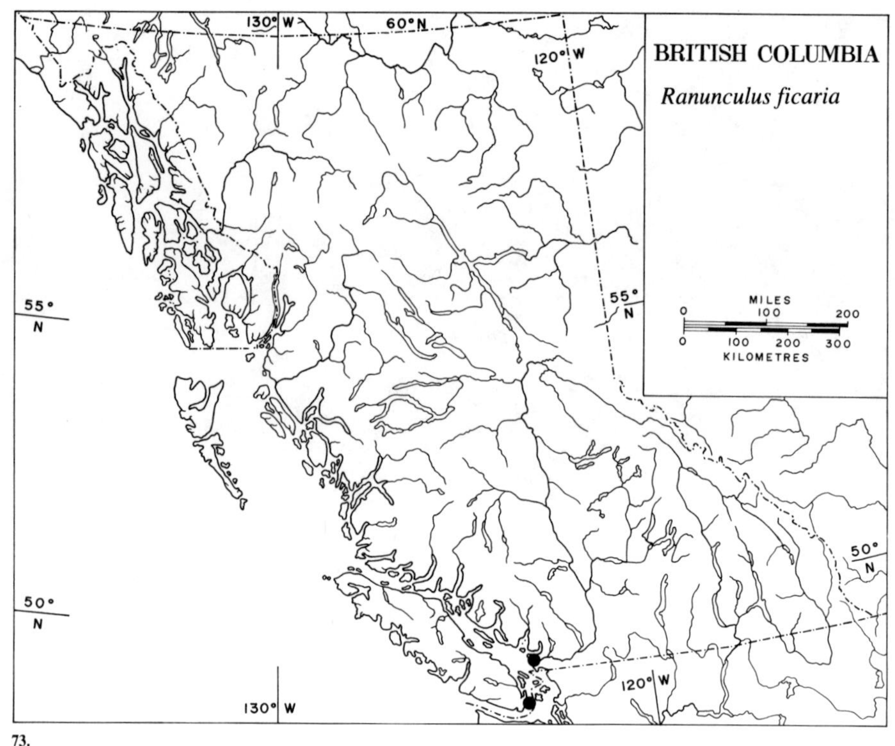

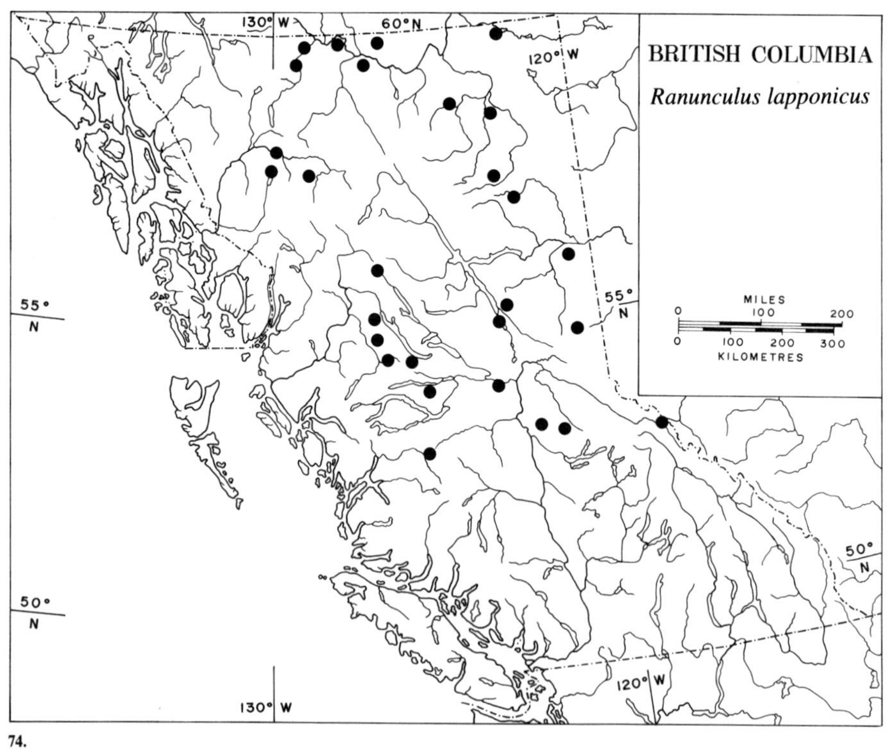

74.

234

BRITISH COLUMBIA
Ranunculus gmelinii
130° W
60° N
120° W
55° N
50° N
130° W
120° W
MILES
0 100 200
0 100 200 300
KILOMETRES

75.

BRITISH COLUMBIA
Ranunculus flabellaris
130° W
60° N
120° W
55° N
50° N
130° W
120° W
MILES
0 100 200
0 100 200 300
KILOMETRES

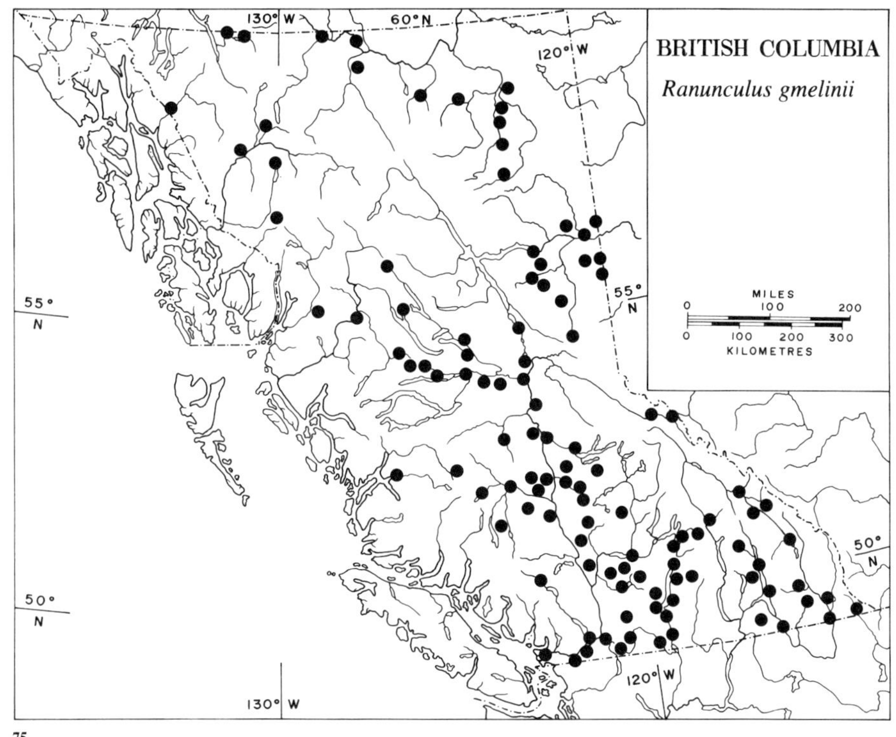

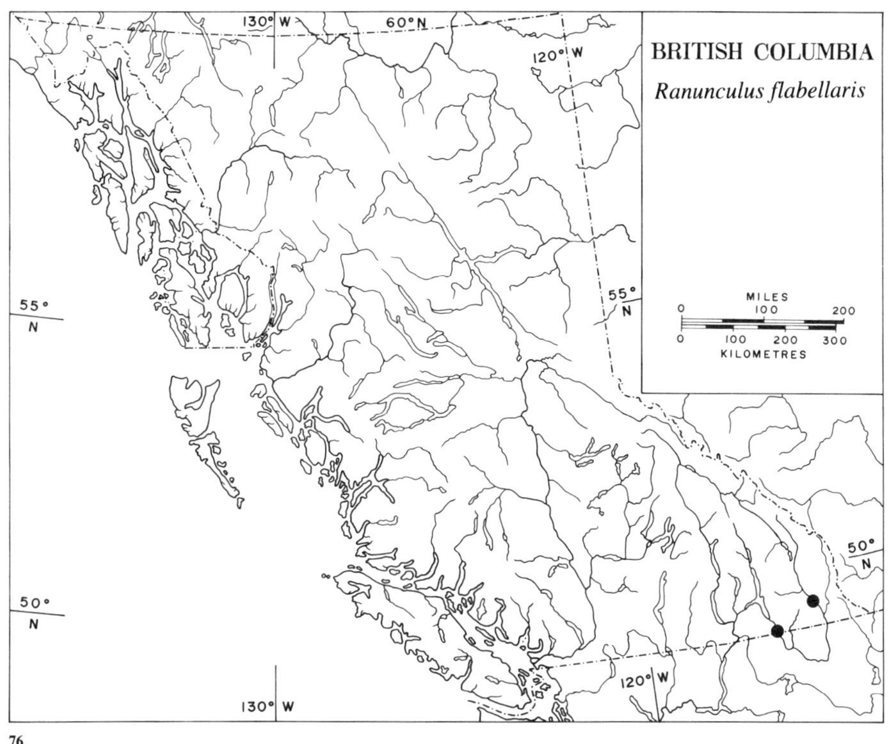

76.

BRITISH COLUMBIA
Ranunculus sceleratus
130° W
60° N
120° W
55°
N
50°
N
120° W
130° W
MILES
0   100   200
0   100   200   300
KILOMETRES

77.

BRITISH COLUMBIA
Ranunculus
hyperboreus
130° W
60° N
120° W
55°
N
50°
N
120° W
130° W
MILES
0   100   200
0   100   200   300
KILOMETRES

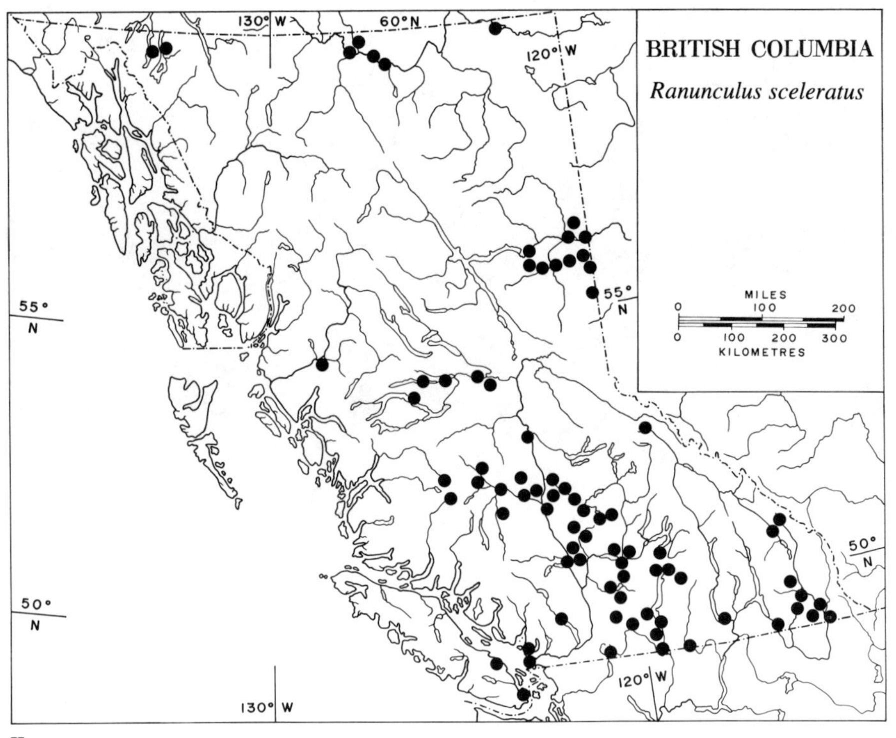

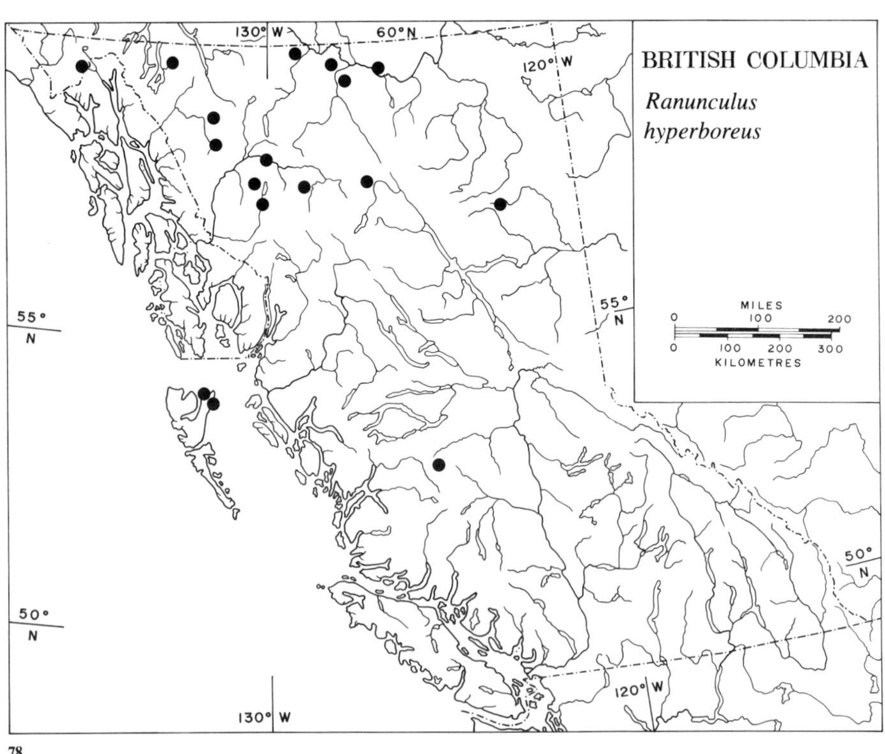

78.

BRITISH COLUMBIA
Ranunculus aquatilis
▲ var. aquatilis
● var. hispidulus
MILES
0   100   200
0 100 200 300
KILOMETRES
130° W
60° N
120° W
55° N
55° N
50° N
50° N
50° N
120° W
130° W

79.

BRITISH COLUMBIA
Ranunculus lobbii
MILES
0   100   200
0 100 200 300
KILOMETRES
130° W
60° N
120° W
55° N
55° N
50° N
50° N
50° N
120° W
130° W

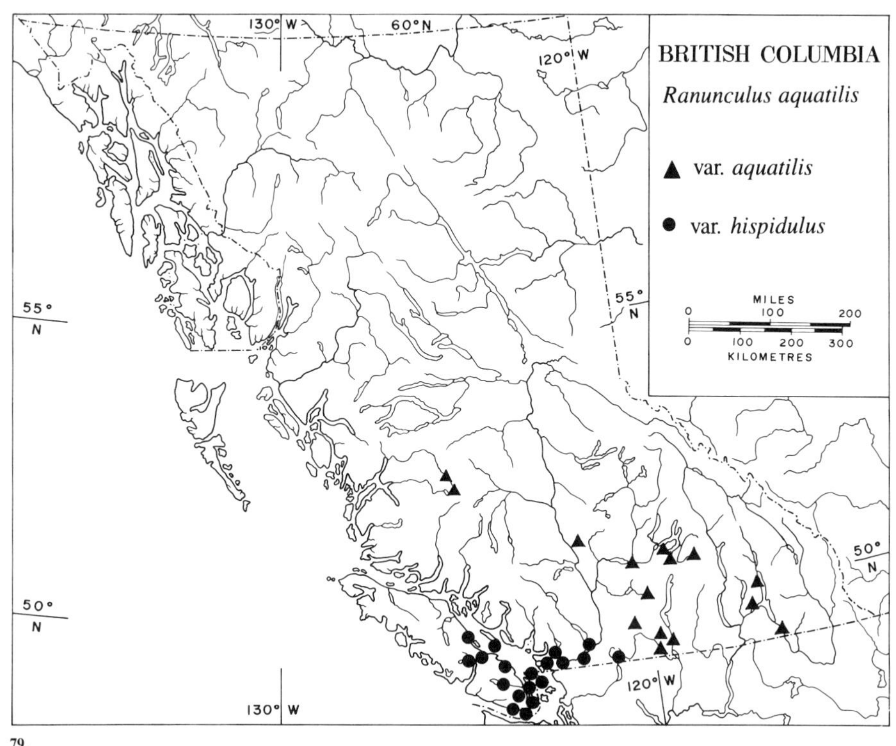

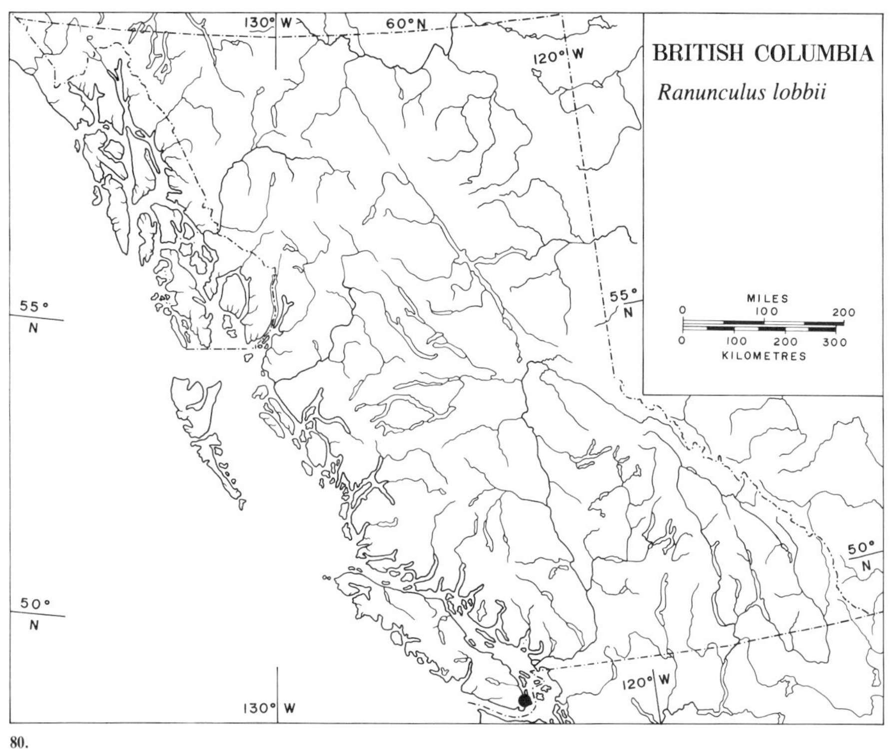

80.

BRITISH COLUMBIA
Ranunculus
trichophyllus
130° W
60° N
120° W
55°
N
55°
N
50°
N
50°
N
130° W
120° W
MILES
0   100   200
0   100   200   300
KILOMETRES

81.

BRITISH COLUMBIA
Ranunculus subrigidus
130° W
60° N
120° W
55°
N
55°
N
50°
N
50°
N
130° W
120° W
MILES
0   100   200
0   100   200   300
KILOMETRES

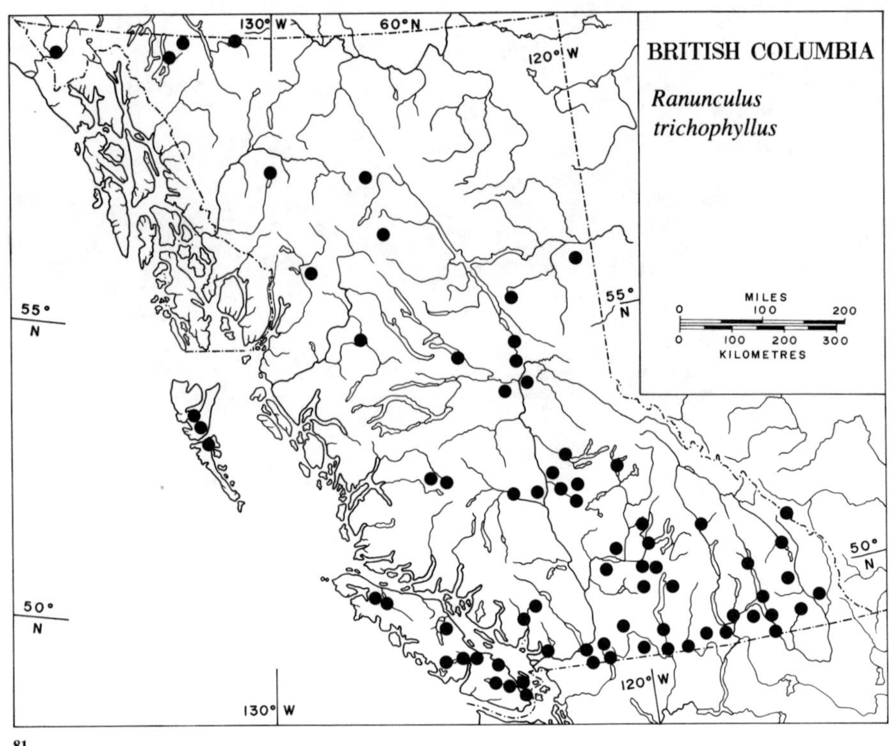

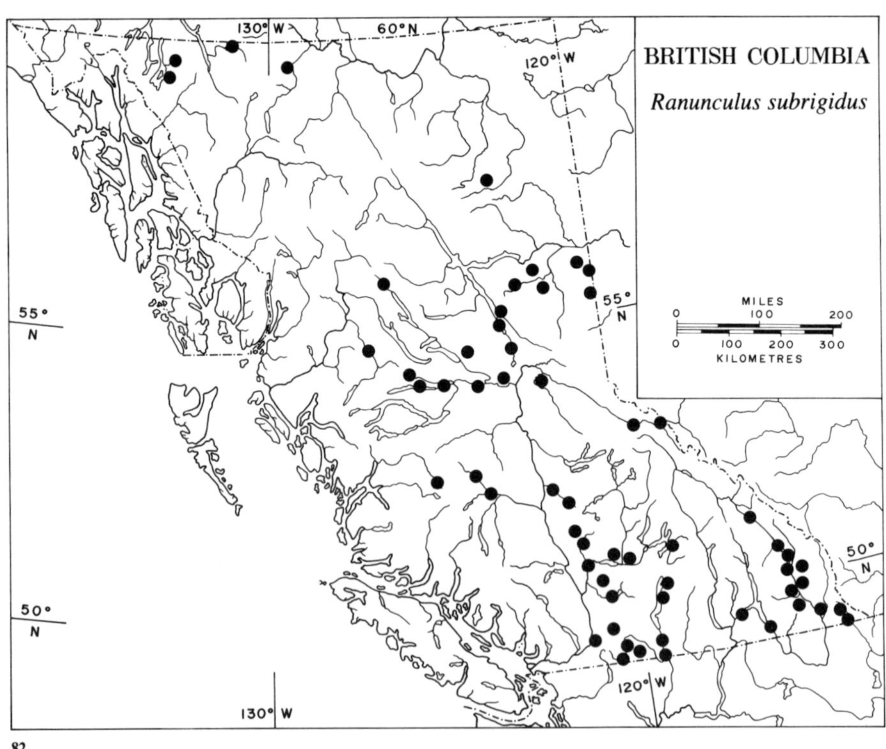

82.

130° W
60° N
120° W
BRITISH COLUMBIA
Ranunculus
cymbalaria
55° N
MILES
0   100   200
0   100   200   300
KILOMETRES
55° N
50° N
50° N
120° W
130° W

83.

130° W
60° N
120° W
BRITISH COLUMBIA
Ranunculus cooleyae
55° N
MILES
0   100   200
0   100   200   300
KILOMETRES
55° N
50° N
50° N
120° W
130° W

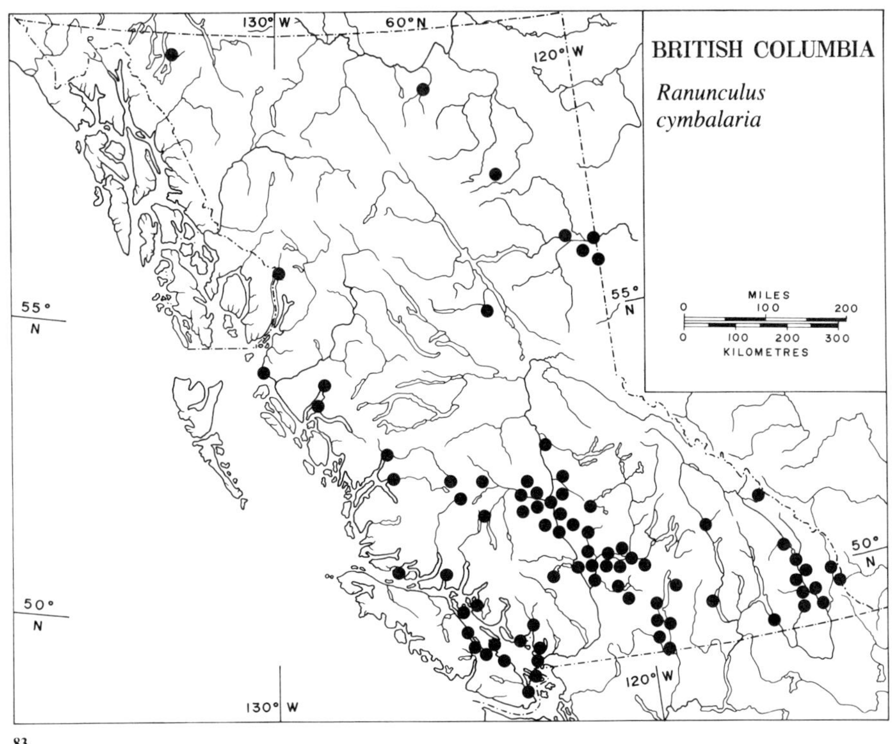

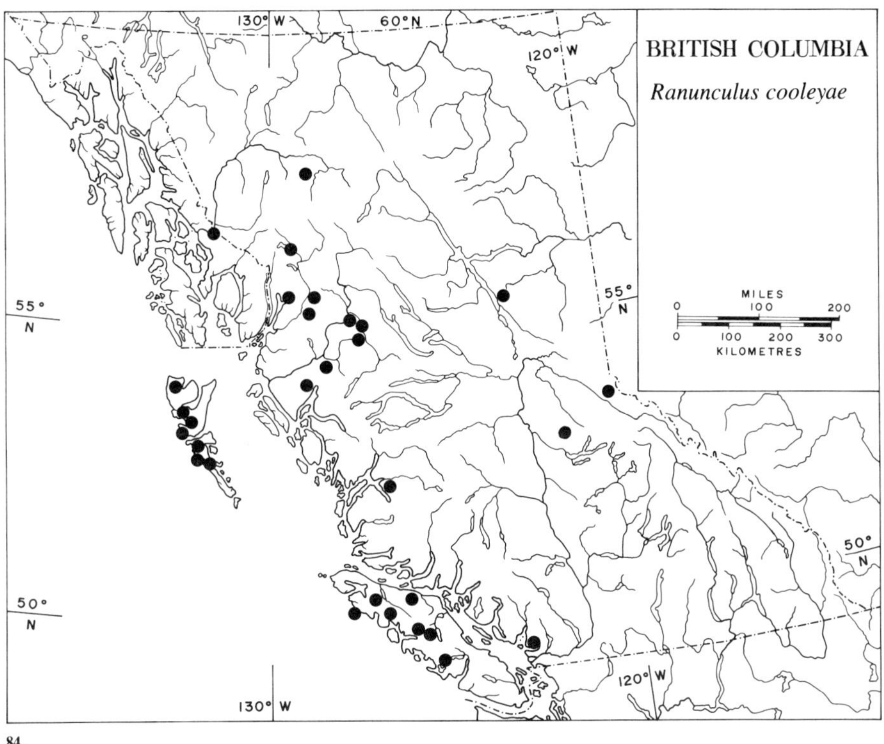

84.

130° W
60° N
120° W
BRITISH COLUMBIA
Ranunculus
testiculatus
55°
N
55°
N
MILES
100
200
0
0
100
200
300
KILOMETRES
50°
N
50°
N
50°
N
120° W
130° W

85.

130° W
60° N
120° W
BRITISH COLUMBIA
Thalictrum alpinum
55°
N
55°
N
MILES
100
200
0
0
100
200
300
KILOMETRES
50°
N
50°
N
50°
N
120° W
130° W

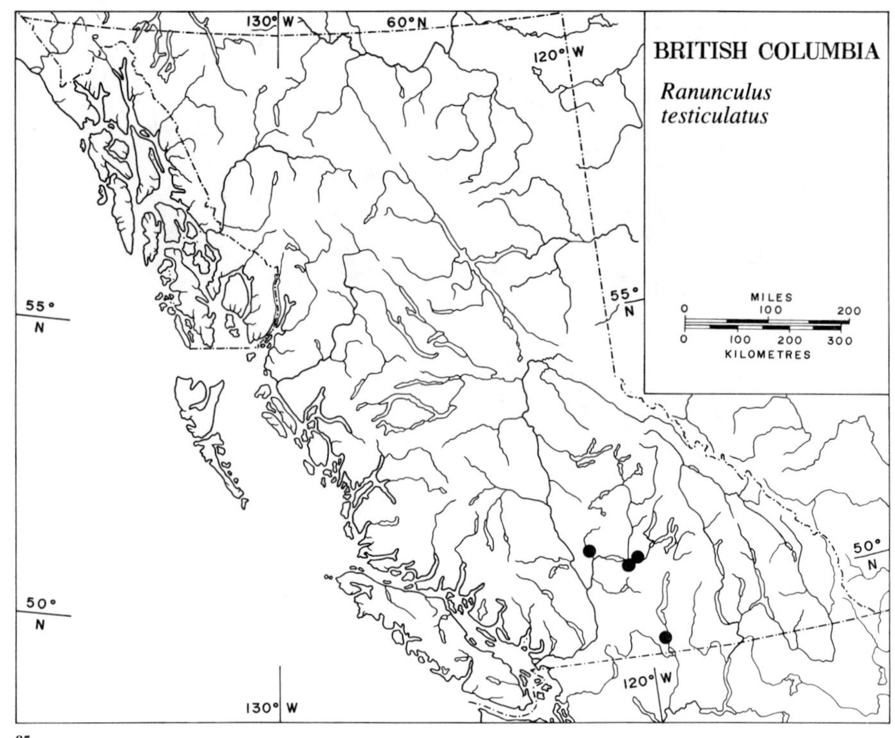

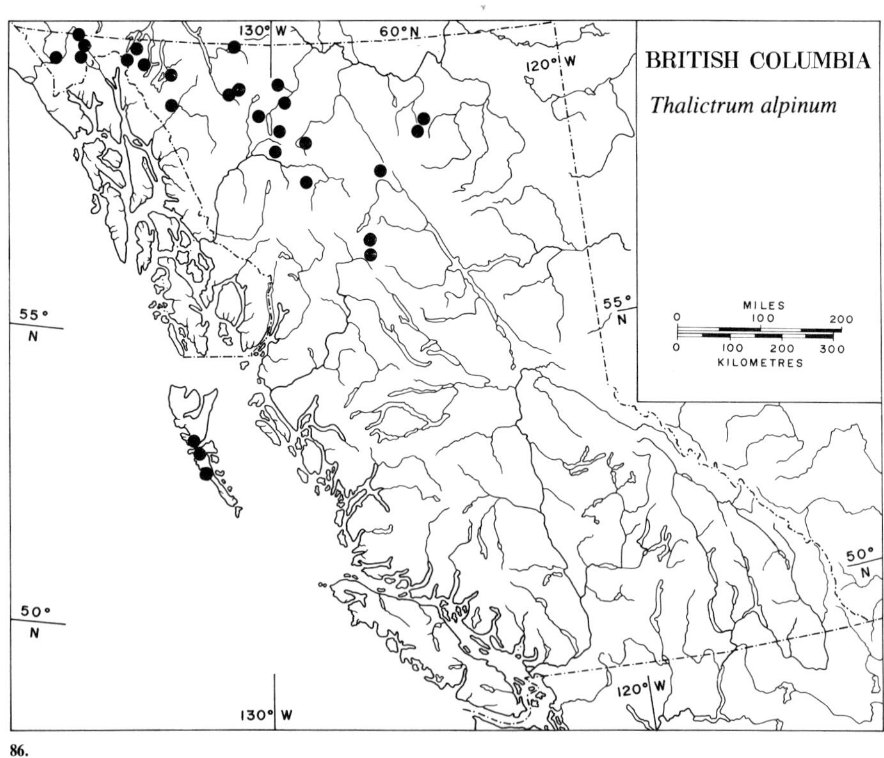

86.

240

BRITISH COLUMBIA
Thalictrum
dasycarpum
130° W
60° N
120° W
55° N
55° N
50° N
50°
N
50°
N
120° W
130° W
MILES
100
0 200
0 100 200 300
KILOMETRES

87.

BRITISH COLUMBIA
Thalictrum occidentale
130° W
60° N
120° W
55° N
55° N
50° N
50°
N
50°
N
120° W
130° W
MILES
100
0 200
0 100 200 300
KILOMETRES

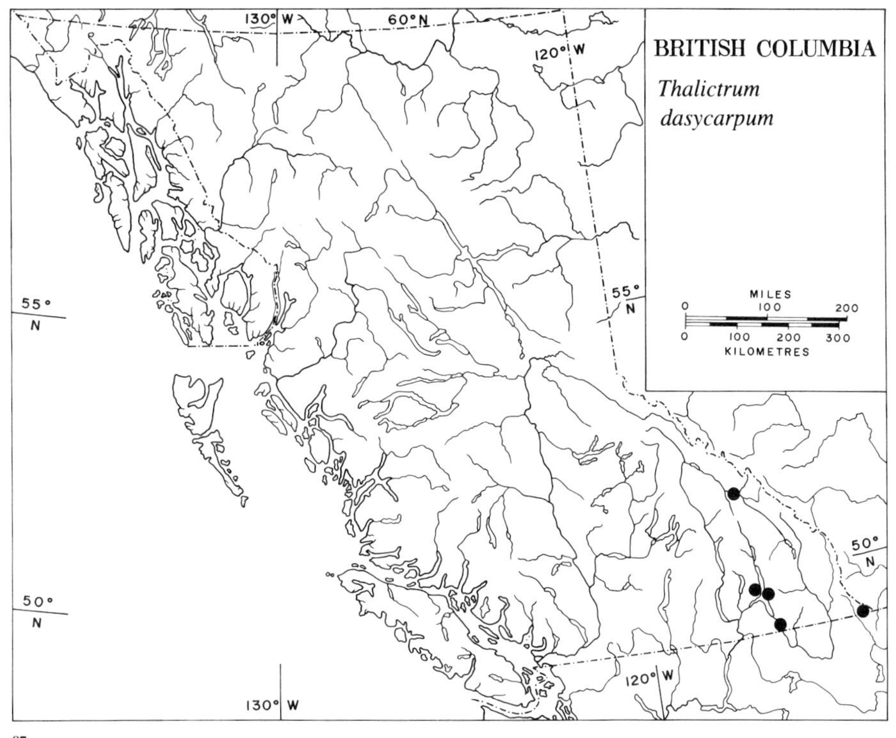

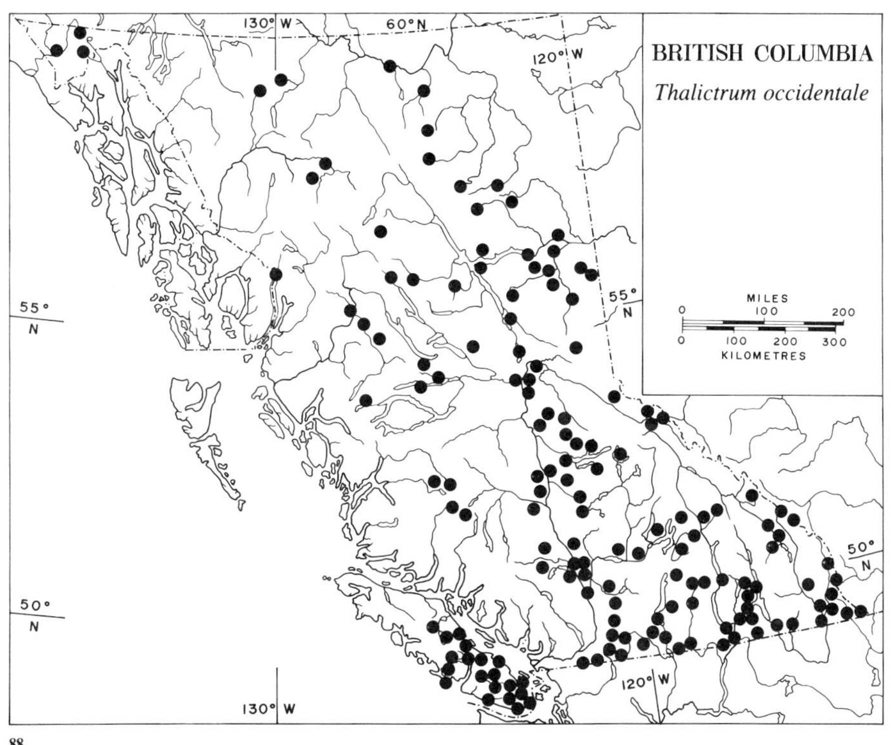

88.

241

BRITISH COLUMBIA
Thalictrum
sparsiflorum
130° W
60 N
120° W
55° N
50° N
130° W
120° W
MILES
100
200
0
100 200 300
KILOMETRES
55°
N
50°
N

89.

BRITISH COLUMBIA
Thalictrum venulosum
130° W
60°N
120° W
55°
N
50°
N
130° W
120° W
MILES
100
200
0
100 200 300
KILOMETRES
55°
N
50°
N

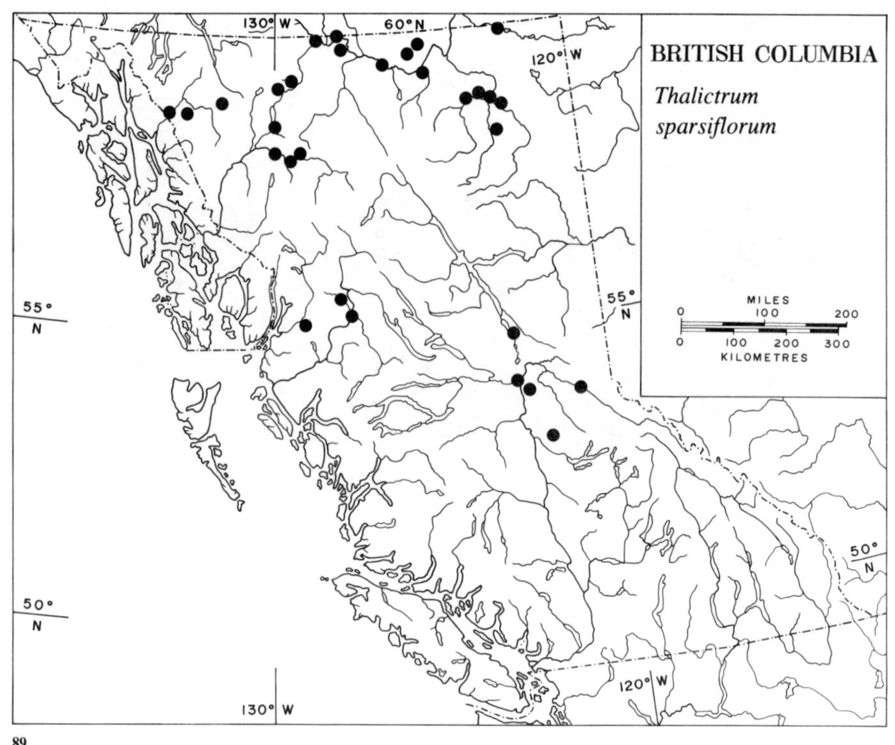

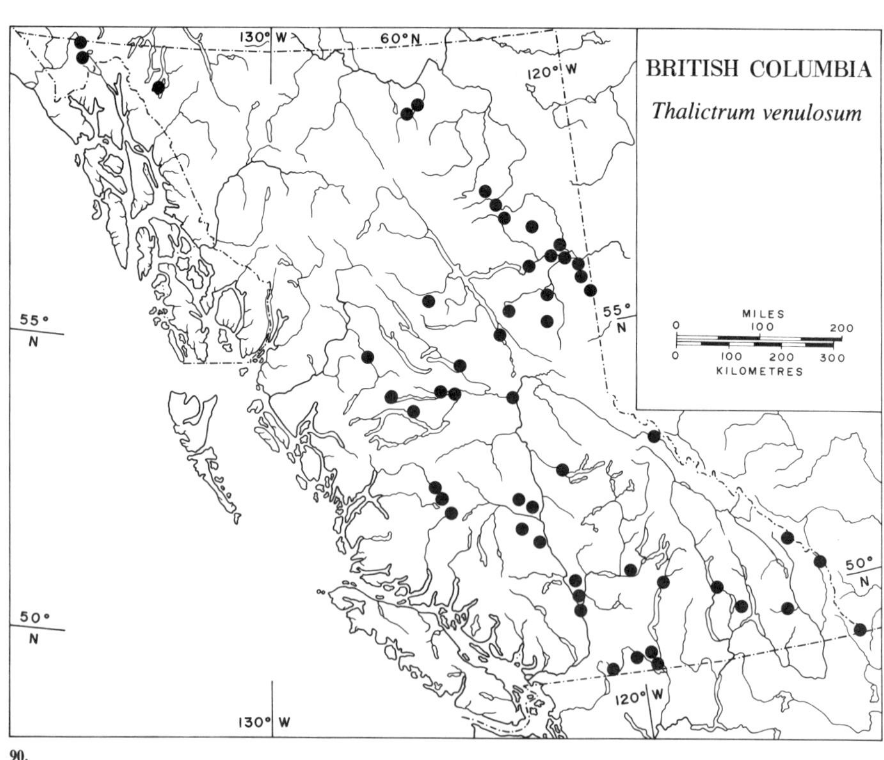

90.

242

BRITISH COLUMBIA
Trautvetteria
caroliniensis
130° W
60° N
120° W
55° N
50° N
MILES
100
200
0
100
200
300
KILOMETRES
120° W
130° W

91.

BRITISH COLUMBIA
Trollius laxus var.
albiflorus
130° W
60° N
120° W
55° N
50° N
MILES
100
200
0
100
200
300
KILOMETRES
120° W
130° W

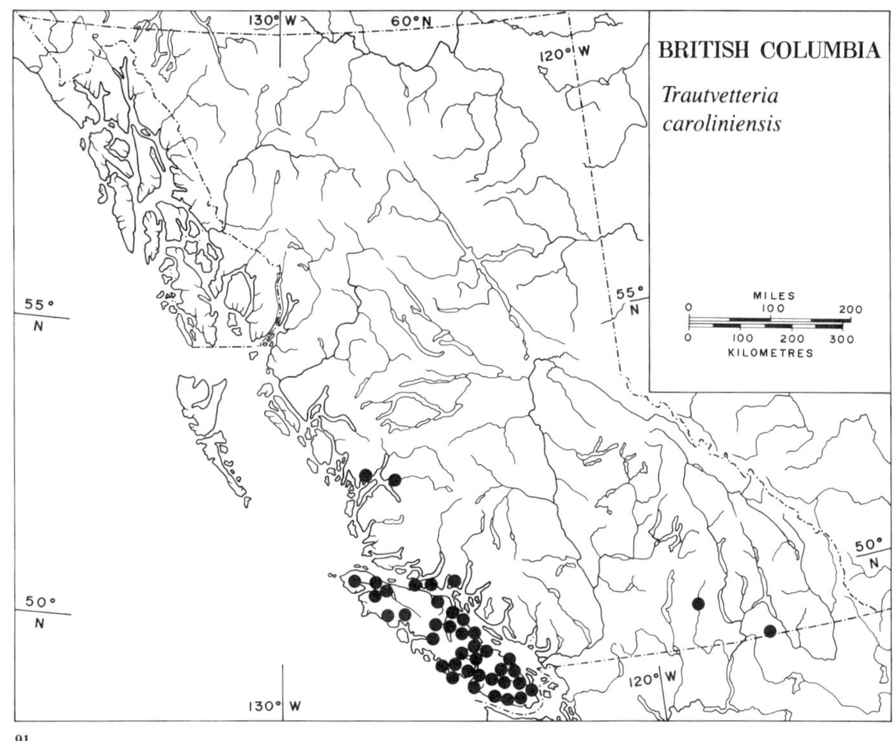

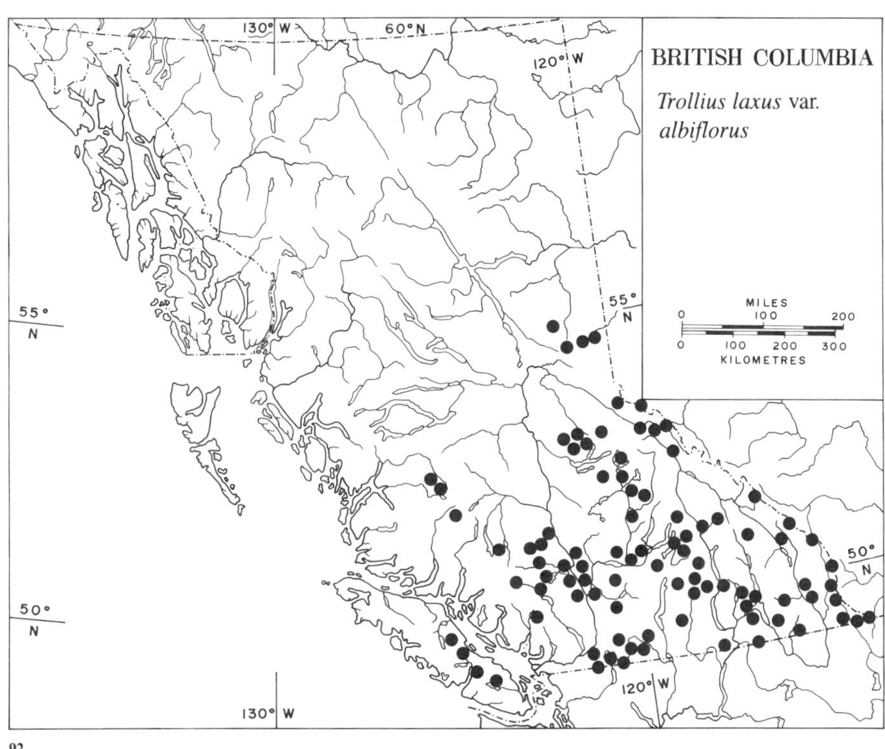

92.

243

BRITISH COLUMBIA
Achlys triphylla
130° W
60° N
120° W
55° N
50° N
50° N
55° N
50° N
130° W
120° W
MILES
0   100   200
0   100 200 300
KILOMETRES

93.

BRITISH COLUMBIA
Berberis aquifolium
ssp. aquifolium
130° W
60° N
120° W
55° N
50° N
50° N
55° N
50° N
130° W
120° W
MILES
0   100   200
0   100 200 300
KILOMETRES

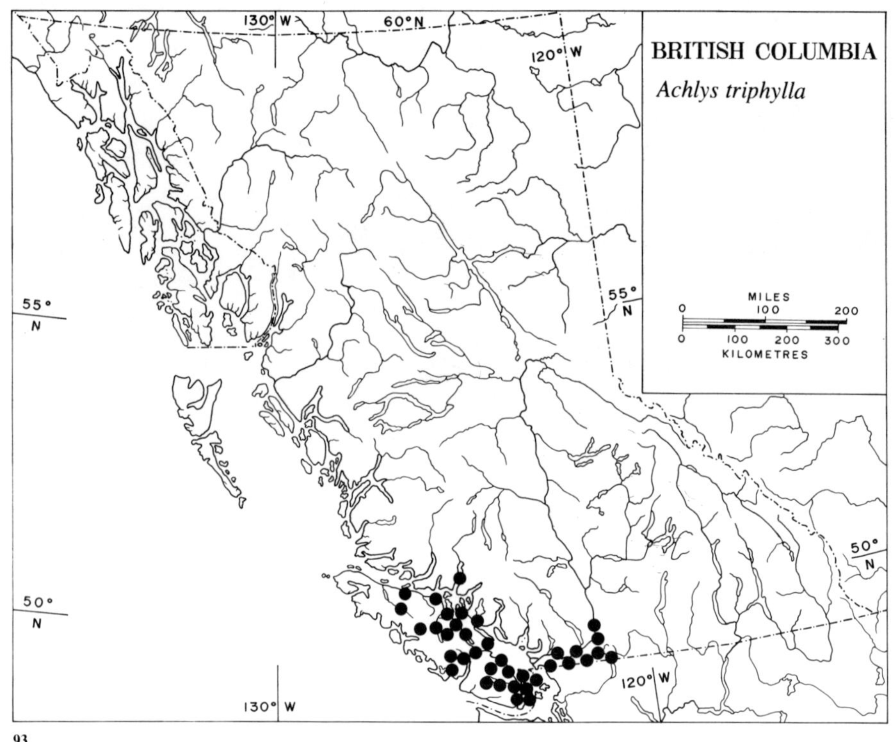

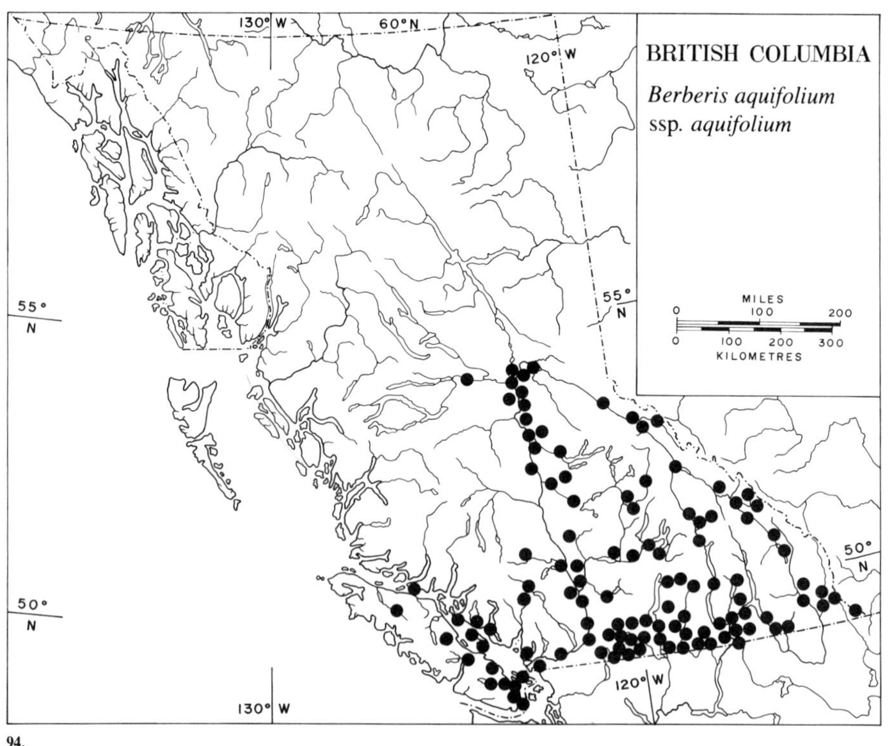

94.

244

BRITISH COLUMBIA
Berberis aquifolium
ssp. repens
130° W
60° N
120° W
55° N
55° N
50° N
50° N
120° W
130° W
MILES
0   100   200
0   100   200   300
KILOMETRES

95.

BRITISH COLUMBIA
Berberis nervosa
130° W
60° N
120° W
55° N
55° N
50° N
50° N
120° W
130° W
MILES
0   100   200
0   100   200   300
KILOMETRES

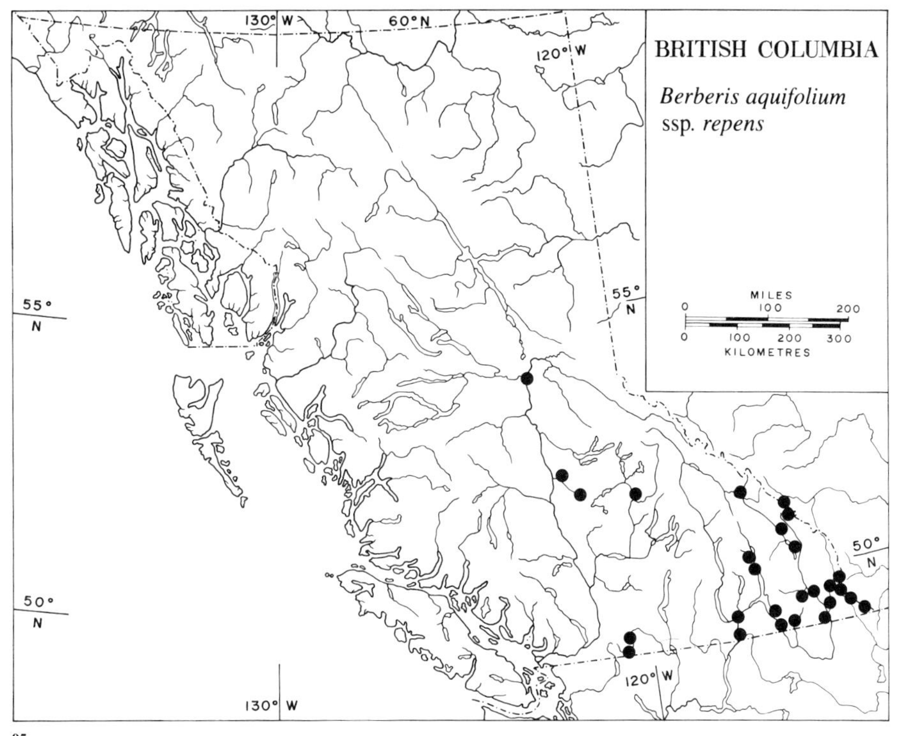

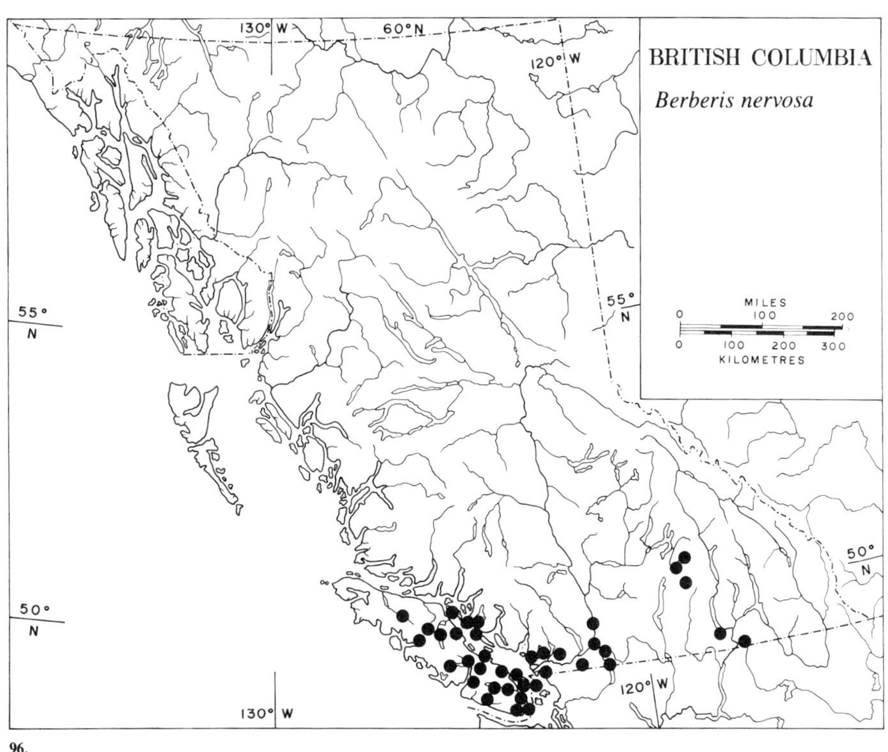

96.

245

130° W
60° N
120° W
BRITISH COLUMBIA
Berberis vulgaris
55° N
MILES
100
200
0
100 200 300
KILOMETRES
55° N
50° N
55° N
50° N
50° N
120° W
130° W

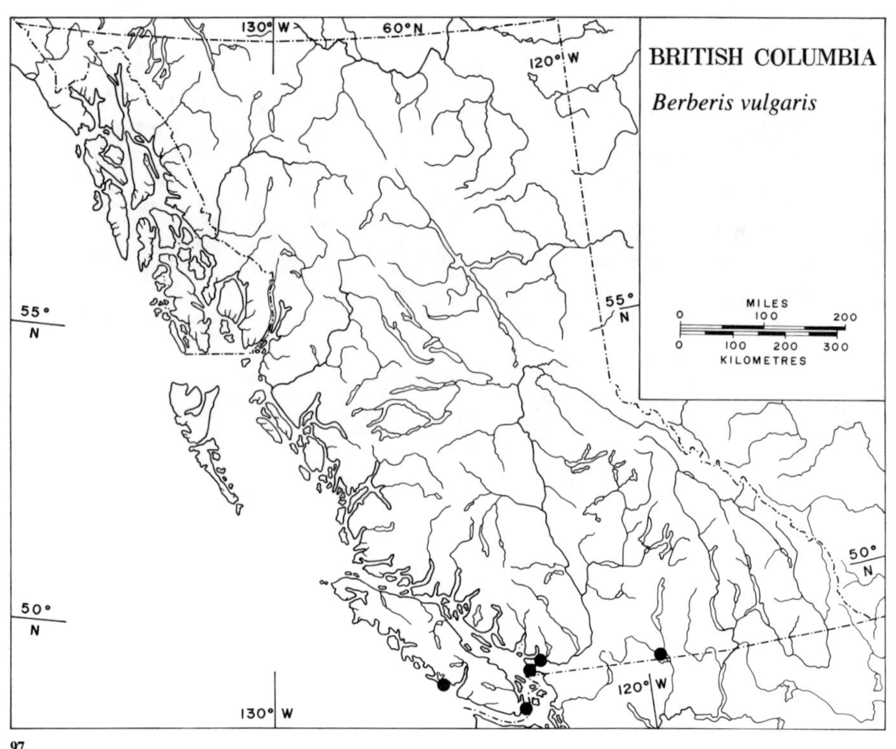

97.

REFERENCES

In addition to those specifically cited in the text, the following list includes additional references that the author has found helpful in a general way in his work on this group.

ARBER, AGNES; 1920.   Water Plants. Cambridge University Press, Cambridge. Repr. 1963 by J. Cramer, Weinheim.

BENSON, LYMAN; 1941.   North American *Ranunculi* — I. Bulletin of the Torrey Botanical Club 68: 157–172.

BENSON, LYMAN; 1948.   A Treatise on the North American *Ranunculi*. American Midland Naturalist 40: 1–264.

BENSON, LYMAN; 1954.   Supplement to a Treatise on the North American *Ranunculi*. American Midland Naturalist 52: 328–369.

BENSON, LYMAN; 1962.   Plant Taxonomy: Methods and Principles. Ronald Press, New York.

BOIVIN, BERNARD; 1944.   American *Thalictra* and their Old World Allies. Rhodora 46: 337–377, 391–445, 453–487.

BOIVIN, BERNARD; 1948.   Two New *Thalictra* From Western Canada. The Canadian Field-Naturalist 62: 167–170.

BOIVIN, BERNARD; 1951.   Centurie Des Plantes Canadiennes — II. The Canadian Field-Naturalist 65: 1–22.

BOIVIN, BERNARD; 1966.   Enumeration Des Plantes Du Canada. Le Naturaliste Canadien 93: (Ranunculaceae: pp. 584–591).

BOIVIN, BERNARD; 1968–1969.   Flora of the Prairie Provinces. Part II. Provancheria 3; reprinted from Phytologia 16–18.

BORAIAH, G. & M. HEIMBURGER; 1964.   Cytotaxonomic Studies on New World *Anemone* (Sect. *Eriocephalus*) With Woody Rootstocks. Canadian Journal of Botany 42 (7): 891–922.

BRAYSHAW, T.C.; 1976.   Catkin Bearing Plants of British Columbia. Occasional Paper No. 18. British Columbia Provincial Museum, Victoria.

BRITTON, NATHANIEL LORD, & ADDISON BROWN; 1913.   An Illustrated Flora Of The Northern United States, Canada and the British Possessions. Ed. 2; Vol. II. Charles Scribner's Sons, New York.

CALDER, J.A. & R.L. TAYLOR; 1963.   A New Species of *Isopyrum* Endemic to the Queen Charlotte Islands of British Columbia and its Relation to Other Species in the Genus. Madroño 69–76.

CALDER, JAMES A. & ROY L. TAYLOR; 1968.   Flora of the Queen Charlotte Islands: Part 1. Systematics of the Vascular Plants. Monograph No. 4, Part 1, Research Branch, Canada Dept. of Agriculture, Ottawa.

CESKA, ADOLF, & OLDRISKA CESKA; 1980.   Additions to the Flora of British Columbia. Canadian Field-Naturalist 94: 69–74.

CLAPHAM, A.R., T.G. TUTIN, & E.F. WARBURG; 1952.   Flora of the British Isles. Repr. 1958 by Spottiswoode, Ballantyne & Co., Colchester.

COOK, C.D.K.; 1966.   A Monographic Study of *Ranunculus* Subgenus *Batrachium* (DC.) A. Gray. Mitteilungen der Botanischen Staatssammlung, München; Band VI, Seite 47–237.

COOK, C.D.K.; 1968.   Phenotypic Plasticity With Particular Reference to Three Amphibious Plant Species. In Heywood, V.H., (ed.) Modern Methods In Plant Taxonomy. (Ch. 8): 97–111. Academic Press, London.

COOK, STANTON A. & MICHAEL P. JOHNSON; 1968.   Adaptation to Heterogeneous Environments I. Variation in Heterophylly in *Ranunculus flammula* L. Evolution 22: 496–516.

CORRELL, DONOVAN S. & HELEN B. CORRELL; 1972 rep. 1975.   Aquatic and Wetland Plants of Southwestern United States: Volume II. Stanford University Press, Stanford, California.

CRONQUIST, ARTHUR; 1968.   The Evolution and Classification of Flowering Plants. Houghton Mifflin Co., Boston.

CRONQUIST, ARTHUR; 1981.   An Integrated System of Classification of Flowering Plants. Columbia University Press, New York.

DALLA TORRE, C.G. DE, & H. HARMS; 1900–1907, re-edited 1963.   Genera Siphonogamarum Ad Systema Englerianum Conscripta. G. Engelmann, Leipzig.

DALLA TORRE, C.G. DE, & H. HARMS; 1958.   Register Zu Genera Siphonogamarum Ad Systema Englerianum Conscripta. H.R. Engelmann (J. Cramer), Weinheim.

DENTON, MELINDA F.; 1978.   *Ranunculus californicus*, A New Record for the State of Washington. Madroño 25: 132.

DOUGLAS, SHEILA; 1983.   Floral Color Patterns and Pollinator Attraction in a Bog Habitat. Canadian Journal of Botany 61 (12): 3494–3501.

DREW, W.B.; 1936.   North American Representatives of *Ranunculus*, Section *Batrachium*. Rhodora 38: 1–47.

DUNCAN, THOMAS; 1980.   A Taxonomic Study of the *Ranunculus hispidus* Complex in the Western Hemisphere. University of California publications in Botany; Volume 77. University of California Press, Berkeley.

ENGLER, A. & K. PRANTL; 1887–1915.   Die Natürlichen Pflanzenfamilien. Edition 2, 1959–1980. Duncker & Humblot, Berlin.

EWAN, JOSEPH; 1945.   A Synopsis of the North American Species of *Delphinium*. University of Colorado Studies, Series D: Volume 2: 55–244.

FASSETT, NORMAN C.; 1940.   A Manual of Aquatic Plants. McGraw-Hill Book Co. Inc., New York.

FERNALD, M.L.; 1917a.   New or Critical Species or Varieties of *Ranunculus*. Rhodora 19: 135–139.

FERNALD, M.L.; 1917b.   Some Color Forms of American Anemones. Rhodora 19: 139–141.

FERNALD, M.L.; 1919.   The Variations of *Ranunculus Repens*. Rhodora 21: 169.

FERNALD, M.L.; 1934.   Some Critical Plants of Greenland. Rhodora 36: 89–97 & Plates 279 & 280.

FERNALD, M.L.; 1950.   Gray's Manual of Botany. Edition 8. American Book Co., New York.

FISHER, F.J.F., J.A. ROWLEY & C.J. MARCHANT; 1973.   The Biogeography of the Western Snow-Patch *Ranunculi* of North America. Comptes Rendus, Societe de Biogeographie 438: 32–43.

FISHER, F.J.F., A. WARNER & E.M. REIMER; 1979.   Anomalous Apetaly: Localized Character Displacement in *Ranunculus eschscholtzii*. Canadian Journal of Botany 57 (20): 2097–2106.

FUKUDA, ICHIRO; 1967.   The Biosystematics of *Achlys*. Taxon 16 (4): 308–316.

GLEASON, HENRY A.: 1952,   repr. 1958; The New Britton and Brown, Illustrated Flora of the Northeastern United States and Adjacent Canada. New York Botanical Garden, New York.

GREENE, E.L.; 1890.   Notes on *Ranunculus*. Pittonia 2: 110–111.

GREGORY, WALTON C.; 1941.   Phylogenetic and Cytological Studies in the *Ranunculaceae*. Transactions of the American Philosophical Society; New Series 31: 443–521.

HENRY, JOSEPH KAYE, 1915.   Flora of Southern British Columbia and Vancouver Island. W.J. Gage & Co., Toronto.

HITCHCOCK, C. LEO & ARTHUR CRONQUIST; 1964 repr. 1971.   Vascular Plants of the Pacific Northwest: Part 2: *Salicaceae* to *Saxifragaceae*. University of Washington Press, Seattle.

HITCHCOCK, C. LEO & ARTHUR CRONQUIST; 1973.   Flora of the Pacific Northwest. University of Washington Press, Seattle.

HULTEN, ERIC; 1968.   Flora of Alaska and Neighboring Territories. Stanford University Press, Stanford, California.

HUTCHINSON, J.; 1959.   The Families of Flowering Plants. (ed. 2). Oxford University Press, London.

JAMES, EDWIN; 1823.   Account Of An Expedition From Pittsburgh to the Rocky Mountains in the Years 1819, 1820. By Order of the Hon. J.C. Calhoun, Secretary of War, Under the Command of Maj. S.H. Long, of the U.S. Top. Engineers. Longman, Hurst, Rees, Orme & Brown, London.

KAPOOR, B.M. & ASKELL LÖVE; 1970.   Chromosomes of Rocky Mountain *Ranunculus*. Cytologia 23: 575–594.

KINGSBURY, JOHN M.; 1964.   Poisonous Plants of the United States and Canada. Prentice-Hall, Inc., Engelwood Cliffs, New Jersey.

LANJOUW, J. *et alia*, eds.; 1952.   International Code of Botanical Nomenclature. Chronica Botanica Co., Waltham, U.S.A.

LEPPIK, E.E.; 1964.   Floral Evolution in *Ranunculaceae*. Iowa State College Journal of Science 39: 1–101.

LES, DONALD H.; 1985.   The Taxonomic Significance of Plumule Morphology in *Ceratophyllum* (Ceratophyllaceae). Systematic Botany 10 (3): 338– 346.

LINNAEUS, CARL; 1753.   Species Plantarum. Facs. ed.; 1957. Ray Society, London.

LÖVE, ASKELL & DORIS LÖVE; 1975.   Nomenclatural Notes on Arctic Plants. Botanisker Notiser 128: 497–523.

MACOUN, JOHN; 1883.   Catalogue of Canadian Plants, Part I. Polypetalae. Geological Survey of Canada.

MADAHAR, CECILY, & MARGARET HEIMBURGER; 1969.   Meiotic Studies in *Anemone multifida*, *A. tetonensis*, and their hybrids. Canadian Journal of Botany 47 (12): 1973–1983.

MOORE, R.J.; 1970.   Index to Plant Chromosome Numbers for 1968. Regnum Vegetabile 68. International Bureau for Plant Taxonomy & Nomenclature, Utrecht.

MOORE, R.J.; 1972.   Index to Plant Chromosome Numbers for 1970. Regnum Vegetabile 84. International Bureau for Plant Taxonomy & Nomenclature, Utrecht.

MOORE, R.J.; 1977.   Index to Plant Chromosome Numbers for 1973/74. Regnum Vegetabile 96. International Bureau for Plant Taxonomy & Nomenclature, Utrecht.

MORRIS, MICHAEL I.; 1973.   A Biosystematic Analysis of the *Caltha leptosepala* (Ranunculaceae) Complex in the Rocky Mountains. III. Variability in Seed and Gross Morphological Characteristics. Canadian Journal of Botany 51 (12): 2259–2267.

MUENSCHER, WALTER CONRAD; 1944.   Aquatic Plants of the United States. Comstock Publishing Company Inc., Ithaca, N.Y.

MUNZ, PHILIP A. & DAVID D. KECK; 1963.   A California Flora. University of California Press, Berkeley.

OGDEN, EUGENE C.; 1974.   Anatomical Patterns of Some Aquatic Vascular Plants of New York. New York State Museum and Science Service bulletin No. 424. Albany, N.Y.

ORNDUFF, R., ed.; 1967.   Index to Plant Chromosome Numbers for 1965. Regnum Vegetabile 50. International Bureau for Plant Taxonomy & Nomenclature, Utrecht.

ORNDUFF, R., ed.; 1968.   Index to Plant Chromosome Numbers for 1966. Regnum Vegetabile 55. International Bureau for Plant Taxonomy & Nomenclature, Utrecht.

ORNDUFF, R., ed.; 1969.   Index to Plant Chromosome Numbers for 1967. Regnum Vegetabile 59. International Bureau for Plant Taxonomy & Nomenclature, Utrecht.

PADMORE, PATRICIA A.; 1957.   The Varieties of *Ranunculus flammula* L. and the Status of *R. scoticus* E.S. Marshall and of *R. reptans* L. Watsonia 4: 19–27.

PELLMYR, OLLE; 1985. Yellow Jackets Disperse *Vancouveria* Seeds (Berberidaceae). Madroño 32 (1): 56.

PIPER, CHARLES V.; 1922.   The Identification of *Berberis aquifolium* and *Berberis repens*. Contr. U.S. Nat. Herb. 20: 437–451 & Plates 24–26.

POLUNIN, NICHOLAS; 1959.   Circumpolar Arctic Flora. Oxford University Press, London.

PRINGLE, JAMES S.; 1971.   Taxonomy and Distribution of *Clematis* Sect. Atragene (Ranunculaceae) in North America. Brittonia 23: 361–393.

RAJU, M.V.S., R.T. COUPLAND, & T.A. STEEVES; 1966.   On the Occurrence of Root Buds on Perennial Plants in Saskatchewan. Canadian Journal of Botany 44 (1): 33–37.

RAUP, HUGH M.; 1934.   Phytogeographic Studies in the Peace and Upper Liard River Regions, Canada. Contribution No. VI; Arnold Arboretum of Harvard University, Jamaica Plain, Massachussetts.

ROWE, J.S.; 1972.   Forest Regions of Canada. Canada Dept. of Environment: Canadian Forestry Service, Ottawa. Publication No. 1300.

SCOTT, PETER J.; 1974.   The Systematics of *Ranunculus gmelinii* and *R. hyperboreus* in North America. Canadian Journal of Botany 52 (7): 1713–1722.

ST. JOHN, HAROLD; 1937.   Flora of Southeastern Washington and of Adjacent Idaho. Students Book Corporation, Pullman, Washington.

SCOGGAN, H.J.; 1978.   The Flora of Canada, Part 3. Publications in Botany No. 7 (3), National Museum of Natural Sciences, Ottawa.

STEWARD, ALBERT N., LA REA J. DENNIS & HELEN M. GILKEY; 1963.   Aquatic Plants of the Pacific Northwest. Oregon State University Press, Corvallis.

TAKHTAJAN, ARMEN; 1961, English ed. 1969.   Flowering Plants: Origin and Dispersal. English translation by C. Jeffrey. Oliver & Boyd, Edinburgh.

TAYLOR, ROY L. & GERALD A. MULLIGAN; 1968.   Flora of the Queen Charlotte Islands, Part 2: Cytological Aspects of the Vascular Plants. Monograph No. 4, Part 2. Research Branch, Canada Dept. Agriculture, Ottawa.

TAYLOR, ROY L. & BRUCE MACBRYDE; 1977.   Vascular Plants of British Columbia. University of British Columbia Press, Vancouver, B.C.

TORREY, J. & A. GRAY; 1838–1840, facs. repr. 1969.   A Flora of North America: Vol 1. Hafner Publishing Co., New York.

TURNER, NANCY J.; 1984.   Counter-Irritant and other Medicinal Uses of Plants in *Ranunculaceae* by Native Peoples in British Columbia and Neighbouring Areas. Journal of Ethnopharmacology, 11: 181–201.

TUTIN, T.G.; 1964.   *Ranunculaceae. In* Tutin, T.G. *et alia*, eds. *Flora Europaea*. Vol. 1. Cambridge University Press, Cambridge, U.K.

VOSS, E.G. *et alia*, eds.; 1983.   International Code of Botanical Nomenclature. Bohn, Scheltema & Holkema, Utrecht.

WELSH, STANLEY L.; 1974.   Anderson's Flora of Alaska and Adjacent Parts of Canada. Brigham Young University Press, Provo, Utah.

WIGGINS, IRA L. & JOHN HUNTER THOMAS; 1962.   A Flora of the Alaskan Arctic Slope. University of Toronto Press.

WILDEMAN, ALAN G. & TAYLOR A. STEEVES; 1982.   The Morphology and Growth Cycle of *Anemone patens*. Canadian Journal of Botany 60 (7): 1126–1137.